Nationwide Praise for
Dreaming Out Loud

"This is a portrait of the music business full of emotional nuances. No one who hopes to make a living from popular music can afford to ignore it."
—*Rolling Stone*

"Any fan of popular music should read this."
—*Denver Post*

"You may think you knew Garth Brooks, Wynonna, and Wade Hayes, but you didn't at all and won't until you read Bruce Feiler's exhaustive profiles. Feiler gets into these artists' lives and under their skin, and along the way presents a compelling look at the heart of Nashville today."
—Chet Flippo, *Billboard*

"Surprising, even shocking, *Dreaming Out Loud* is one of the best books about country music in years…You'll never look at the medium the same way again."
—*Charleston Post and Courier*

"An important, honest commentary on today's country music—sort of like a backstage pass to some of the greats. *Dreaming Out Loud* is significant to our times."
—*Wichita Times Record News*

"Insightful."—Michiko Kakutani, *New York Times Sunday Magazine*

"A penetrating look at the workings of the music business in Music City USA."—*Atlanta Journal Constitution*

More Praise for *Dreaming Out Loud*

"Intimate and insightful...A fascinating look
at the changing face of Nashville."
—*Dallas Morning News*

"A remarkable book...vividly accurate.
It's Feiler's rapport with Nashville and his reciprocating
honesty about himself that makes this book work."
—*Music Row Magazine*

"*Dreaming Out Loud* is an entertaining,
perceptive assessment of the business."
—*Charlotte Observer*

"Riveting...Feiler's a born storyteller, a craftsman
and a cut up—in short, he's the genuine article."
—*Salon*

"Nashville may finally have its *Bonfire of the Vanities*."
—*Nashville Life*

"With a Fellini-like support cast of gossip columnists,
obsessed groupies and high-flying promoters,
Dreaming Out Loud is as quirky and entertaining
as Robert Altman's 1975 satiric film."
—*The Ottawa Citizen*

"Like a great country song, *Dreaming Out Loud*
will become a touchstone for future stories about
its subjects. With a researcher's eye for detail and a fan's
forbearance, Feiler puts human faces on Brooks' ambition
and the Judd family's dysfunction...All the while,
Feiler's casual first person narrative gives
Dreaming Out Loud the feeling of intimacy country fans
expect from their performers. It never forgets that it's
not only about dreamers—it's for them."
—*USA Today*

Andrew Feiler

About the Author

BRUCE FEILER is the *New York Times* bestselling author of *Walking the Bible*. He is a contributor to NPR's *All Things Considered* and writes for the *New York Times Magazine, Condé Nast Traveler,* and *Gourmet.* He lives in New York City.

Also by
Bruce Feiler

DREAMING
OUT LOUD

GARTH BROOKS,
WYNONNA JUDD, WADE HAYES,
AND THE CHANGING FACE OF NASHVILLE

BRUCE FEILER

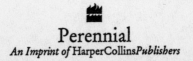

Perennial
An Imprint of HarperCollinsPublishers

Grateful acknowledgment is given to Sony / ATV Music Publishing for permission to reprint lyrics from the following songs: "Don't Make Me Come To Tulsa," by Don Cook, copyright © 1994 by Sony / ATV Songs LLC / Don Cook Music; "Old Enough To Know Better," by Chick Rains and Wade Hayes, copyright © 1994 by Sony / ATV Songs LLC; "Our Time Is Coming," by Kix Brooks and Ronnie Dunn, copyright © 1992 by Sony / ATV Songs LLC/Buffalo Prairie Songs/Showbilly Music; "On A Good Night," by Don Cook, Paul Nelson, and Larry Boone, copyright © 1996 by Sony / ATV Tunes LLC/Sony/ATV Songs LLC/Don Cook Music/Terilee Music; "From Hell To Paradise," by Raul Malo, copyright © 1992 by Sony/ATV Songs LLC/Raul Malo Music   All rights administered by Sony / ATV Music Publishing, 8 Music Square West, Nashville, TN 37203   All Rights Reserved. Used By Permission

The excerpt by Peter Kinder on page 118 appears courtesy of the *Southeast Missourian* .

Photograph captions and credits are listed on pages 405-406 .

A hardcover edition of this book was published in 1998 by William Morrow.

First Spike paperback edition published 1999.

Reprinted in Perennial 2002.

Designed by Kellan Peck

The Library of Congress has catalogued the hardcover edition as follows:

Feiler, Bruce S.
Dreaming out loud : Garth Brooks, Wynonna Judd, Wade Hayes, and the changing face of Nashville / Bruce Feiler.
p. cm.
Includes index.
1. Country musicians—Tennessee—Nashville—Biography. 2. Country music—Tennessee—Nashville—History and criticism. I. Title.
ML.394F42   1998   97-49638
781 642—dc21   CIP

ISBN 0-380-79470-5 (pbk.)

02 03 04 05 06 RRD 10 9 8 7 6 5

•

May the circle be unbroken

For the memory of
George Alan Abeshouse
and
Ellen Abeshouse Garfinkle

•

I hear down there it's changed you see
They're not as backward as they used to be.

-Bob McDill,
"Gone Country"

CONTENTS

VERSE IV

VERSE V

CODA

NOTES

PRELUDE

THE PLAYERS

At what should have been a crowning moment in his career, Garth Brooks made a rare misstep.

He was wearing his black Stetson that evening and his lace-up ropers as well—the signature accessories of American myth. As usual, his black Wranglers were three sizes too small—the better, he confessed, to conceal his often unruly weight. ("Truth is," he told me, "I don't mind making fun of my body. That way we all know straight out of the box: 'Don't worry, guys, I don't have a hangup about how I look. I'm not a sex god.' ") All together, with his puffy eyes, his chipmunk cheeks, and his dumpling chin, he looked a bit like a high school football coach who had gotten all dressed up to help chaperon the prom.

"And now, to present the award for Favorite Artist of the Year, please welcome, once again, superstar Neil Diamond . . ."

The tuxedoed crowd in the red velvet seats of the Shrine Auditorium in Los Angeles applauded politely as Neil Diamond stepped to the Lucite podium fondling the trademark triangular envelope of the American Music Awards.

"Thank you so very much," said the onetime pop idol who himself had just cut an album in Nashville in an effort to capitalize on the exploding interest in all things country. "It's my privilege to present the final award of the evening, Favorite Artist of the Year. The five nominees represent the wide range of music, and they are: Hootie and the Blow-fish, TLC, Green Day, Boyz II Men, and Garth Brooks . . ."

As the cameras panned the faces of the nominees, Garth Brooks, sitting spread-eagled on the front row, bowed his head and clutched the arm of his wife Sandy. Having already claimed two awards that night—Favorite Country Album and Favorite Male Country Artist—he was having another banner evening, coming off another record-breaking year, and climaxing what had been the fastest rise to stardom of any artist in American history. In a little over six years, the thirty-three-year-old former college javelin thrower from Tulsa, Oklahoma, had captivated America with his modern-day version of an old-fashioned singing cowboy—Gene Autry with a social conscience. His songs, many of them self-written, were an irresistible fusion of James Taylor-esque you've-got-a-friend-in-me bear hugs and John Mellencamp-like let's-get-rowdy anthems that had people all across the nation saying, "Hmm, I didn't know country could be cool." His shows were an eye-popping mixture of honky-tonk raucousness and arena rock pyrotechnics that coaxed even die-hard country music haters to call their friends and declare, "If he's the captain, I'm playing on his team." And play they did. In the span of half a decade, Garth Brooks sold an astonishing 60 million albums—one for every household in America and more than Michael Jackson, Madonna, Billy Joel, even Elvis Presley. Along the way, he was certified as the bestselling country artist in history, the bestselling artist of any genre in the 1990s, and the bestselling solo artist in North American history, behind only the Beatles in total sales.

Garth Brooks, in other words, was big. But like Tom Hanks in the movie of that name, Garth had achieved his status by acting like an innocent kid who accidentally struck it rich. It was all part of his charm. He read the Bible, thanked his mother, and, for all appearances, drank his milk as well. His imagery was so simple, so elementally American—hat, guitar, mama's boy, a touch of evil—that in no time Garth had become America's you-get-what-you-pay-for hero, our national Eagle Scout-at-Large. At a time of nihilism and unfettered cynicism, here was someone prepared to stand up with a straight face and deliver an unapologetically upbeat message. As he sang in one early benchmark song, "The River": "Don't you sit upon the shoreline/And say you're satisfied/Choose to chance the rapids/And dare to dance the tide."

One reason Garth Brooks became so big was because of people like me. Though I was born and raised in South Georgia, I grew up hating country music. Nashville (and by that I mean the country music side of

Nashville) embodied everything about the South I abhorred—hay bales, banjos, overalls, bigotry. Most of all, as a teenager who was eager to escape the region and its choking stereotypes, I disliked what I viewed as the music's lingering attachment to outdated images and ideals. Four years at Yale, followed by three in Japan and one as a graduate student in England, only increased my distance from what I perceived as the voice of the obsolete South.

Garth Brooks, by contrast, embodied the New Nashville and, by extension, the New South. Born in Tulsa, where the South, West, and Midwest collide, Garth was smart, college-educated, and savvy—as hip to Hollywood as he was to West Virginia; one part movie star; one part fraternity brother; three parts "Home on the Range." No part hillbilly. Garth Brooks's *The Hits* was the first country album I bought, and I loved it—its fun-loving pop ditties ("Papa Loved Mama"); its soaring, sensitive-guy ballads ("If Tomorrow Never Comes"). I urged everyone I knew who mocked country music to buy it. I put the winding fiddle intro from one of its songs, "Much Too Young (to Feel This Damn Old)," on my answering machine. I even made love to the album on the floor of my apartment in Washington, D.C. (Actually, I *tried* to make love. My partner needed a dose of Mary Chapin Carpenter to finish the deed. Then she dumped me.) In short, I became obsessed with Garth Brooks in the way that only a recent convert can feel. In particular, I relished the promise he held for fallen Southerners like me—you *could* go home again and not feel embarrassed. That feeling only intensified when high-minded friends of mine, including a diplomat from Singapore, sat me down and told me politely that I was losing my mind.

I didn't care. As I reached my thirties, country music became a passion and, for someone who had spent so many years abroad, a way back into America, a real America, far removed from the cynicism and manipulation of New York, Los Angeles, or Capitol Hill. I decided to learn more about this world, which is where Garth Brooks came in. When I first visited Nashville in 1995, I was told that Garth was just coming out of a period of semi-retirement and was interested, if the circumstances were right, in opening up to a writer. Once I passed several rounds of screening (which included breakfast with his handlers at the famed Pancake Pantry and some heavy breathing on my part about my then-favorite Garth song, "Unanswered Prayers"), I was invited to meet him. In time we developed a friendly, intimate, though always professional relationship, and during the course of the next several years I was with him on a regular—at times weekly—basis in what I later realized was a rare, and perhaps unprecedented, opportunity to witness a major American entertainer at the height of his stardom. Also, as it turned out (and though neither of us anticipated it), the time I spent with Garth coincided with a dramatic turning point in his career.

When I first met Garth, I was captivated by his magnetism. In person he was just as cordial as he appeared on television, with an aw-shucks charm and effortless good manners. It was almost as if he had stepped out of an episode of "My Three Sons." "Nice to meet you, sir," he said upon our first meeting. "Yes, ma'am, no, ma'am," he replied to questioners on a radio show. And while being hurried to a press conference, Garth brushed off his handlers to greet a girl who was cowering in the corner. "Hi, I'm Garth," he said, locking onto her with his spellbinding eyes. "What's your name?" Instantly the girl began to cry. Later, when prompted by a reporter to tell where he hoped to be in ten years, he offered an impossibly humble reply: "Hopefully I see myself as a father, as a husband. I would still like to be a son ten years from now. And that's about all I know. If it's music, then God has blessed me more than I could ever imagine, and if it's not, I hope it's something where I still get a chance to make a difference." The crowd erupted in applause.

The focal point of Garth's charisma was his eyes. When he first burst into public view in 1989, Garth's eyes were a constant topic of conversation—among friends, fans, and colleagues alike. They were dreamy, cerulean blue eyes with a hint of martial green. They searched the sky constantly for inspiration, and when they found it they locked on the spot with near-religious intensity. It was as though he was following a spirit that no one else could see, but Garth made it so believable that everybody else wanted to follow. And when he turned those eyes on you—in person, onstage, or even in photographs—you couldn't help but be transfixed. As Jimmy Bowen, the head of Garth's label and later his chief antagonist, wrote of the first time he saw Garth perform: "Garth's eyes got so wide and fiery, you could tell what color they were from twenty rows back. He looked out over four hundred people in one area of the crowd, and every person thought he was making eye contact with him or her." Garth Brooks had the kind of eyes that made you feel important, but also made you feel as if he was looking into the depths of your fantasies.

Over the years, though, Garth's eyes began to change. They became less innocent, and more calculating. Instead of looking out at the world, they began to peer inward. The puffy-cheeked kid with the too-big baseball cap and the dream of becoming a singing cowboy became increasingly overshadowed by the hard-nosed mogul with the degree in advertising and the desire to extend his empire to all corners of the globe. Garth, of course, always had both traits in his personality, but all during his rise he had brilliantly straddled the line between the two. Sure, he strategized, but he did so in a way that made you root for him. Caught cheating on his wife early in his career, Garth and Sandy went on TNN (The Nashville Network), confessed, and discussed their commitment to rebuild their relationship. It was a brilliant act of self-effacement. When Garth's video for "The Thunder Rolls" was pulled from TNN and CMT (Country Music Television) for its depiction of spousal abuse, Garth rallied support in battered women's shelters. Again fans backed him. Every month he seemed to be in the news for a different cause (saying he might retire after the birth of his first child; announcing his sister was a lesbian on Barbara Walters), and though some accused him of being a crass manipulator, more supported him. After all, here was a man who talked openly of old-fashioned dreams and throwback heroes. After winning country music's prestigious Horizon Award in 1990, Garth thanked four people: his father, George Jones, George Strait, and John Wayne.

Now, five years later, Garth was clearly beginning to strain under the pressure of trying to live up to his own image. "I've thrown up, cried, and passed out," he told *Billboard* magazine of his new album, *Fresh Horses*, his first in two years, "everything physically possible you can do with a record." With another interviewer Garth went even further. "It just got too loud in my head," he said of fears that became so intense that one day he blacked out on his farm. "It just got . . . too . . . loud. People were screaming, and I was one of them inside there: 'What are you gonna do? *What are you gonna do?* WHAT ARE YOU GONNA DO? If it doesn't work.' " He spent the first three days after turning in his album in a fetal position, worrying. "I know this artist's worst nightmare is being forgotten," he said to me in the odd manner he developed of speaking of himself in the third person. "I'm going to have to come face-to-face with that. I've never been a fan of the backside of the bell curve."

All of which raised a series of questions that haunted his mind as he waited to hear the results of the American Music Award for Favorite Artist of the Year, questions that would play themselves out in a vivid and sometimes disturbing way in the months to come: Could it be possible that the reigning sovereign of American music, having almost singlehandedly brought Nashville to an unprecedented primacy in America, had overstayed his welcome? Might it be true that his efforts to transform himself into a modern-day John Wayne would fall short? And, more importantly, what would happen if the public (especially those in Nashville) sent him a message that they no longer believed his humble, normal-guy image?

"And now . . ." proclaimed Neil Diamond with a dramatic pause as soon as the clips of the nominees' videos were complete, "the Favorite Artist of the Year is . . ." He carefully slid open the triangular envelope as the fans in the balcony began to scream out their votes—Hootie, TLC, Boyz II Men—until Diamond's own voice cut them off: "Garth Brooks!"

Sitting in his seat, Garth lurched backward and his eyes bugged out in surprise. "Everything suddenly went white," he told me later. Sales of his new record had been disappointing. He had not performed publicly in over a year. Surely he didn't deserve this award. His mind was spinning. He felt sick to his stomach. Finally, after a moment, he leaned over and kissed his wife. He rose, hugged the other (startled) nominees one at a time, and finally scampered to the stage. "I just kept saying to myself, 'Man, I don't deserve this. Man, I don't deserve this. Man, I don't deserve this . . .'" Once in the light, he removed his hat, bowed, and collected the trophy from Neil Diamond. Arriving at the podium, he paused, looked directly into the camera, and delivered the speech that for months afterward would be viewed as the signal that Garth Brooks was becoming so consumed with own image—and with creating media stunts to enhance that image—that he risked losing his effortless, common-man touch.

"Thank you very much," Garth said in his familiar, small-town mayor tone of voice. "So you'll know right off the bat, I cannot agree with this . . . Music is made up of a lot of people, and if we're one artist short, then we all become a lesser music. So, without any disrespect to the American Music Awards, and without any disrespect to any fans who voted, to all the people who should be honored with this award . . ." He set the trophy down on the podium. "I'm gonna leave it right here." And with that he left the stage empty-handed.

The audience, stunned, sat still in its seats.

In Nashville, tired of such speeches, people clicked off their sets: Garth had already accepted two awards that evening. What was wrong with this one?

And by the following day word had spread around town: Garth Brooks seemed to be on the verge of a breakdown.

                              ⚬   ⚬   ⚬

The first dime-store cowgirl with a camera and flashbulb slid on her knees to the front of the stage even before anyone appeared in the doorway. The second and third ones followed soon after. By the time the announcer turned on the microphone and unleashed his generic bass radio voice, nearly two dozen feathered-haired fans in creased Western shirts and slickly stitched boots were bobbing like schoolgirls in front of the cramped corner stage at Bronco's Lounge, "Central Virginia's Number One Country Showplace."

"Welcome to Bronco's, ladies and gentlemen. My name is Cal, and we're glad you're here." The squeal, even then, could not have been louder. Cal's voice was as low as could go. "And now, what you've all been waitin' for . . ."

The roar in the room grew rowdier with each word. A few of the three hundred people overcrowded in back started to bang on the red-and-white tabletops. "Is he going to come out of that door?" a young woman asked as I emerged from backstage. When I nodded, she yanked her mother from her seat and took her picture in front of the closed door.

"His first single, 'Old Enough to Know Better,' has been on top of the K95 charts. It's been on top of the *R&R* [*Radio & Records*] charts. It's been on top of the *Billboard* charts for the last *two* weeks as the number one country hit in our nation. You've heard it before. You're going to hear it again. Ladies and gentlemen, give it up for Columbia recording star, Mr. Waaaaaade Haaaaaayes!"

With a blaze of fiddle fire and a flash of Kodak light, the backside of Nashville's newest hunk in a hat went streaking through the side door of the lounge and paused, momentarily, in the center of the stage.

"One, two, three, four." The drummer, a blond, rapped his sticks together. "Five, six, seven, eight."

His friend, still hidden, turned to face his new life.

"Don't make me come to Tulsa
You know I hate to fly
I ain't been back to Oklahoma
Since you said good-bye."

Two weeks shy of his twenty-sixth birthday and sixteen weeks after starting from scratch, Wade Hayes—guitarist, mama's boy, sex object— suddenly awoke from the density of his dreams one night in the spring of 1995 and found himself in the unexpected role of Nashville's start-up superstar of the moment. Standing before his first sold-out crowd—on the wrong side of Interstate 95 north of Richmond—the soft-spoken Oklahoman was startlingly tall in his creased blue Wranglers, dark flowered shirt, and standard-hunk-issue black cowboy hat. He was startlingly awkward, almost foallike. He could pluck his Telecaster guitar with admirable esprit de stardom. He could flick back his shoulder-length brown hair with insouciant sex appeal. But in his stare—his dark callow stare—he looked deeply unprepared. "I just couldn't believe," he said later, "that what they all wanted was for me to turn around, so they could take a picture of my butt."

Following the path newly widened by Garth Brooks, whose poplike success paved the way for a new kind of superstar to emerge out of Nashville, a sudden influx of young men and women began flocking to Music City in recent years with nothing but range-fed expectations and a handful of half-written love songs. Out of this pack, few would ever see the inside of a record label, even fewer would see the inside of a recording studio, and fewer still would have any music released to the fickle public. Those who do survive this gauntlet and achieve even a modicum of success are then forced to confront one of the most exhausting, and wrenching, personal transformations awaiting any young artist in American life. One day Wade Hayes was nervously standing before a microphone on his first-ever visit to a recording studio; six months later, a song from that session was the most-played record on 2,600 radio stations, reaching 70 million sets of ears in North America alone.

Stories such as these, with their timeless echoes of small-town heroes triumphing over big-city realities, are one of the principal reasons that millions of Americans have been transfixed by the stories coming out of Nashville recently. Wade Hayes was born in Bethel Acres, Oklahoma (population 2,505). His father moved his family to Nashville in the early 1980s after being promised a record deal himself. But when that deal proved fraudulent, the elder Hayes was forced to sell his home in Nashville and, with his family, slink back to Oklahoma. Ten years later when Wade dropped out of college and moved to Nashville, he was carrying not only his own ambitions, but those of his entire family. Like many young singers, Wade began his hunt for vindication by singing at Gilley's nightclub in the evenings and working construction during the day—the kind of one-two punch-and-hope that nurses many a failed ambition. Wade, however, emerged from that jam of hopefuls through an impromptu meeting with Don Cook, one of Nashville's premier producers, responsible for boot-scootin' hotshots Brooks & Dunn and ultrahip crooners the Mavericks. At the five-minute audition, Wade took out his guitar and sang a self-written ballad, "I'm Still Dancin' with You," about a young man still in love with his ex-girlfriend. The following day he signed a letter of intent to be the first artist on Cook's new label at Sony Music.

"Any of you guys ever heard of Merle Haggard?" Wade asked after his second song. Blinking back shock in response to their cheers, he licked his lips and smiled at the front row. "Well, gosh, he's always been my favorite, too. And if y'all don't mind, I'd like to play you a song of his."

Far from easing his burdens, Wade's overnight success only heightened the pressures. If anything, his speedy start—the fastest and most promising of any artist in the Nashville Class of 1994—raised the stakes considerably: Suddenly millions of dollars were on the line, lots of people were watching, and legions of rivals were hoping he would fail. Having first met Wade in early 1995 on the day his debut single reached number one, I spent countless hours with him over the next several years, going into the studio, out onto the road, and back home with his family, as he struggled first to define his ambition, then to achieve it. As a window into Nashville, not to mention the larger world of American celebrity, the view was unrivaled. Also, as a counterpoint to Garth's ongoing battles with his success, Wade's struggles to *achieve* success were, in many ways, even more revealing.

Unlike many of his contemporaries, Wade studied the marketplace around him, constantly reexamining his appearance (Should he get plastic surgery on his broken nose?), his reputation (Should he *try* to get in the tabloids?), and his competition (What should he do about his clean-cut counterpart, Bryan White, coming up fast?). Above all, his experience raised a dilemma: How does a young man, full of so many dreams, fresh from one of the tiniest towns in Middle America, who had never even been on an airplane before he had a record on every country radio station in America, possibly navigate all the opportunities and hazards of being an artist in the age of conglomerates? In Nashville alone, the examples of early success gone bad are legendary. For every Garth Brooks or Dolly Parton who persevere through early trouble to achieve acclaim, there are legions of artists like George Jones, who nearly squandered his talent on drugs, or Keith Whitley, who drank himself to death.

For Wade the questions were less apocalyptic, but no less pressing: Could he build on his early success to develop a long-lasting career?

Could he withstand the corporate pressures that seem to overwhelm many young artists? And, most importantly of all, could he overcome his considerable personal insecurities to become a dynamic, full-fledged star?

"I'd like to tell y'all a story," Wade said after finishing another song or two. "Y'all might know this song. It's our brand-new single. You might have heard it on the radio or seen the video on CMT." His voice was canyon-deep and rich. His smile was wry, with an occasional chuckle. He was clearly warming to the hour. "I was sittin' at home one night right before I moved to Nashville, Tennessee, from Oklahoma. And this girl and I had just broken up. I had dated her for three years." Two girls in front stood up to cheer. A few older women behind them nodded knowingly. They would want Wade Hayes as a son-in-law.

"I want to tell you I was heartbroken, I was in bad shape. I set down on the couch one night and this song started coming out. I moved to Nashville and this song got me my record deal. Later it got me my publishing deal. Got me here in Richmond, Virginia, pickin' for you guys tonight." He paused. "And guess what: I'm kinda over it by now, if you know what I mean."

The moment was solid gold. Wade Hayes was on a roll. And by the time he reached the end of his ninety-minute set, through stately, if predictable, tributes to Merle, George Jones, and Keith Whitley, as well as most of the songs on his first album, Wade handled the moment the crowd had come to see with precise and admirable grace.

"Before we go, I'd like to the thank all of y'all out there in the audience for buying my record. I'd like to thank my record label for getting behind me. But above all, I'd like to thank my Heavenly Father for giving us our first . . . *number one song!*

> 'Neon lights draw me like a moth to a flame
> Mama raised me right
> That just leaves me to blame
> When I get a little sideways on a honky-tonk tear
> I'm old enough to know better
> But I'm still too young to care.' "

Even before he reached the end of the first verse, all three hundred people in Bronco's—including the men, the chef, even the jaded owner—were standing on their feet and singing along. And when he came to the final verse, tossed his pick into the room, and slithered from the stage, the audience continued to stand and cheer. "Wade. Wade. Wade. *Wade!*" And wait for his return.

But Wade did not return to the stage that night. The audience continued to stand and cheer, but the stage door never opened. After a moment, the owner put on the house tape, and the road manager—older, brash, knowing that Wade's new number one would mean he wouldn't play overcrowded honky-tonks like this much longer—insisted his charge not denigrate himself by returning to the stage after the house lights were on. Wade, frozen in indecision, capitulated. He took off his hat and returned to the bus. It was, under the circumstances, the wrong decision.

"I hate to say it, but I got a bad feeling about him," the owner said as the rented bus pulled out of Bronco's and returned with Wade to the road. "He's real nice. He can sure sing. But who's in charge of his career? Who is making the decisions?"

After a two-hour delay, they were glad to see her.

"You're bad, girl," someone in the back shouted. "You're bad."

After a two-year absence, she seemed scared to see them.

"Oh, Lord," she uttered. "Oh, Lord."

Back they came, in revival style.

"We love you, Wy. We love you!"

And then she replied, in her own kind of prayer.

"Oh, my God, I forgot how hard this can be."

With a strobe of blue light and shock of red hair, Wynonna Judd walked onto the stage of Nashville's Sunset Studio on a Saturday evening in early 1996, flashed an utterly horrified expression in front of three hundred and fifty special guests and eleven swooping television cameras, and abruptly terminated in spine-tingling fashion an eighteen-month isolation from her fans, her music, and the epic oddness of her life. At thirty-one, with an infant son, a newlywed husband, a famous mother, a starlet sister, a father who turned out not to be her own, and a child in her belly conceived out of wedlock and soon to be delivered—legitimately—in the midst of a two-year world tour, Wynonna Judd was arguably the most famous woman in America about whom it could be said that everything known about her comes from somebody else: her mother.

Naomi Judd, née Diana, and her elder daughter, Wynonna, née Christina Ciminella, burst onto the mostly male country music scene in the flaccid-sounding early 1980s and rode a wave of dulcimer melodies, sweet-tongued interviews, and cloying *Ladies' Home Journal* covers to seemingly overnight fairy-tale success. A high school dropout, occasional fashion model, and battered woman, Naomi Judd was the steely will behind the duo. She pounded the pavement, sweet-talked executives, and steered the stunning rise of her family from barefoot poverty to the glitzy heights of American celebrity. When at the height of their glory, Naomi suddenly came down with a mysterious liver disease—took time off, returned to the stage, got sick again, announced her retirement, then led her family on the Judds' interminable Farewell Tour—focus shined even more brightly on Mama Judd. That was followed by a best-selling book, a multicity book tour, a magical recovery, an NBC miniseries, another PR tour, and, as always, a drumbeat of cover stories documenting her every emotional high and low. Naomi Judd became a poster child for the Oprah Era, a Chinese menu of talk-show topics, ready and willing to be plumbed.

While Naomi was doing all the talking, meanwhile, Wynonna was doing all the singing. Though little noted at the time, the Judds' meteoric rise—six Grammy Awards, at least a dozen Country Music Association (CMA) Awards, and over 5 million albums—was based on the spunky vocals of Wynonna, softened by the harmonies of her mother. When Naomi decided to retire, Wynonna (that's "WHY-nona")—mimicking Madonna—dropped her surname and went out on her own. Her self-titled debut album, released in 1992, became the bestselling studio album by a female country artist in history. It was also universally acclaimed by critics as the work of one of the country's greatest singers—a startling combination of Dolly Parton and Aretha Franklin, a sort of hybrid country soul. Wynonna, pundits proclaimed, was finally realizing her potential as the "Female Elvis." A second album, released in 1993, was equally heralded, but sold less well. By then the tabloids began to catch up with Wynonna. Her mother's book portrayed Wynonna as a rotten brat. The man her mother claimed to be Wynonna's father admitted he really wasn't. And in a final, self-inflicted coup de shame—at least in country—Wynonna got pregnant out of wedlock and announced she was keeping the baby. Exhausted, she withdrew from public life in May 1994. This night, a year and a half later, was meant to be her carefully midwifed return to form.

"I want to do that song again," she said at the end of the first number, "just because I can." The audience laughed. A handful were friends and industry insiders; the rest were members of her fan club who had waited in line over thirteen hours to participate in this taping. The fan club members, naturally, were fawning, while the industry people sat on their hands and watched with skeptical curiosity, itching to pounce.

"Now, don't you worry," Wynonna said to the audience as she paraded off the stage. "I can do this. I know that I can do this." She stopped and brought her hands together as if in mock worship. "I'm a professional. I'm a Judd. I can do anything."

At the moment that hardly seemed true. In addition to being over two hours late (the result, she told me later, of a crippling panic attack), Wynonna's opening song had been out of key. Her body—dressed in satin trousers, a teal shirt, and a long golden Captain Hook coat—was rigid. Her face—porcelain smooth, geisha white—was visibly frightened. She looked like a ghost who had scared herself. The fans didn't mind. They would have stayed there half the night (and eventually, in fact, did; though most of the industry brass went home). But this debut was revealing a different side of Wynonna. Far from the reckless young woman her mother portrayed, she came across as a more fragile person, a deeply unsure performer. "Settle down. Settle down," she said to herself over the microphone. "I'm just so nervous I can't stand still."

On second try ("Now, everybody, you have to give the same kind of standing ovation or it won't match up on tape . . ."), Wynonna made it through the first song and into the next. Between each take, a team of hair curlers, blush applicators, and lipstick daubers would scurry onto the sprawling stage and prepare her for the next song.

"I don't know what's wrong with me tonight," she announced three bars into her third song. "When I do that at home, I'm just moving all over."

"Don't worry, we're just like family," a woman called from the audience.

"But you always want to do good for your family," Wy pleaded. To which a guy replied excitedly from the back, "In that case, when's the barbecue?"

The whole event was beginning to sound a lot like a twelve-step support group: Artists who Lost Their Mothers and the Fans who Support Them. This was country music at its mawkish best and the kind of story that Nashville has long told best. Patsy Cline, Dolly Parton, and Tammy Wynette have, at one time or another, all been the cover girl for troubled women in America. In the nineties, Wynonna inherited the legacy of those women, and how she chose to handle her afflictions—from a dysfunctional family to a stubborn weight problem—would determine her future as a viable artist. Moreover, as a very public woman,

Wynonna faced a challenge over the coming two years that Garth and Wade could only imagine: trying to develop her own identity while battling her dress size, her hair color, her parentage, and her pregnancies on the pages of the *National Enquirer*.

These pressures were written all over this night. If the key to Garth's psyche was his eyes, the key to Wynonna's was her body. It was a big body, but one that was surprisingly malleable, even expressive. Particularly when she sang, Wynonna would sway back and forth as if part of a huge church choir. You could see the tension escape from her arms as the music consumed her being. It was in these moments that Wynonna seemed to reach out and embrace everyone around her: I see you, I hear you, I love you. Her comments between songs only echoed this theme. "This one's for all of you out there who are totally in love . . ." "Now, listen to me, if it doesn't work out, this is what you do . . ." "Don't worry, you can survive. Just let me be your inspiration. I want to be your healer . . ." Wynonna was a modern version of the traveling preacher. Not a singing cowboy, but a singing talk-show host.

But as soon as the music stopped, Wynonna lost her composure. She became tense and unsure. Her body seemed to lose its shape. It was particularly noticeable when her mother entered the studio that night, dressed to perfection in a plum red suit and black pillbox hat. Wynonna bristled. Many in the audience, sensing the transformation, gasped. "Good God, who could compete with that?" Naomi was clearly the main ghost haunting Wynonna and the one demon she had yet to surmount. And for those around Wynonna, indeed for anyone with even a modicum of sympathy for the difficulty of contending with a perfect mother, it was in these moments that they felt the need to reach up onstage and embrace *her*: We see you, we hear you, we love *you*. And, most of all, we need you to keep singing.

"You don't know how much this means to me that you're here," Wynonna, sensing the embrace, said after several more songs. "This is a big fat hairy deal to me. Things are starting to feel better now."

But no sooner did she reach the comfort of that moment than the unsteadiness of her situation returned. After finishing several songs from her new album, *revelations*, Wynonna sashayed to the center of the stage and introduced her special guest. This was meant to be the climactic moment of the evening. "Now I want to introduce you to a *personal* friend of mine," she said with her familiar sass. "I met her last spring. I had the opportunity to work with her. I flew out to Hollyweird and together we sang. Since then I have been really blessed to know her as a friend. But you all know her as the wacky, the wonderful Bette Midler!"

The entire audience quickly rose to its feet in the most spontaneous applause of the evening. The I LUV JUDD license plates and WY'S MY GIRL signs started shaking deliriously. Even the industry snobs who knew it was coming looked around in admiration that Wynonna could rake in such pure star appeal. "Man, she looks *fabulous* tonight," one normally cynical writer cooed. Nashville didn't need to go Hollywood; Hollywood was coming to middle Tennessee. Bette Midler strode up the three stairs and paused at the lip of the stage long enough to let the full impact of her shimmering lemon pantsuit and picture-perfect blonde curls linger in the air. Then she launched into the opening verse of her signature song, "The Rose." Arriving at the chorus, she began to sashay herself in an elegant Broadway strut right toward the center of the stage, where Sister Wy was supposed to join her in a show-stopping duet.

When Wynonna opened her mouth, though, the voice that came was not the powerful vibrato she had been using for much of the previous hour. There was none of the lustiness of Aretha or the purity of Dolly. There wasn't even a hint of Elvis. Instead there was only a shadow of a former self, a whimper, a painful reminder of Wynonna's browbeaten past. Daughter Wy, not Mama Wy. As if to fill this breach, Bette immediately started talking when the song was finished. Unlike Wynonna, she was confident, brash, and eminently in control. "Wynonna's supposed to chat with me," she said. "Go ahead, Wynonna. Chat. I dare you."

"I dare *you*," Wynonna repeated meekly.

"The sign does say 'Chat,' " Bette continued. "So what should we chat about? How about children? I hear you're going to have another. My advice is breast-feed as long as possible." The pace of this so-called chat was almost overwhelming. Bette didn't stop talking. She didn't wait for answers. She just kept going. Line after line. Punch after punch. "Well, what are we supposed to do now . . . ? I can't remember . . . I've been in that trailer so fucking long I've forgot what I'm doing . . . And don't go tell the papers I swore in front of you . . ." Eventually she proposed that Wynonna play the ukulele. Wy didn't want to. "Come on," Bette said. "I'll teach you how to do the hula."

"No," Wy said meekly, "I've been down this road before."

But Bette persisted, just like Mother, and Wy relented, just like in the past. Bette started singing "Ukulele Ladies," surely one of the hokiest songs ever written, then started doing the hula. Eventually the two women were kicking up their heels right next to each other in the middle of a soundstage, on the west side of Nashville, in front of television cameras that would beam this cornpone scene, as if straight from "Hee Haw," to sweeps-month televisions around the country.

Until she blew.

Four steps into the dance, Wynonna had finally had enough. "I'm not going to do it anymore," she said. "I had to do this stuff for ten years with my mother and I'm not going to do it again."

"Why don't I sing my solo now?" Bette suggested.

"You're the boss," Wynonna replied and headed off the stage, more in regret than disgust. "This was supposed to be my show," she said. "I had so many things I wanted to do. I had so much I was going to say."

VERSE I

ONE

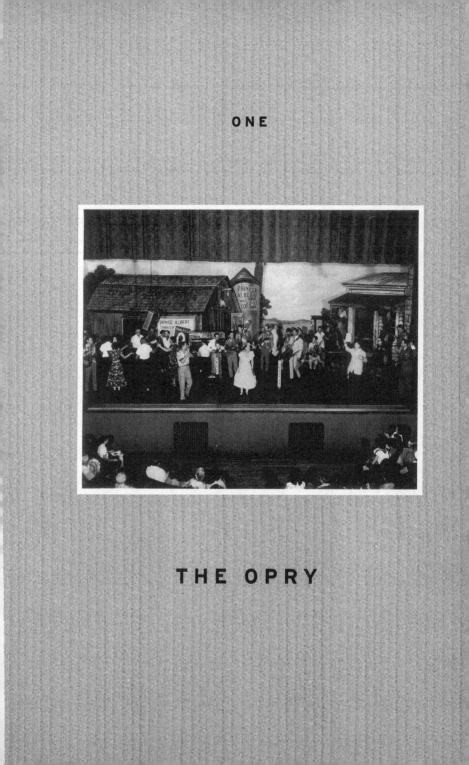

THE OPRY

The sheets of rain were falling so hard and the rush of headlights was so expectant that it was easy to miss the cloverleaf exit that banks hard to the east off Briley Parkway and deposits the driver right into the heart of what the neon hails as MUSIC VALLEY, U.S.A. Turn here for Shoney's. There for Cracker Barrel. You don't even have to turn at all to veer into the WORLD-FAMOUS NASHVILLE PALACE, which appears to be a Denny's with an overly ambitious Vegas Strip sign attached. All these establishments, with their blinkety beacons and boppity billboards promising extra-fluffy biscuits and the BEST CATFISH ANYWHERE!, are to the east off McGavock Pike, which itself is ten miles east of downtown Music City, not far from the blue-collar outposts of town, and just up the road from the real countryside. In short, in the middle of nowhere. A fine place to recreate a small corner of Everywhere, circa Anytime At All.

WELCOME TO OPRYLAND, U.S.A.
TURN RIGHT FOR THE MAGNOLIA LOBBY.
LEFT FOR OLD HICKORY LOUNGE.

Unlike the ones across the street, these signs are written in soothing script. They are painted in white on handsome red boards that remind you of fairy-tale wholesomeness. And they unfold before the driver's eyes on a faux country tableau of fluffy green grass, a white picket fence, and dozens of sculpted storybook trees wrapped in thousands of twinkling lights. On this night, uncommonly cold and rainy, the normally idyllic entrance to the Opryland Hotel, with its churchlike steeples, plantationlike columns, and county courthouselike red brick grandeur, has been overtaken by an even more idyllic constellation of candy cane light poles and plush fir wreaths in a holiday vision out of Currier and Ives. Even the guardhouse has been rebaked as a two-story peach-colored gingerbread house complete with giant plastic gumdrops and bouncing marshmallow ladies who wave at every car that makes it through the valley and into the parking lot of Gaylord's fantasy: WELCOME TO A COUNTRY CHRISTMAS. ALL SELF PARKING $4.00

Gaylord, in this instance, is Edward Gaylord, the Walt Disney of Southern culture and the aging, legendarily frugal proprietor of Gaylord Entertainment Co., which controls the dominant institutions in country music, the Opryland Hotel, Opryland Theme Park, CMT, TNN, and the gemstone in this hillbilly tiara, the Grand Ole Opry itself. All of these he has gathered in a 406-acre entertainment complex that draws visitors to an undistinguished plot of land not far from the airport with the paradoxical promise of unimagined wonder and old-fashioned small-town wholesomeness. Ironically, the hotel alone is larger than many small towns. With 2,870 rooms—the seventh-largest in the country and the largest outside Las Vegas—the hotel is an air-conditioned biosphere-cum-theme park complete with giant murals, dancing fountains, and a 6-acre, glass-covered interiorscape with over ten thousand tropical plants. The newest extension, known as "The Delta," has a fifteen-story-high glass dome covering 4.5 acres and featuring a 110-foot-wide waterfall and a quarter-mile-long "river." Move over, Mickey Mouse. Tom Sawyer is the new American icon here, and he has gone uptown.

Driving around the facility (24 MPH SPEED LIMIT), past the Conservatory, alongside the Cascades, under several neo-Georgian overhangs, I finally arrive ten minutes later at the backdoor entrance to a modest, unornate, redbrick theater that since 1974 has been home to America's most beloved radio show, the Holy See of Country Music, the Grand Ole Opry.

"Hi. May I have your name please, sir . . . ?" The officer retrieves a portable stop sign that has blown over in the wind.

"And who are you here with . . . ?" He starts flipping through his clipboard.

"Ah, well, Mr. Brooks is expecting you. He's not here yet. Just drive up to the canopy and park wherever you can. The show starts in about an hour."

The Grand Ole Opry is as old as country music itself and is one of the few institutions in American life to survive the transition from radio to television to global satellite intimacy. Begun in November 1925 as one of a burgeoning breed of down-home variety shows popping up on newly formed commercial radio stations, the "WSM Barn Dance" was simply a marketing tool for the National Life and Accident Insurance Company to sell products to rural listeners. Broadcast from the studios of clear-channel WSM-AM (an acronym for the company's slogan, "We Shield Millions"), the show mixed banjos, fiddles, harmonicas, and string bands, along with homespun advertisements and a flamboyant host, George D. Hay, the "Solemn Old Judge." It was Hay, in 1927, who gave the show its indelible name when he followed a network broadcast of opera classics by saying, "You've been up in the clouds with Grand Opera, now get down to earth with us in a performance of Grand *Ole* Opry." It's never missed a weekend since, moving to the legendary Ryman Auditorium in downtown Nashville in 1943, and then decamping to its current suburban roost in 1974.

Since moving to the suburbs, though, the Opry has lost much of its clout. Garth Brooks, like many new stars, may identify his induction into the elite cast of seventy-five performing Opry "members" as the highlight of his career (a position that is wise marketing, if nothing else), but his career no longer depends on the Opry. Simply put, he has outgrown it. As a result, it was something of a surprise that Garth chose the Opry this night—just a month after the much-heralded release of his new album, *Fresh Horses* and still a month before his gaff at the American Music Awards—to break his year-long hiatus from public view. Under the circumstances, it was clearly designed to send a message. Having flirted with rock 'n' roll for years, Garth had long been viewed with suspicion among some traditional country fans. "MORE POSSUM [George Jones]," one station in St. Louis had announced on a billboard, "LESS GARTH." Those fears had only intensified in recent weeks after he released his new single, a manic version of Aerosmith's heavy metal song "The Fever" that tanked on the radio after only a month and became his lowest-charting single. Suddenly, it seemed, the fans were sending him a message: Don't get too high above your raisin'.

"I just don't understand it," he told me. "I thought radio would jump all over that song. That's why I released it. Nobody else wanted to. Not the label. Not anybody. Maybe I should start listening to the people around me." In particular the episode added to his growing fears that the public was tiring of him. Maybe he *was* losing his Midas touch. Maybe his forthcoming tour, slated for the spring, would also flop. As odd as this anxiety seemed for a man who had so completely rewritten the history of popular music, it was fast becoming the defining question Garth Brooks carried around in his head: "What are you gonna do? *What are you gonna do?* WHAT ARE YOU GONNA DO . . . ?"

If they don't love you anymore.

Once inside the building, I gave my name to a woman sitting underneath a blue and green acrylic painting of Minnie Pearl, Hank Snow, and Porter Wagoner ("The Grand Ole Opry is as simple as sunshine," the painting quoted George Hay as saying), and she gestured me backstage. Instantly I was lost. The manicured opulence and carefully ordered charm of the hotel grounds gave way at once to casual pandemonium and random misdirection. People of all size, shape, and accent were milling about the halls with what appeared to be nothing better to do than gossip with their neighbors and nibble on deviled eggs. Three men with cowboy boots held fiddles above their toupees as they edged through the crowd. A woman with balloons of blonde hair and Wal-Mart crystal heels wiggled past some ogling teenagers. A group of high school tap dancers wearing lime green gingham skirts went clickety-clacketing toward the ladies' room. Were they performers, background singers, or waitresses? The whole atmosphere reminded me of a small-town water-melon festival, except it was being held not outdoors in front of the hotel in what would be an ideal setting for such an event, but inside a cramped, fluorescent maze of hallways with linoleum floors and orange ventilated metal lockers that looked as if they came right out of the backdrop of "Harper Valley P.T.A." The Mother Church of Country Music may have had stained glass and maple in its original downtown cathedral, but its current suburban cathedral-lite was decorated with vinyl burlap on the walls backstage and peanut-butter-cracker orange carpeting on the three stories of mock church pews out front.

No sooner had I paused to orient myself than the bustling crowd suddenly parted like two halves of a backlot sea, and a navy blue raincoat and black cowboy hat came shuffling through what appeared to be a cheerleading funnel. At first I mistook the body for a prop, but then I realized that the person was very much alive, shaking hands, and, even though I could not see his face, drawing considerable respect from the crowd. He pivoted back and forth with each step forward, shook the hand of each admirer, and mumbled something indistinguishable under his brim. I watched the scene with great admiration—(A secret Opry ritual perhaps?) until I realized that this tradition was heading rapidly in my direction and that I would have to do my part to keep the whole spectacle moving. The shaking went on—back and forth, left then right—and I surveyed the movements with nervous anticipation. Surely there was no way I would recognize this person. The Opry has dozens of elder luminaries, most of whose names I didn't know, and most of whose faces I couldn't recognize. Just the other day I had sat in a crowded room with the greatest songwriter in Nashville history—a retiring man with no pretense at all—and didn't know it until he had left.

Still, like a test, the moment arrived. The shuffling body advanced through the line and finally reached where I was standing. All eyes were following his every step. People around me were snapping pictures. I stepped back to let him pass. Instead of moving on, though, the daunting vintage hat slowly lifted to reveal a set of shocking white sideburns, a gentle nose, and the most benign, almost luminous face that I had ever seen. It was the face of peace, of total consolation, of a man who had lived a life of such exalted dreams that he floated above the worldliness around him. He knew who he was. He knew where he came from. And he inhabited his status as lightly, and as comfortably, as any luminary could.

"Good night," he said, raising his hand first.

"Good night," I repeated. We shook hands.

Though I had been coming to Nashville for several years, though I had met plenty of superstar artists before, it was not until that moment—on that dreary night in winter—that I truly understood the almost spiritual effect that country music artists can have on their fans. I had been at the Grand Ole Opry for five minutes. I had just met the "Father of Bluegrass," Mr. Bill Monroe.

By the time Mr. Garth Brooks, the "Father of New Country," arrived backstage, the broadcast was already under way. Seventy years after it debuted, the Grand Ole Opry is still, at its essence, a radio show, performed live before up to 4,400 fans and broadcast simultaneously over dozens of AM stations to a listenership estimated at 3.3 million people. There are two shows on Saturday nights and, depending on the season, one or two on Fridays as well. (On this Friday night, there was only one.) Each show lasts three to four hours and is divided into half-hour segments. As in the old days, each segment has a sponsor, whose advertisements are performed live from the Opry stage. The props these advertisers use all litter the offstage areas and look, quite frankly, like ramshackle props from an elementary school production of "Goldilocks and the Three Bears." To the right was a yellow ranch bell on wheels for Cracker Barrel restaurants and to the left a gray signboard several feet tall painted to look like a bag of Goo Goo Clusters, the signature candy item of the Grand Ole Opry, a cloyingly sweet yet irresistibly chewy hamburger-sized concoction of marshmallow, milk chocolate, and peanuts. (Though the name of this candy mimics the acronymous nickname of the Grand Ole Opry, GOO, the Standard Candy Company insists that it's only a coincidence.) To a newcomer, what was most remarkable about these props is that most of them represented companies I thought no longer existed: Dollar General Store, Martha White Flour, Goody's Headache Powder. One segment was even sponsored by a product called Jogging in a Jug, a mystery elixir that reminded me of stories I had read about miracle unguents and shady cure-all balms that advertised on early Opry broadcasts. The only difference now is that every advertisement for Jogging in a Jug began *and* ended with a grave caveat: "There is no scientific evidence to support medical claims of Jogging in a Jug."

Garth Brooks, making his first appearance at the Opry all year, was scheduled to be in the fifth segment, sponsored by Dollar General Store. Like other members of the Opry "family," he is technically required to perform at least twelve times a year (it had been as high as twenty-six times a year), though he, like other newer stars, rarely meets that obligation. As a result, most weekly shows at the Opry seem like the Seniors Tour in Country Music, with long-forgotten stars singing once-begotten hits for mostly octogenarian audiences. (Bill Monroe, at eighty-four the oldest member of the Opry, was inducted in 1939.) Garth (1990) was scheduled to appear during his segment along with Jean Shepherd (1955), the fiery traditionalist from Oklahoma, and Hank Snow (1950), the legendary Canadian crooner known for his flashy suits and squirrelish toupees. Appearing in the same segment, however, does not require coordination. In fact, by the time Garth strolled into the backstage area a little after nine o'clock, he had no idea of the schedule, nor what his band would play. If anything, he was more interested in boasting about the Christmas present he had just received from his wife: a brand-new, six-wheel 4x4 Chevy truck. Its color, he bragged, was autumn wood.

"Did you see that baby?" Garth cooed to the band, several members of his staff, and the manager of the Opry, all of whom had gathered in a private office to discuss Garth's set for the night. "It's a dualie, with lights on top and this great stereo. Man! I slipped in a CD of George Strait, and I was happy as could be."

Garth was wearing blue Wranglers and one of his freshly pressed navy blue and cardinal Mo Betta cowboy shirts that he helps design. His Stetson was tan tonight.

"Why not wait until Christmas?" someone asked.

"We couldn't hide it," Sandy said. "My mother drove it over this afternoon."

"It has that extended cab," Garth continued. "Real nice seats."

"With three kids we're going to need the extra room," Sandy said as she sat down on the sofa. Two days earlier, Garth had hinted on "CBS This Morning" that Sandy might be pregnant with their third child.

"Three kids!" Garth exclaimed. "What are you talking about?" He walked over and started massaging her neck. "I want seven."

Everybody laughed . . . except Sandy, who dropped her head.

"I've already been pregnant every year this decade," she said. "I don't want to go into next decade, too."

"Mr. Brooks . . ." The call came from just outside the door. "Mr. Brooks, you're on in five minutes."

As soon as the call came, Garth snapped into work mode. The members of the band started hurrying to the stage as Garth hooked his guitar across his shoulder. Sandy got up and started fidgeting with the collar of her husband's shirt. She looked nervous, as did he. He started licking his lips. She gave him a quick kiss and headed through the door. "I don't know what to do," he said, partly to himself, partly to the room, which was empty now except for me. "It's traditional at the Opry, if you've been away for a while, to play your next single and the one that got you here in the first place."

"So what is your next single?" I asked.

"I don't know," he said. "After what happened with the 'The Fever' . . ." He shook his head. "You spend a lot of time trying to surround yourself with people you trust, but I'm not sure if I can trust *anybody* these days. I'm still trying to figure that out." He picked up a bottle of Evian and walked through the door.

"Ladies and gentlemen . . ." Porter Wagoner stepped to the center of the Opry stage and grabbed a portable microphone. He was wearing his fabled magenta rhinestone jacket with the flashy inner lining that says HI! The audience usually loves it when he flashes the lining, but at that moment the crowd couldn't care less. The fans had already begun to squeal. "He's the biggest star in country music today and one of the most decent men I've ever met. Please welcome my friend Mr. Garth Brooks . . ."

The stage of America's longest-running radio program is unlike any stage in the world. Its slick floorboards are hardly extraordinary. Its sound system is not unusual in the least. Along its base sit sixty-three 40-watt lightbulbs and at the rear stands a sixty-by-thirty-foot painting of a bright red barn. But what makes the atmosphere of the stage so electric—so unexpectedly exhilarating—is its casual informality. When Garth walked onto the stage for his first time in a year, the audience burst to its feet. Old men. Young women. Mothers and grandmothers all spontaneously left their seats, hobbled, hopped, or skidded toward the stage, took photographs, waved, offered up little envelopes and gifts, and generally acted as if the reigning master of American music was coming directly into their living room. It wasn't the applause that was a surprise or even the cameras flashing. (Though how many theaters still allow that?) It was the complete lack of pretense or even showmanship. Garth took several steps toward center stage, removed his hat, and bowed to the audience. He collected up several of the gifts, waved to the balcony, then slowly made his way toward the microphone.

Meanwhile behind him another and, in many ways, more extraordinary aspect of the Grand Ole Opry was being played out. When Garth walked onstage from the sizable offstage waiting area, about a dozen of the people who had been waiting with him—including the manager of the Opry, Sandy, and even me, pulled along unknowingly—walked onto the stage with him. The manager just stopped behind the drummer. Sandy, who effortlessly transformed herself from sassy equal to dutiful wife, sat on a stool behind the backup singers. And I, feeling awkwardly out of place, hurried to sit down on one of the four old pews from the Ryman that had been placed for hangers-on in front of the red barn backdrop. An elderly couple welcomed me with a nod as we all watched Garth make his way to the honorary six-foot circle of maple that had been transferred from the Ryman and embedded in the floor. It was as if we were settling in for an afternoon hayride to no particular destination. Finally, after what probably wasn't very long but seemed like forever, Garth arrived at the microphone with the call letters WSM on the placard around its stand, hunched his shoulders in that earnest way that only serves to heighten his pent-up energy, leaned forward on his Justin toes, and began to sing his own sermon for the night, the tale of a woman whose husband dies on a rodeo bull and who then kills herself on the "Beaches of Cheyenne."

Garth wasn't halfway through the song, though, before the members of the audience slowly built up the courage to resume their own intimate exchange with their leader. Some did listen to the song. Others closed their eyes and nodded along. But a surprising number just continued calling out greetings, snapping pictures, and generally ignoring this much-ballyhooed "only Nashville appearance of the year" by the "bestselling artist in American history." The truth is, they didn't appear to want to hear him sing at all. They certainly didn't want to listen to his band. They wanted to talk, to tell him things, to chat, and to call out declarations of love that were deeply meaningful to them at that moment. "I love you, Garth!" "Nice shoes, buddy!" "Do you like my coat?" "Say hi to Taylor." All in the middle of one of the most desperate songs he has ever written.

Finally the song ended and the declarations could be unleashed even more. Garth, naturally, soaked in all this affection like a sponge. He waved to the audience, giggled at some of the questions, and picked up a few more of the stuffed animals (for some reason, they all seemed to be green). Then, in an extraordinary exchange that long afterward came to epitomize, for me, the odd intimacy that country music superstars share with their fans, a man halfway up the orchestra section in the most expensive ($17) seats, lifted his voice above the crowd.

"My wife loves you, Garth!" he called, and the people around him giggled. Garth, who was just preparing to launch into his second song, stopped, smiled in the man's direction, and called out, "What's your wife's name?"

"Karen!" the man replied with gusto.

"I love you, too, Karen," Garth said, and you could hear the women swoon. A moment passed, then Garth continued. "And by the way, what's your name, sir?"

There was a pause. "Brian," he said. Now you could hear laughter; he had the men as well.

"Well, I love you too, Brian."

He nodded for the drummer to start.

Though it seems anachronistic today, it would be hard to overstate the importance of the Grand Ole Opry to the identity of the American South. While not the first Saturday night barn dance, nor initially the most prominent (that honor belonged to WLS in Chicago), by the late

1930s the signal of the Grand Ole Opry reached over thirty states, helping it to become the defining cultural program for millions of rural
Americans, many of whom walked miles to their neighbors' houses to
hear their favorite show or rigged radios up to truck batteries. Those
performers fortunate enough to be invited to join the broadcast—Roy
Acuff, Ernest Tubb, Minnie Pearl, Bill Monroe—saw their careers rise
with the success of the show. By the mid-1940s, Roy Acuff had become
so well known as a result of his appearances on the Opry (as well as a
few movies that spun out of the show) that Japanese soldiers trying to
psych out American soldiers fighting in the Pacific shouted, "To hell with
Franklin Roosevelt! To hell with Babe Ruth! To hell with Roy Acuff!"

Central to the success of the Opry, and to the music it helped promote, was an overwhelming sense of place. Especially after the show
moved to the Ryman Auditorium in 1943, with its upright church pews
in front, cramped dressing areas in back (which spawned the habit of
wandering onstage), and internationally acclaimed lack of air-condition-
ing, the Opry oozed humid huckleberry tradition through every staticky
receiver it reached. "You listen to the Opry, and pretty soon you have
a place in mind," wrote Garrison Keillor in a famous *New Yorker* profile,
"a stage where Uncle Dave sang and told jokes and swung the banjo,
where the great Acuff wept and sang 'The Great Speckled Bird,' where
Hank Williams made his Opry debut with 'Lovesick Blues' . . . where
Cousin Minnie Pearl calls out, 'Howdee! I'm just so proud to be here!'
And eventually, you have to go and be there, too."

Even as late as the early 1950s, when Paul Hemphill, a Birmingham
teenager and later a Nashville chronicler, made his first trip to the Opry,
it was a moment of transcendence for the lifelong baseball and country
music fan. "Mama and my sister made up fried chicken, potato salad,
and beans on the night before," he wrote, "and all during the five-hour
ride to Nashville the next day in our new Dodge Coronet sedan—over
twisting asphalt back roads my father knew like the back of his hand,
Daddy keeping the car radio tuned to country stations, so as to heighten
the anticipation—I tingled. 'Wonder who'll be there. Hank? Naw, they
had to fire him for something. Little Jimmy Dickens really that little?
Wish I had an autograph book.'" And when he finally arrived on Fifth
and Broadway: "That imposing shrine of fierce redbrick that is the
Ryman Auditorium might as well have been Yankee Stadium."

But within months of Hemphill's visit to Nashville, the importance
of the Opry started to erode. Elvis Presley, told to return to truck driving
after his only Opry appearance in 1954, landed the first blow with his

upstart rock 'n' roll revolution. The rise of television (so key to Elvis) further undermined the centrality of radio, not to mention the front-porch tradition it represented in the South. "Within a year after that visit to the Opry," Hemphill recalls, "my father was buying our first television set. That black-and-white nineteen-inch screen brought prom-ises of another world into our living room in Birmingham—California beaches, New York skyscrapers, Yankee Stadium, the Rockies (much more impressive than our Appalachians), the World Series (So why go watch the Birmingham Barons play?)—and for the next twenty years, I would try to cover up my roots and engage in a quixotic search for those promises."

A decade and a half later, when the Opry finally succumbed to reality and sought relief in the suburbs, die-hard fans proclaimed the end of an institution. "As I awoke, slowly, on Saturday morning," Garrison Keil-lor wrote of the day the new facility opened, "it dawned on me that I didn't know where Opryland was, or how to get there. By midafternoon, I knew that Opryland was eight miles from the hotel by freeway and six by river, either way as practical as the other. There was no bus to Opryland; a rented car could be returned on a Sunday only at the airport, and I was leaving on the Sunday train; the route by road was not direct, one would have to hitch two or three rides to get there; and my boots weren't made for walking."

The Opry, and with it an important part of the Old South, seemed to be waning. And though many lamented this passing, others—myself included—did not. As a child growing up in the Television Era South, the Opry, and the country music it represented, embodied a South I didn't like. Far from anchoring me to a timeless past, to me it symbolized the tyranny of place, a set of shackles that beautifully rendered a bygone region. Like Garth, I was a rock 'n' roll or at least a Top 40 kid, waiting up on New Year's Eve to see which sentimental seventies hit would be number one for the year. I can still remember the seminal day I received a copy of Elton John's "Philadelphia Freedom" for our junkyard jukebox. It was my older brother's favorite and thus mine as well. As a fifth-generation Georgian with parents educated outside the South, living in a neighborhood with carefully cut lawns and driving to school in a station wagon, I found Billy Joel more relevant than Lefty Frizzell. Hank Wil-liams was not my guiding light; Patsy Cline was hardly my mother's milk. If anything, like Paul Hemphill, I couldn't wait to escape the South after high school and seek my destiny in the larger world.

It didn't work. In fact, no sooner had I left the South than I realized

I was bringing more of it with me than I was prepared to admit. It wasn't that I had a secret hunkerin' to eat pork rinds and drink Jack Daniel's. It wasn't even that I had a deep-seated desire to reembrace my childhood days of honeysuckle wine and BB guns. That wasn't my desire because that wasn't my South and because that South doesn't particularly exist anymore anyway. My desire was for something more complex and more contemporary. I was holding on to the thinnest of narrative threads in my life. The part of me that wants, despite the headlong advance of cynicism and despair, to hold on to a storyline that seems somehow bedrock. It's that part of me that brought me home to the South, and, I believe, it's that part of the country that has also brought it home to the South at this time in our history.

It was almost a decade after leaving home that I first started listening to contemporary country music. At first it was just curiosity calling. Like millions of Americans, I couldn't help noticing that a new kind of sound had started coming from the radio—a little louder, a little faster, with familiar drums, less twang, and, for the first time, the hint of an edge. Some friends dragged me to see Mary Chapin Carpenter perform in Washington, D.C. I saw Garth Brooks swinging from a rope on television and he reminded me of a rock star. The Judds seemed to be everywhere, and at least one of them (Was it the mother or the daughter?) sure could sing. The emergence of this new breed of country music was interesting to me for another reason: It seemed to lack the dusty old icons that made the Opry so offensive to Southerners like me, especially those of us who worked on computers, watched "Saturday Night Live," and tried our hands at using chopsticks. "I try to think about Elvis / Memphis / Oprah in the afternoon," Patty Loveless sang in one song, ". . . Sushi bars and saxophones." Say what you will about old-fashioned country music, but it's hard to imagine Hank Williams, Sr., watching Ricki Lake and nibbling on pickled ginger. The Land of Cotton was at last forgotten; the Confederacy would not rise again.

I, for one, was delighted. I love the South. I love the attachment to home, the sticky weather in summer, the storytelling, the slightly ribald humor. But I also love sushi bars, saxophones (I tried to play one once, but wasn't very good), as well as, on occasion, Oprah in the afternoon. To me the best of contemporary country music speaks, at its heart, to both these elements. With its playful wit and unflagging directness ("If you have to explain a country song," an executive in Nashville once told me, "it's not a good song . . ."), the strongest music coming out of Nashville these days captures the new hybrid that has been emerging across America out of the combination of rural and urban motifs—the mix of apple pie and neon, if you will; the blending of dirty fingernails with designer shoes. More to the point, I loved that this ideal was being created not in Hollywood or Greenwich Village, but in a relatively isolated creative community in a six-square-block area of middle Tennessee, which only a few years earlier had still been offering up "Hee Haw" to the world.

Once I started exploring country music, I found that I was not the only one who had been attracted by its changing face. By 1993, 42 percent of Americans were listening to country radio every week, twice the number of a decade earlier. According to the Country Music Association, these listeners included a third of all Americans who owned a home, 40 percent of single adults between twenty-five and thirty-nine, and half of all Americans who owned a snowmobile. With 2,642 radio stations programming Nashville's latest, country had become the dominant radio format in the United States, reaching 20 million more people a week than its closest competitor: adult contemporary. Moreover, these new fans were spread out in all regions. In 1994, country radio was the top-rated format in 55 of the nation's largest 100 cities, including Baltimore, Buffalo, Milwaukee, Seattle, San Diego, and Washington, D.C. A million people a week were listening to country radio stations in New York, Chicago, and Los Angeles.

Country fans, meanwhile, were more educated than either adult contemporary or rock audiences. According to the *Simmons Study of Media and Markets*, 36 percent of country music fans had postgraduate degrees, as compared to 30 percent for adult contemporary and only 22 percent for rock. They were also wealthier. Forty percent of individuals with annual incomes over $40,000 listened to country music, as did a third of individuals who earn over $100,000 a year. This money had completely changed the record business. In 1995, country album sales, which accounted for 18 percent of the market, surpassed $2 billion a year, three times their total in 1990 and twenty times what they were in 1970. In 1985, ten new country albums went gold—sales of half a million units—while seven went platinum, meaning they topped 1 million. In 1995, fifteen albums went gold, twenty-six went platinum, five reached 2 million, four reached 3 million, one passed 7 million, and one, Garth Brooks's *No Fences*, topped 10 million units.

America, as Bob McDill wrote, had "Gone Country." And though McDill's song, which Alan Jackson took to number one in early 1995, actually poked fun at country's trendiness, the trend was still real. In less than a decade, the once ridiculed world of Nashville had become an unlikely focal point of American pop culture, a new patron city for the American dream. What had been a quaint, listless corner of the entertainment industry overnight exploded into *the* mainstream American music. Moreover, just as rock 'n' roll foreshadowed many of the changes in gender and race relations that followed in the sixties, country music in the nineties—with its themes of family and renewal—became the clearest reflection of many of the conservative ideals that were just beginning to surface in American life. In short, country music, once the voice of a distinct minority in America—working-class Southerners—had become the voice of the new American majority: middle-class suburbanites. No longer about one particular place, country music had become about *every* place.

Eventually I decided that the best way to understand this community would be to move to Nashville and try to become as much a part of it as I could. As a Southerner and, by that point, a frequent visitor to Nashville, I knew enough not to be surprised by the absence of hay bales and corncob pipes. I also knew enough not to be shocked when I met more people from North Hollywood and the West Village than from Southern Alabama. But what did surprise me was the magical and welcoming way in which the city opened up to me. Before coming to Nashville, I thought that the stars and the industry would remain steadfastly aloof to an avowed outsider. What I found was the opposite. Once people realized that I did not have an agenda, that I was not some disgruntled rock critic trying to infiltrate the community and serve up another hatchet job about the industry and its fans ("You be the judge," *Rolling Stone* wrote several years ago. "Garth Brooks is at the top of the pop music charts. At the same time, the nation's verbal SAT scores plummeted to an all-time low. Can anyone save our children?"), the town embraced me. Sure, they wondered what I would write. Certainly they feared an exposé. But still they talked with me, openly and honestly. They invited me backstage. They invited me to travel with them on the road—in the bus or on the jet.

In the course of making those travels, I tried to answer the question of what the changing nature of country music said about the changing nature of America: our values, our culture, our shifting sense of place. Nashville these days is exhilarating. It's certainly rapidly changing, but it's also, I believe, vitally important. It is impossible to understand the maturing, tradition-craving, socially conservative America of the moment without understanding the chief expression of many of those views: country music. And, as I discovered, the best way to understand that music is to understand the people—colorful, intense, at times calculating, even gothic—who have made the music the centerpiece of their lives.

After Garth's standard two-song set, the audience refused to be quieted. They bathed the stage again with applause, strobing flashes, and a veritable menagerie of stuffed animals (presumably for his children). Porter Wagoner appeared back onstage and asked the crowd if they would mind if Garth sang another song. Reluctantly, they agreed.

"Before I do that," Garth said to the crowd—he was warming to the role now; less consumed with his own insecurities—"I would like to introduce you to someone. She's my best friend and my wife, and I don't know what I'd do without her . . ."

Sandy stepped forward from her perch at the back of the stage and waved politely to the crowd. She was wearing a blue dress with lace around the collar and her shoulder-length blonde hair was draped around her face. She looked like a small-town Sunday school teacher.

"Somebody asked me recently what I wanted for Christmas," Garth continued. "I said I think I've already got it. Sandy's not been feeling too well in the mornings . . ." The crowd erupted in cheers. "So, if you don't mind, I'd like to dedicate this next song, 'Little Drummer Boy,' to Sandy, my two girls, and the third one on the way . . ."

As he reached his final "ra ba ba bum," Garth stepped to the base of the stage one last time, collected the animals, the flowers, and the chocolates, and headed back to his makeshift dressing room. He was sweating profusely when he arrived. With his hat removed, his hair clung in matted strings against his scalp, and his eyes still had a distant, aghast glow of someone stunned by a near-religious encounter. He grabbed a bottle of Evian and started gulping it down.

"Wow! That felt great," he said and, for the first time in several months, seemed genuinely uplifted. In particular his decision to announce Sandy's pregnancy had been a masterful stroke. "I thought this was as good a place as any to confirm it," he said. Indeed, the Opry seemed made for such announcements, which is why, I now realized, he was there to begin with. On my first trip to Nashville, long before moving, I had purposely planned to leave town on Sunday so I could visit the Opry. Its existence was part of my basic understanding of life in the South: football on Friday night, Opry on Saturday night, church on Sunday mornings. What I soon discovered, though, was that nobody on Music Row cared that much about what was happening in front of the red barn backdrop out by the airport. *That* was about nostalgia, certainly not business. It certainly wasn't the least bit important in determining what artists to sign, what music to cut, or what songs to release to country radio. Radio is still king in country, but it's FM radio now. And since most FM country stations refuse to broadcast the Opry (its listeners are too old, its ratings too small), the Opry is out of the loop.

Though shocking to me at the time, I changed my ticket to Friday afternoon and never even went to see the Opry. Now, visiting the place for the first time, I was beginning to realize how telling my decision actually was. One of the central challenges facing country music is how to exploit the unprecedented opportunities of the present while still maintaining a link with the past. If anything, Garth Brooks's current problems—becoming so obsessed with his own empire that he seemed to be forgetting his roots—were a reflection of this problem. Sure, Garth Brooks didn't need the Opry: His forthcoming three-year world tour would earn him close to $1 million a week. By contrast, his stint on the Opry that night would earn him a mere $221. But his appearance on that stage still bought him a badge of legitimacy that he, more than most people in Nashville, needed. The Old South may be dead—the Opry itself may be little more than a living museum—but country music fans, more than those of most other genres, still hold on to the one remnant of the past that remains foremost to them: its old-fashioned values.

Within minutes, Garth was ready for the reunion part of the evening. After handing his gifts to his road manager, meeting a terminally ill teenager who had been flown in from Milwaukee, and kissing his wife good-bye, Garth stepped outside the office. What followed was a stunning display of celebrity endurance, not to mention public relations. Fans, neighbors, elders, youngsters, Opry staff, and other performers from the show all gathered in an informal throng outside the door to the manager's office backstage, just in front of the men's room. Though the show was still continuing onstage, few here seemed to care about that. No one shoved. No one screamed. And Garth stood there on the linoleum floor and began to sign autographs. If anything, he was defying orders by being there. A special notice was posted around the facility urging guests not to pester performers for autographs or pictures. Garth not only ignored the order, but also asked the manager if it had been placed there on his behalf. (It was for security concerns, the manager explained. Earlier that year, two runaway girls from Ohio, ages sixteen and twenty-two, had initiated an ill-fated scheme to meet Reba McEntire by taking all 4,400 fans at the Opry hostage with semiautomatic machine guns and demanding that authorities produce Reba as ransom. When the girls' landlord, an eighty-six-year-old grandmother, refused to give them bus fare, the younger girl hit the woman twenty-five times over the head with a porcelain clock, then smothered her to death with a pillow.)

Garth not only flouted the manager's order, but went out of his way to test the propriety of his fans. The fans, meanwhile, went out of their way to push the limits of propriety as well. They tickled him, hugged him, tried on his hat, and pinched him on the seat of his Wranglers. Though most maintained a respectable degree of distance, what was more striking was how close they actually came. Not in a physical sense, but in a personal sense. They acted like family and were treated as such. Garth, for his part, seemed to recognize most of the people in line: "How you doin', girl!" "Weren't you here last year at this time?" "How'd that picture we took in Dallas in 1989 come out?" "How'd your daughter do on that geography test she took after Tampa in '92?" The fans knew this was coming, and they were prepared for it. They brought those pictures from Dallas. They brought the report cards from their children's tests. And, even more surprising, they brought phone numbers, which he asked for and which he then gave to his road manager, who then promised that Garth would call them next time he was in their neighborhood: Tampa, Cleveland, Houston, or Phoenix. It's an absurd notion to believe that Garth Brooks could possibly know every one of the 60 million people who had purchased one of his albums, but I actually spent several minutes that night trying to contemplate the mathematical possibility.

"There is something about my mind," he told me later, "that if I see a face, I'll remember seeing that face. If I sign somebody's name, I can go back to when I signed it, look down at the picture, and read the name."

"But how do you maintain that?" I asked. Most people who achieved his level of fame, I suggested—Michael Jackson, John Lennon, Elvis—descended into isolation and oddity.

"But none of them are country music," he said. "I think that's part of the whole rock 'n' roll image. Inaccessibility. Having weird things. It ain't country. The thing that I love the most about country music is that it can give you the sales you can have in every format, and, at the same time, when you come off that stage, they treat you like an average Joe."

Garth Brooks, the Dale Carnegie of popular music, had built a fortune based on this principle: You can be ambitious and still appear accessible. Indeed, it might even be in your best interest to do so.

"You know how I could have sold a hundred million records?" he once asked me in a tone that indicated he wished he could have. "Sign every autograph I could have. After a while, though, I just couldn't. I was becoming late for my next dates. One morning after a show in Dallas I looked up at the end of a night of signing and realized that the sun was coming up. I went home that night—I remember it was my birthday—and cried. The next week I had to stop signing."

After almost two hours on this Friday evening, Garth had made his way through most of the crowd. The rest of the Opry was empty by now. The other performers had all gone home. The staff had cleaned up the popcorn boxes from the seats. Even Rosa, the seventy-seven-year-old nurse who distributes coffee and pink lemonade backstage in front of the dedication plaque signed by President Nixon, had packed up her swizzle sticks and gone home for another night, her latest in an unbroken span stretching more than twenty-six years. When the hallway seemed to be clear, Garth straightened his hat, gestured to his road manager, and rounded the corner to the entrance desk, where the woman still sat underneath the blue and green portrait of Minnie, Porter, and Hank. Almost instantly there was a roar. Without realizing he had been just around the corner, several dozen people were camped out at the entrance. Though it was after eleven by now, over two hours since he had last sat down, Garth instinctively plunged into the crowd. It was almost as if he was drawing life from the people.

Once again he stood. And once again they came. Each couple, or mother-daughter, or group of high school friends having him sign as much as they could, then running to their cars to retrieve more belongings. Another hour passed. Midnight came and went. It was just before one o'clock in the morning when he hugged his last fan and stepped into the chilly night, where another fifteen or so people had waited since the end of the show. Fifteen more minutes. Then thirty. "Doesn't he ever go to the bathroom?" one of the women asked. At this point the ten or twelve people still awake in Tennessee were standing in an informal clump around the bestselling bladder-control expert in America. "So much has changed for you," one man said, nearly weeping, as he finally arrived for his audience. "But you haven't changed one bit." Garth embraced the man. And by this point, it was becoming clear that this marathon session was serving a much different purpose than it appeared. Though perhaps originally intended to send a message to the world that he was still accessible, Garth's effort was actually sending a message to himself: that not only did the public still care about him, but that he, more importantly, still cared about them.

In that sense, Garth reminded me of Bill Clinton. I often thought it was no coincidence that Bill Clinton was the dominant political figure in America during the period when Garth Brooks was the dominant musical figure. Both were children of the New South, who had larger-than-life appetites (food, women, power) and great marketing savvy. Also, both were challenged on the motives of many of their actions, which were often criticized as being manipulative and insincere. Regardless, both men succeeded because they were able to communicate a profound sense of empathy and shared purpose with the American public. That feeling of camaraderie ("I feel your pain," Clinton said to voters; "I *am* you," Garth said to fans) was vividly on display this night, when Garth, like Clinton, seemed to gain strength by coming out of isolation and pressing the flesh with his constituents. If anything, as he set out to relaunch his career, it was becoming apparent that Garth still needed his fans in a way that was much more visceral—much more desperate—than they needed him.

Finally, at 1:42 in the morning, three hours and forty-five minutes after he finished his three-song set at the Opry, Garth Brooks was finished for the evening. He unlocked the door of his new 4x4, hopped behind the driver's seat, and started up his Christmas gift for the brief ride north to his home. As he backed out of his parking space, he rolled his window down. I was now the only person standing in the cold.

"Can I offer you a ride to your car?"

"No, thank you," I said. "It's just around the corner."

"Well, then, have a safe drive home," he said. "And don't forget. We'll be back here again tomorrow."

TWO

THE STUDIO

The men gathered in the soundproof room on Division Street, just east of Music Row, were legends. They had come, on this day, to help create another.

It was just before two on the first Monday in December when the Toyotas and Chevys, the pickups and hatchbacks, started squeezing into the cramped parking strip alongside the Soundshop Studio, a somewhat dreary, single-story, cedar-sided studio down the street from the Country Music Hall of Fame. The first few cars to arrive were dented and bruised, the men who emerged from them less than pristine themselves. With stringy hair, faded jeans, and bleary, late-night barroom bags under their eyes, these middle-aged men hardly seemed the stuff of the white-bright smiles of America's newest glamour industry. And, in many ways, they weren't part of that smile. They were the Frankenstein pieces behind the smiles: the hands, the ears, the strumming arms, the pedaling feet, and most of all the fingers behind the music that made those smiles ring true. They were the session players, heard but not seen.

The last person to arrive drove the nicest car of all. It was a power red Mercedes coupé with a license plate that monogrammed its owner as DKC. The initials claimed the parking spot of honor, directly in front of the studio door, and the man behind the initials—perky, preppy, a jolly roger on this cruise—strode through the glass door, down the short hallway, and directly into the seat of power, an aqua blue rolling office chair on the upper level of a split-level room.

"Okay, gentlemen," he said upon arrival. "What do you say we make some music?"

The first day of a recording session is invariably a moment of some circumstance in Nashville. The producer is expectant, the musicians abuzz. And the artist is both of those—expectant and abuzz—and also a little bit scared. The artist in this case was a twenty-six-year-old former construction worker turned reluctant stud when the debut single off his debut album stunned everyone in Nashville (including most of the people in this room) by steadily climbing the country charts a year earlier and peaking for two triumphant weeks at number one. More than just a song, "Old Enough to Know Better (But Still Too Young to Care)" was that rare gift in popular music, a clever lyric, a catchy tune, and— that indefinable—an irresistible spirit that together helped set up a young man's dream. The single, "Old Enough to Know Better," went on to become one of the most-played records of the year. The album, of the same title, went on to become the bestselling debut record of the year. And ultimately the phrase itself helped create a certain aura around a still-lanky young man from Oklahoma, *Billboard*'s Debut Artist of the Year, as he sang it often, discussed it even more, and, on this day, arguably his most important in a year, even wore it around his waist in the form of a giant silver and gold belt buckle.

Wade Hayes was a new kind of country star, with the looks of a dashing video hunk, the voice of a deep-canyon echo, and the somewhat old-fashioned sensibility of a young man trying to do the right thing even as his life spins out of control. He was, for starters, extremely well mannered. Even after knowing him for over a year, when I attended a small luncheon in honor of his new album some months after the re-cording session, I received a handwritten note from Wade two days later thanking me for attending. In three years in Nashville, including dozens of such luncheons, Wade's was the only personal note from an artist I ever received.

As cordial as he was in private, in public he was awkward. Shy to the point of being taciturn, Wade almost never spoke in a group (includ-ing his own band) and when called on to make brief remarks—at a reception at his label, say, or at a ceremony awarding him a gold rec-ord—he would offer abject apologies for his inability to express himself. Once, several years into his career, after being introduced to a crowd of well-wishers at the Wildhorse Saloon in Nashville, all of whom had come to hear of his participation in a new tour, Wade became so flustered that he sat down without finishing his remarks. Minutes later, after his colleagues on the tour offered reams of effortless repartee, Wade returned to the podium, confessed he was so nervous he was sick to his stomach, and politely finished his thank-yous.

Because he was so uncomfortable with speaking, music was how Wade communicated with the world. As he put it in a television interview at the start of his career: "Music is about the only way I feel comfortable expressing things. I've always been a real emotional person, but I've never felt comfortable about being emotional in front of people. I've always had these feelings inside of me going crazy. My mind's always going a million miles an hour, but I try to stay quiet and stay reserved in front of people. Sometimes it drives me crazy. Writing a song or playing an instrument is a pretty good outlet for that. Maybe that's why it means so much to me."

On this day, Wade was even more anxious than normal. Dressed in deep blue Wranglers, a long-sleeve teal T-shirt, and brown cowboy boots, he looked lean and fit. Without his normal hat, his shoulder-length hair was pulled back from his head, drawing more attention to his flat, globular nose, thin lips, and deep-set, dark and mournful eyes. Despite his height, broad shoulders, and widely acknowledged powerhouse voice, he looked at this early stage in the session like a sapling just waiting to be toppled by a gust of wind. He wasn't glamorous. Not even sexy. More like "adorable," in the words of one female friend of mine.

"I guess I slept a little more last night than I did a year ago," he said as the musicians gathered around the console. Even his speaking voice was deep, but pleading: pure Oklahoma, a rich cousin of twang. "Back then I didn't sit down for weeks. I was *so* nervous."

"That was your first time, though," offered DKC.

"I know, and I almost didn't make it through."

Callow as he appeared, Wade Hayes was rapidly maturing into a full-bodied star—perhaps even more than he realized himself. His shooting status was no more apparent than in the roster of country music legends who dropped by the darkened studio this afternoon to pay their respects. First there was Bill Anderson, Opry luminary, one of the few singer-songwriters in Nashville to have hits in every decade since the 1950s and coauthor of "Six Feet Tall," a song on the album. With a quick smile and standard-issue baseball cap (Atlanta Braves), he was one of the handful of men who settled in for the afternoon on the baby blue sofa in the back of the control room. Next to him sat Ronnie Dunn, the red-bearded half of the superduo Brooks & Dunn, headliners of the most recent tour on which Wade was the opening act and cowriters of the first song to be cut, "Our Time Is Coming." (Kix and Ronnie, along with Don Cook, alias DKC, had also written "Steady as She Goes" on Wade's first album.) Out of respect for the nature of the occasion, Ronnie removed the cellular phone from his belt, turned it off, and put it in his lap.

Despite the presence of all these stars, the person in the room who received the most deference was a gentle, gray-haired man with a Jay Leno jaw and slight stoop who sat quietly in the corner while the other men in the room exchanged introductory dirty jokes. He watched silently as the first song meeting came to order, listened as they played the demo of the cut and discussed possible directions for the instrumentation to take, and about a half an hour into the session—during a Diet Coke break—slowly rose to his feet. Immediately the conversation ceased. Everyone in the room rose as well. They all turned in his direction. One by one he wished them good-bye. ("I love you, man," he told Don Cook, before repeating the same comment to Ronnie Dunn.) Then Harlan Howard, the most prolific hitmaker in Nashville history, the "Cole Porter of Country Music," shuffled over to the youngest person in the room.

"Wade, son, I just want to tell you: You've got a real great team around you. A real A-plus group." His voice was quiet, genuine Kentucky bourbon. It was the soothing spirit behind more than four thousand songs, among them over fifty number ones, including Patsy Cline's "I Fall to Pieces." "Everything in your life is just great right now. I'm so happy for you."

Speechless, Wade smiled, took a short step back, and imperceptibly bowed his head. He had been anointed. And he knew it.

By just after three, the musicians were in place and the green lights on the control panel started to dance. There were forty-eight columns of green lights in all, each a slender six inches tall, and each one reaching for the bar of red lights above it in a seemingly neverending strobe that was a constant reminder of both the technological underpinnings and the glamorous potential of making a record in the Digital Age. Music may be art, but it's also science.

"I'd like a little more groove than on the demo. A little action on the guitar and drums. You know, U2 does country." Don Cook spoke from the upper deck of the control room. A native of San Antonio,

Texas, he had dreamed as a teenager of becoming the next Bob Dylan. When that ambition proved too lofty, he moved to Nashville in 1971 to try his hand at writing country. He had his first Top 5 hit in 1977, Barbara Mandrell's "Tonight," and the following year had eleven chart singles. Over the next decade, he became one of Nashville's most prolific songwriters, with twenty-five Top 10s and ten number ones. In 1991, he produced his first record, Brooks & Dunn's smash debut, *Brand New Man*, and overnight became coveted as a producer for his keen song sense and strong commercial instincts. By 1994, riding a wave of popularity, he talked Sony Records into giving him his own boutique imprint (DKC) under its Columbia division. Wade was his first artist.

Despite his power, Don was known as one of the most affable men on Music Row. He had a quick smile and a sandy walrus mustache that served to distract attention from his balding head. He came to the studio on this day in blue jeans, loafers, and a generous-bellied, white button-down shirt with a monogrammed tennis racket on the pocket. He was relaxed and in effortless control: the Skipper to Wade Hayes's Gilligan. "I'm pretty much of a working guy," he said. "I don't make such a big deal about myself. I'm not the artist. I'm the producer." A lot of producers, he noted, are so insecure about what they contribute to the process that they create personas around themselves that make them seem like gurus, "the only way to God." "That's just patently not true," Don said. "I'm just one of a bunch of guys who do the same thing. When you get me, you get whatever skills and talent I have. You pay a lot of money for it—" He chuckled. "And you get it. But I'm not a magician. If anything, I'm a hired hand."

But what about the halo of power that seems to hover around the most famous producers in Nashville: Owen Bradley, Jimmy Bowen, Tony Brown? On Music Row they are often more mythologized than the artists.

"I once played golf with this golf pro," Don said, "and of course he was a phenomenal golfer. But I expected that when he played fire would come off the end of his club, that the rest of us would stand at the tee box and just sigh every time he hit the ball." It didn't happen that way, though. "He dressed like we did," Don said. "He interacted with the rest of us. He ate a tuna fish sandwich with us when the round was over. In our midst, though, he was just a fuck of a lot better than we were." The pro shot a sixty-five; Don shot a ninety. "And it impressed the hell out of me," Don said. "That a guy could just quietly be that thorough and that good. I would like to be thought of like that. Real power is just quality and what you can actually bring to the table when all the bullshit is done with."

Sitting in the studio that afternoon, the absence of bullshit was pronounced. Far from an artiste, Don, a native of Texas, acted like the most popular kid in class, the kid who grew up to become the unanimous chairman of the social committee at the country club. He would rock back in his chair and listen to the musicians, then lean forward and press a red button labeled THE VOICE OF AUTHORITY that carried his words to the other side of the window where the musicians were gathered in the dimly lit studio, about the size of a doctor's waiting room. "Any of you guys know why you can tell the toothbrush was invented in Oklahoma?"

Several of the band members, speaking into their microphones, chortled their nonresponses.

"Because anywhere else it would be called a teethbrush."

Wade, the token Okie, laughed the loudest.

Considering the stakes, as well as the time and money that goes into writing, pitching, and selecting material, the process of actually recording a song in Nashville is suprisingly haphazard. First of all, the song itself is never really written down. There are no notes, no melodies, no little measures with tempos and crescendos jotted helpfully throughout. Instead, in the five minutes before a song is recorded forever, to be played on radio stations and home stereos potentially millions and millions of times, a representative of the band—in this case, the acoustic guitar player—sits down in front of a piece of white copy paper and pencils a series of numbers—1, 1, 3b, 2—that refer to the chords and that he then Xeroxes and distributes to his colleagues. These cryptic numbers, known collectively as the Nashville Number System, are part of the pidgin system of communication developed on Music Row in the 1950s. Though it has the obvious disadvantage of not capturing every note for posterity, the Nashville Number System does have several distinct advantages: It frees songwriters from having to notate different instruments, it frees musicians to improvise creative flourishes, and it frees self-taught artists from the embarrassment of not knowing how to read music. And ultimately it gives everyone involved in the recording process leeway to do whatever they want to a song—change the key, add a bridge, slow it down, speed it up.

It is this feature of recording—its chance—that makes the process even more remarkable. If you're going to be dazzling, one wants to say, at least break a sweat. (Some pop music is recorded in this manner, though the bulk of recorded music is still done with the aid of notated music.) Instead, all those fanciful openings, those catchy licks between choruses, that soaring touch of fiddle in the middle of the second verse that makes you stop and stare at your radio are all made up *entirely* on the spot. Impromptu. A few might be included on the songwriter's demo of the song, but most are invented by the studio musicians. "I think we need a little more Vince Gill in that opening," somebody will say. "How 'bout some Alan Jackson energy on that fiddle bridge?" "Now, *that's* George Strait." For better or worse, these studio performers, having played on so many other records (musicians good enough to become session players rarely travel on the road; road musicians, like those in Wade's band, are rarely invited to play on an album), bring the entire history of the genre into the room with them. That makes a recording session a little like "Afternoon at the Improv."

"Doesn't this haphazardness make you nervous?" I asked Don as the musicians started warming up. "Aren't you worried that you won't get it right in the few minutes you all gather to invent it?"

"Of course I'm nervous." His voice had a casual Southwestern aw-shucks quality to it. "But that's why I live. If you're not anxious, you're not alive." He paused. "And don't forget, it's also quite expensive to have these people here."

And so it is. Each of these union musicians, part of the two dozen or so elite studio players in town called the A-list (on this day, one each for fiddle, drums, electric guitar, acoustic guitar, bass guitar, piano, and pedal steel guitar—the last being the free-standing instrument, sort of a guitar on a table, that gives country records their distinctive whine), was earning double union scale, or $500 for a three-hour session from 2–5 P.M. Then they received a catered dinner (lasagna on Monday, barbecue on Tuesday, steak on Wednesday), followed by another $500 each for an evening session from 7–10 that night. Add to that the price of the engineer—around $750 a day—an assistant, a rental fee for the studio, backup singers, and of course all that food (veal parmigiana on Thursday, salmon on Friday), and the cost of recording just the background instruments on an album in Nashville these days can top $20,000 a day. Then multiply that times eight or ten days.

Even for that cost, astronomical compared to Nashville budgets of the past (as late as 1980, a country album that sold 100,000 units was a phenomenal success; now it takes three times that just to break even), the productivity was impressive. By half past three, this de facto band, most of whom had played on Wade's first album, as well as albums by

Brooks & Dunn and the Mavericks (all produced by Don Cook), had tinkered with their instruments long enough to have devised a course of action. Don, staring into a Macintosh PowerBook with the list of songs on the screen, gave the signal to start. The engineer, Mike Bradley, who was seated on the lower level in front of the $250,000 cockpit panel, started rolling around in his chair like a mad scientist flicking switches and punching buttons. And suddenly—one, two, three, *four*—a deafening burst of music came pouring from the studio.

Don tapped along with his finger on the table. Mike hurriedly adjusted some levels. And Ronnie Dunn, sitting in the back, proudly swaggered up and down to the song he had written with his partner, Kix Brooks.

"We need more bass. More fiddle, too." The instructions come at the end of the song. First from Don, then from the studio. "Hey, Bradley!" the fiddle player calls. "Can you give me more drums in my ear?" "How about some lyrics, too!" calls the drummer. "Do you want to sing higher?" Don asks. "No, that feels good," says Wade. "Good, then give me more power in the choruses."

After several more practices, they were ready to record. The instruments only—the tracks—were being laid at this stage. The lyrics would be rerecorded later. "Okay," said Don, "let's roll the tape." The assistant disappeared into the back room. The red light went on above the door. And they recorded their first track of the day—pounding, rocking, heavier drums than the demo cut, a deeper, almost darker groove. Twice more they did it, on different strips of tape. Once a little faster. Once in a higher key, just to make sure. Each time Wade sang the lyrics, just to set the pace. "Times are hard / And the money's tight / Day to day / We fight that fight / Nothing new / It's the same old grind / Uphill all the way. . . ." When they finished, Don called the entire team back into the control room.

"Was that swingin' or what?" crowed Wade, picking up a carton of chewing tobacco and tucking a wad under his upper lip. "That sounds like a hit to me."

"You can make Kix and Ronnie some more money," said Don, to great laughter.

"Yeah," Wade countered, "somebody needs to do that."

"We were worried about Christmas," Ronnie said.

"Okay," Don announced when the laughter died down, "what do you say we play it back and listen?"

◇   ◇   ◇

For most of the rest of the afternoon session, the musicians tinkered with that first track alone. They paced back and forth between the control room and studio, practiced their parts, recorded additional takes, then shuffled back behind Don's chair to hear their work replayed on two of the largest speakers I'd ever seen. In general the mood was Sunday afternoon subdued. "I think people do their best work when they're calm and things are peaceful," Don explained, "and when the anxiety-provoking parts of their life are locked outside the door." Indeed, with no windows, only one faint clock, and a giant steer skull hovering on the wall, the session had an oddly timeless feel, except for the technology, which anchored it irretractibly in the present. Elvis, in his day, didn't have computers, nor million-dollar recording equipment. Wade Hayes had all of that and more.

When the recording age dawned in earnest in Nashville in the mid-1950s, studios were constructed like padded rooms in sanitoriums, with thick insulation in the walls, carpet on the ceiling, and foam rubber stuffed into every nook and cranny. The reason, Mike Bradley explained to me, was that all the instruments were placed into an open room and it was important that sound from one instrument not interfere with the others. "You needed things dead, so the room didn't ring," he said. (Like many studio veterans, he constantly peppers his speech with the studio musicians' own private lingo: "It's out of pocket"; "It's overgrooved"; "Open it up"; "Bring it on down"; "There's not enough pulse"; "There's too much meat"; "Put the wang on it"; "Take the vibe off"; "That'll stick 'em to their seats"; and my favorite from the week, from Wade himself, "That'll make 'em rub some monkeys.")

Beginning in the 1980s, though, the sound coming out of Nashville started to change: less dead, more alive. The padding was replaced with cloth wallpaper, the carpeting was removed in favor of tile, and isolation booths were constructed around the perimeter of the studio, enabling the acoustic instruments, like the guitar and piano, to be separated from the harder instruments, like the electric guitar and drums. "Back then drums weren't nearly as predominant in records as they are now," explained Mike, himself a roly-poly, soft-spoken man who had been working as an engineer on Music Row for twenty years and was recently awarded the first-ever Engineer of the Year Award from *Music Row* magazine. "You wanted them to sound good, but they weren't as much of the overall sound. We are looking now for a big bright roomy drum sound, where you can actually use the room. When I record the drums these days I actually put room mikes out so that I can record how they make sound in the room, not just the way they sound six inches away from the microphone."

Other changes were made as well. The bright orange and red lights of the past were replaced with softer bulbs—75- and 100-watt—and subtler colors—blues and grays. ("With softer light, I think you can stay in here longer without feeling you're in a Kroger," Mike said.) A series of RPG defusers—small panels of wood arranged at different depths, like children's blocks—were suspended from the ceiling to disperse sound. And the dense exterior walls were rebuilt using three sheets of common drywall separated by fiberglass sound-deadening boards. Though much more effective in keeping out street sounds, these walls are still not completely impenetrable to low-grade rumbling, which explains why many studio owners in Nashville rose up in protest when Reba McEntire announced her intention to build a helicopter landing pad on top of her new management palazzo on Music Row. Faced with such pressure, Reba backed down and the biggest threats to pristine sound nowadays are the chokings and honkings of tourist RVs that drive the wrong way down Music Row's one-way maze. "Harry, look, there's MCA." "Oh, my God, there's Wynonna's truck!"

The net effect of all these changes is that studios these days yield a more urgent, snappier sound. "If you go into the hallway out there and holler against the wall, it's going to bounce back and you're going to hear a slapback," Mike said. "If that wall, instead of being flat, has different angles on it, like those in the studio, you're going to get reflections bouncing back in different directions, at different times, making what you hear sound less blunt, like when you're in your shower."

"Is that why shower acoustics are so good?" I asked him.

"I wouldn't say they're good. It's a square chamber, and parallel walls are the biggest dangers in acoustics because anything that's parallel allows waves to bounce back and forth. Instead of being defused, they just keep going. It produces flutters that are hard to use."

"So, in fact, people think they sound good in the shower when actually they sound worse?"

"It sounds good in the sense that you have a natural reverb, but if you try to record in the shower you're going to hear what we call 'flutter tones.' This studio, because the walls are *not* parallel, gives you something more usable."

After several takes spread out over several hours, Don Cook was satisfied that they had a usable version of the song, a basic track of all the instruments that could then be slowly improved. Here's where the bene-fits—and perhaps the drawbacks—of modern technology became appar-ent. With the basic track laid, the musicians were released from the studio and wandered into the lounge, where they nibbled on Chex, sucked on cherry fireballs, and watched endless loops of "Headline News." From there they were summoned back to their places one at a time, where they replayed their parts—over and over—until each note was perfect and each lick unblemished. The steel guitar player in particu-lar, Bruce Bouton, took so long at this stage that a few of his colleagues snuck out to run errands. "If a doctor told me I had six months to live," Don Cook joked, "I'd spend it with a steel player 'cause it'd feel like two years."

This degree of perfection is possible nowadays because of the highly touted, yet still controversial, digital recording systems now standard in upscale studios in Nashville. In essence, the system works like this: All the musicians set up their instruments in the studio. The engineer then places multiple microphones on or around each instrument—twelve for the drums, two for the piano, one for the fiddle, and so on. The microphone, essentially a plastic diaphragm sprinkled with gold dust, functions like a high-tech eardrum, converting the acoustic energy from the instrument into electric energy. That electric signal then moves through the wire and passes into the console, or "board," where it is equalized (frequencies balanced), compressed (squeezed into a narrower band), or simply left alone, before proceeding on to what appears to be an old-fashioned reel-to-reel tape machine, so big that it's stored in a room unto itself. Though it looks old-fashioned, this machine, a Sony 3348 Digital Multitrack Re-corder, actually records not on conventional audio tape but on digital audio tape, otherwise known as DAT.

First introduced into Nashville in the early 1980s by L.A. Über-producer Jimmy Bowen (Reba McEntire, Vince Gill, Waylon Jennings), DAT technology marked a dramatic breakthrough in the history of studio recording and opened a still-raging debate about what constitutes genu-ine music. Unlike traditional analog recording, in which the sound actu-ally saturates the tape like, say, water poured onto a paper towel, in digital recording the electric signal literally sits on top of the Mylar in separate sections, like water poured into an ice tray. This breakthrough allows a great number of tracks to be available, up to forty-eight, instead of the previous maximum of twenty-four. That's forty-eight different sounds that can be isolated from the others, tinkered with, toned down, retuned, or just plain recut, before being seamlessly reconnected to the whole. Compared with the early recordings of Elvis at Sun Studios in Memphis, for example, in which everyone had to play their part perfectly at the same time, the improvements are staggering. Two tracks followed that system, then three (Patsy Cline recorded on three), rising quickly to eight (the Beatles pioneered this), and finally ballooning to twenty-four tracks by the 1970s. The arrival of forty-eight tracks in the mid-1980s has meant a much wider array of options, but has also left a nagging concern among some that technology is overtaking the artistic process.

In analog recording, because the sound actually *saturates* the tape, sound from one instrument often bleeds into sound from another, giving the recording what some believe is a warmer feeling. Some of Nashville's leading producers, among them Tony Brown (Wynonna, George Strait) and Kyle Lehning (Randy Travis, Bryan White), still use analog recording occasionally. By contrast, in digital recording, because each sound is isolated from the others and is converted into electronic ones and zeros, the music tends to be crisper and the overall feeling more precise.

"How you feel about this tends to reflect how you feel about technology," a techie friend of mine explained. "Would you rather drive a twenty-year-old Porsche 911 and get the organic feel of the car or would you prefer to drive a Lexus, which is snappier and has all the latest gadgets?" For Don Cook, the answer is clear. "Digital sounds better from top to bottom," he told me. "I've cut Brooks & Dunn records with both, done blind tests with them, and they picked digital thinking they were picking the analog because they'd always heard that analog is warmer than digital. That's bullshit. Digital's just better. Better because the frequency response is more complete. Better because it's just cleaner. And better because it doesn't deteriorate when you overdub. Every time you run analog tape through heads, it deteriorates. Digital doesn't."

As a neophyte, discovering how much such technology has altered production was shocking. This was especially true when it came time to record vocals in week two. These days if a lyric simply reads "I love you," it's not only possible, but also likely that a singer will sing "I" on one day, "love" on the next, and then even go on vacation to Mexico, have throat surgery, get married, get divorced, fall in love again, lose a parent, have a child, and in the process completely change his conception of what love means, before returning to the studio to sing the final "you." George Jones's 1980 seminal hit "He Stopped Loving Her Today," which won a Grammy for Country Song of the Year, was notoriously done in this way, taking nearly a year and a half to complete. Since most consumers have no idea this occurs, I couldn't help wondering if the whole process isn't a bit fraudulent: People think they're getting a sublime batch of brownies from Garth Brooks, when in fact instead they're getting the best brownie from seventeen different mediocre batches.

"But you don't think it's fraudulent the way automobiles are manufactured or movies are made," Mike Bradley countered when I raised this suggestion. "In the old days, everything used to be handmade; nowadays everything is machine-made. We spend a lot more time trying to achieve perfection than we did back then because a lot of times we couldn't get perfection then. Time and budgets didn't permit it. And if you listen to those records today, even the ones we thought were great turn out to be full of mistakes. There are places where the tempo shifts or where somebody's out of tune. By today's standards, you say, 'Boy, is that sloppy.' But at that time it didn't matter to us. Today, we're trying to make a perfect record."

Perhaps that's the essence of Nashville's transformation: Perfection is the coin of the realm these days, rather than authenticity. A Hank Williams record is a performance of a song; a Wade Hayes record is more like multiple performances of a song, the Platonic ideal of that song, a perfect rendition that exists *in digitas*, if not in reality. If anything, recording an album has become more analogous to writing a novel—especially in the age of word processing. First you get the idea, then you lay down the plot, then you slowly build the characters. "Are people buying whole sentences from you and asking, 'Did you write all the words at one time?'" Don Cook asked me in defense of this system. Fair enough, but especially in country music, where artists represent themselves as being real people singing about real emotions, the irony is almost too painful to accept: The same technology that allows songs to be recorded perfectly now almost guarantees that no singer will ever be able to perform them that way again.

Four days of recording in the Soundshop yielded eight complete tracks, including "Our Time Is Coming," "The Room," "Six Feet Tall," and "Hurts Don't It." Four more songs would be added in February. Several more in March. That would leave Don, Wade, and assistant producer Chick Rains the freedom to choose the ten or so cuts that would make it onto the final album. For the moment, though, it was time for Wade to move on to vocals. This painstaking process began exactly one week after the opening day and, for the first time, offered a hint of the emotional struggle of recording an album, especially for a newcomer.

"Afternoon, everybody. How y'all doin'? Sorry I'm a little bit late."

It was just after two on Monday afternoon when Wade Hayes strolled into the studio alone, dressed in his familiar tall, tight jeans, a BUZZ 107.9 T-shirt, and his now quite pronounced one-week-old beard. In his back pocket he packed a Primo del Ray eight-inch cigar, which he occasionally gnawed on for nicotine in lieu of his normal chew. And in his hands he carried a low-fat banana oat bran muffin and a bottle of Evian natural spring water.

"I bet Hank Williams never brought that to the studio," Don Cook cracked. "He always came to record with Menatrol and whiskey, like you're supposed to. Hell, even the real singers today say, 'Bring me Robitussin and a blonde.' "

"Darn," Wade said. "Sometimes I don't know if I'm a real singer at all."

Don winced. He propped his tan loafers up on the desk. "He's not quite as scared as he was last time," he said when Wade disappeared into the cutting room to begin his vocals, "but he's still real pressured. The problem is, he has much less time now. It's the age-old saying, 'You have your whole life to make your first album, and you have a year to make your second.' "

"It's true," Wade elaborated later. "The first time I did this I was so frightened I couldn't sing. My whole body tensed up and I couldn't function for days. This time I tried to get a lot of sleep. I sucked down vitamins for days. I tried to tell myself I could do it."

They got down to work. For the vocal session, the assistant engineer had prepared a small blue spiral-bound pamphlet containing the lyrics to each of the songs, followed by eight empty spaces. As Wade made different passes through the song, each of the several people in the room would mark his copy of the grid, indicating which track they believed contained the best version of each line.

Once in place, Wade asked that the lights in the studio be turned off completely (leaving only the control room lights, filtered through the glass panel, to illuminate his lyric book). A small Oriental rug was placed under his feet to reduce reverberation. And a special pop filter—the round, saucer-sized piece of nylon mesh that prevents robust p's and t's from disrupting the balance of sound—was placed in front of the microphone, in this instance an old-fashioned, capsule-shaped Neumann tube mike, a deliberate throwback to the nontransistor age when mikes were thought to capture warmer sound.

He started to sing. "Times are hard/And the money's tight . . ." With each phrase he swayed from side to side. "Day to day/We fight that fight . . ." The mood, though hopeful, was notably tense. "Nothing new/ It's the same old grind . . ." And the rhythm, for some reason, wasn't quite right. "Uphill all the way."

"Let's start over," Don suggested when the song was finished.

"That wouldn't hurt," Wade agreed.

"I don't like the way your voice broke up."

"Yeah, I think I'm not in the groove yet."

The afternoon proceeded in a similar fashion—starting, stopping, rethinking, redoing. The day became evening. The evening became the following day. There was a slow, almost deliberate massaging to the whole process. First they would address the general issues. "Just whip it," Don would say. "Really feel the passion." He would offer periodic criticisms—"We need more emotion in the verses," then "The chorus doesn't grab me"—mixed with regular notes of encouragement—"Did I tell you you were a genius today?" And at times he would even stop the session and summon Wade back to the control room for a frank coaching lesson. "You have to remember that this is a song, not a blueprint," he said during one such encounter before dinner on day two. "You can't make it up. You must sing what's on the page." Eventually, once they had a general pass on each song, they would move on to fixing particular lines. "Give me more voice in 'street of gold,'" Don would say. "I need to hear your anguish on 'good Lord only knows.'" Finally, when the overall rhythm was right, they moved on to individual words. "Don't get off 'close' so fast." "'Time' had too much scoop in it." Every now and then, the whole session would break down in frustration and Wade would return to the control room almost in tears.

"Man, I can't hear anything through those headphones," he said near the close of the fourth and most rigorous day when he was recording "The Room," the most emotional ballad on the album.

"Don't worry," Don said. "The same thing happens to everybody. Hell, I can't tell pitch through headphones at all."

"Well, I can't tell anything at all. I just don't even know what I'm supposed to sound like anymore." Wade sat down at the console and ran his fingers through his hair. Don stood up and pushed the door closed behind him. He waited in silence for several minutes, then wheeled his chair beside his pupil.

"Well, if you ask me, I think you're pretty damn good," Don said. He put his hand on Wade's shoulder. "I think you can flat sing."

"Shoot," Wade replied. He looked up and stared straight ahead. "Sometimes I think I can. Sometimes I think, 'What am I doin' here?' " He turned to face Don. It was a look of total vulnerability. Don had seen it before. When they had finished recording Wade's first album, Wade listened to the mastered cuts of the songs and announced he wanted to sing everything again. "It's the producer's job," Don told me, "and hopefully the producer has enough experience to deal with that kind of problem, to say, 'We've got to stop somewhere. This is it.' In his case, we did sing a couple of things again, but we didn't sing the whole album. Young artists who have never done albums and never had songs on the radio tend to get real freaked out about their perception of the quality of the product. Everyone else's album sounds finished, but theirs doesn't. Somebody just has to say, 'We're done.' "

Don was totally calm. "You're not quite as insecure as Ronnie Dunn," he told Wade, "but close. . . ." He chuckled. "Just the other day, we were in the studio and a call came in from Tim DuBois at Arista and Ronnie said, 'Oh, shit, what did I do?' It was like the principal was calling to tell him he fucked something up, when in fact all Tim was doing was calling to tell him how much he adored him. The reason is, Ronnie Dunn is the embodiment of quality, only he just doesn't believe it. And you're the exact same way. You're so good you don't even know it." There was a pause. Don offered a laugh again. "And I, for one, hope it stays that way. God help us if you ever discover what you've got."

The grimace on Wade's face slowly eked into a grin. "I hope it goes away."

By the end of the week, the first eight songs were cut. Later they would be mixed and then mastered, the laborious process by which an engineer arranges the tracks in perfect balance. The album, for the first time, was beginning to take shape. They had strong mid-tempo songs, "Our Time Is Coming" and "Oughta Be Over You by Now." Emotional and fairly dark ballads, "The Room" and "Where Do I Go to Start All Over?" But only one up-tempo number, "Six Feet Tall." They would try to add more later, though a tight release schedule might prevent them from having time.

Now, though, the time had come to play the cuts they had for the head of Sony Records, Paul Worley, himself a prominent producer hired to head the label several years ago, who came to the studio late Friday afternoon just before leaving town on vacation. Like many executives in Nashville, he was wearing blue jeans and a company leather jacket. He leaned back in the engineer's chair and listened to the cuts with a visible—and growing—smile on his face. At the end he didn't even pause before speaking. "That's great," he enthused. "That's really great! I want to thank you from the bottom of my heart. That's an excellent Christmas present and a great New Year's gift. I can't wait for people to hear it."

"Well, I'm glad I brought you over," Don said, barely containing his glee. "Nobody believes me around here anymore."

"Certainly not me," Paul said with a grin that hardly belied the hint of rivalry that hovered over two of the most powerful men in town.

He stood up to leave. Don bounded to his feet. The two of them spent several minutes glad-handing each other and discussing their plans for the week ahead—skiing in Colorado for Paul, boating in Florida for Don. Paul reached into the half-empty tin of Danish cookies. Don Cook finished his umpteenth Diet Coke. And all the while Wade Hayes sat motionless at the desk, dribbling an occasional drop of tobacco juice into a cup and quietly smiling to himself.

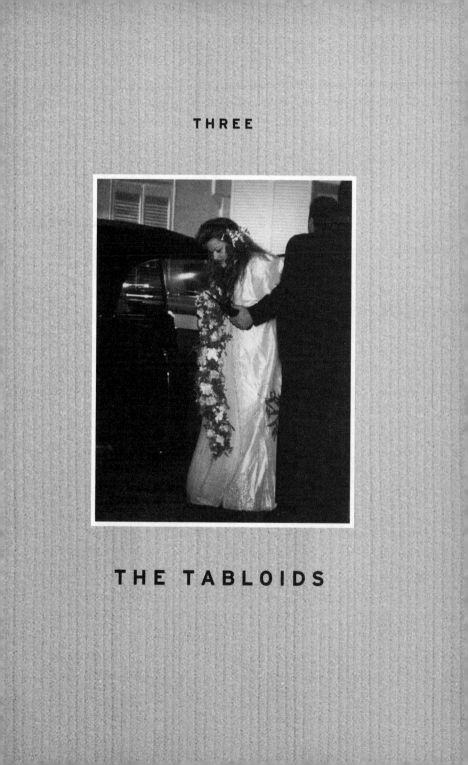

THREE

THE TABLOIDS

The telephone rang in Hazel Smith's kitchen at 8:32 A.M.

"Hello, honey," she said, picking up the receiver. "You're late this morning." Her voice was uncommonly cheery for the hour, the result, she said, of over two decades of waking up two grumpy boys by herself. Though the boys were both grown now and had boys of their own, Hazel, a single grandmother, a lifelong songwriter, and for two and a half decades the leading gossip columnist in country music, still awoke every morning at daybreak. By the time of her weekly call from a syndicated radio show in Indianapolis, she had already downed a cup of coffee, eaten an orange ("I hear it's better for ya than juice . . ."), and scanned the morning paper, paying special attention to page three, where onetime hard-nosed reporter turned tart-tongued gossip columnist Brad Schmitt parsed out a daily dosage of music industry innuendo, breathless musings about the romantic progress of Tim McGraw and Faith Hill, and constant drooling over his true obsession, the love life of Ashley Judd. "That marvelous Ashley Judd is at it again! Wait till you hear which Hollywood hunk got a handful of her delicious derriere." On this winter morning, Hazel had also managed to pull a purple turtleneck with purple sequins over her upswept gray hair.

"Okay, listen," she continued now on the telephone. Her bare feet were turning pink from the cold. A box of uneaten Krispy Kreme doughnuts sat on her counter. "I am going to talk about Wynonna this morning. But I'm not going to tell you what I've got 'cause I don't want you to know until I go on the air . . ." She paused to wait for a response from the producer on the other end. With her Southern biscuit spirit and sweet cane syrup accent, Hazel could have been a stock caricature in a Hollywood sitcom: Ethel Mertz meets Martha White. But she was real. The framed picture of George Strait on her wall was autographed: "To Hazel, I love you. I couldn't have done it without you."

"Good, baby," she said when she heard what she wanted. "Now, I've also got a story about Faith Hill, but I call her 'Faithless Hill,' you know. And a story about Dolly singing at her nephew's wedding. I don't think anybody knows this. And I have to tell you where Garth and Sandy were shopping on Wednesday. You're going to think that was very funny." Her voice seemed to gain speed and titillation with every item she ticked off her yellow pad. She was bobbing up and down like a dashboard grandmother. "Oh, honey, I can't wait till we get on the air. I'll talk to you in fifteen."

Perhaps the most distinguishing feature of the country music community is its unabashed feeling of clubbiness. With the music and its stars long subject to derision, some country fans keep their passion hidden away. I was amazed when I first announced my intention to move to Nashville by the number of people in places like Washington, D.C., or Manhattan who pulled me away from the earshot of others and confessed their secret obsession with Trisha Yearwood, Alan Jackson, or George Strait. "What did it for you?" one publishing sales rep asked me. "For me it was Garth Brooks's 'Friends in Low Places.' " "It's like therapy," one U.S. Senate staffer said. "I know my problems aren't as bad when I listen to Patty Loveless." Others, meanwhile, have responded to the sense of shame by flinging open the attic doors with glee. It's these fans who force colleagues to listen to country radio at work or pull out a Tanya Tucker album when friends come to visit. (Gotcha!) It's these fans especially who make up the backbone of the Elks lodge of country music. For twenty years, the den mother of that lodge has been Hazel Smith.

Hazel Smith need make no pretension to being country. The future doyenne of country music gossip—the Liz Smith or Army Archerd of Music Row—was born in a house with no electricity off a dirt road in Caswell County, North Carolina, not far from the Virginia border. Her father was a tobacco farmer and deputy sheriff.

"Did you go to any college?" I asked as we waited for her syndicated radio broadcast to begin. I had come to see what she had to say about Wynonna and the most hotly anticipated event of the year.

"You must be kidding!" she squealed. "The way I talk. You know I ain't been to no college. I finished high school, that's all. I got married when I was nineteen, and everybody thought I was an old maid because they were all married and making babies. That marriage lasted fifteen and a half miserable years."

Her husband worked at the Lorillard Tobacco Company making Kent cigarettes while Hazel raised tobacco on the side. "At the time I was a pretty good-lookin' woman," she said, "and let me tell you, my friend, I got top dollar for my crop." On the weekends, her husband, a fiddler, played square dances and weddings, and Hazel wrote songs. "I was always attracted to the music," Hazel said, "but I hated being married. I never liked somebody who was not as smart as I was trying to tell me what to do. And I don't mean to say that to sound like a female chauvinist piglet. I simply say it because it was the truth. Fifteen and a half years is a long time to live with someone who's telling you you're never going to amount to nothin'."

Finally, in a scenario repeated countless times across the South and immortalized by Loretta Lynn (a coal miner's daughter), Dolly Parton (also a tobacco farmer's daughter), and Naomi Judd (a gas station attendant's daughter), Hazel Smith left her husband, took her children, and followed the music in an almost blindly self-destructive way until it led to the threshold of the Grand Ole Opry. "It was 1969," she said. "I was staying in a motel down by the Ryman. I had no idea where to go or what to do. I just fell down beside the bed and started praying, and when I come to my senses the bedspread was wet I had cried so hard." Noticing a phone book beside her on the table, she opened it up. Her eyes fell on the Reverend John Christian, Goodlettsville, Tennessee. "I called him and said, 'I've got two children, three hundred dollars, and I've just got me a job at an employment agency.' And he said, 'I know where there's a place that I can get for you for seventy-five dollars a month. It's in the middle of a cow pasture.' I said, 'I don't care, I'm a country girl.' I was just so happy to be there. It was freedom. It was *free*-dom!"

In time Hazel got a songwriting contract at a local music publisher that paid $100 a week as an advance on future royalties. There she met Tompall Glaser and Kinky Friedman (leader of the Texas Jewboys), two of Nashville's more colorful characters. They hired her to do publicity. "We went to New York once and I wore this cowboy hat that said

'Kinky' on the front and didn't know why this girl said, 'Do you want to go out with *me*?' That's how dumb I was." It was the early 1970s. Country music was about to be turned upside down by Glaser, Friedman, and their pals Willie Nelson and Waylon Jennings, all of whom worked out of a small, windowless stucco building on 916 19th Avenue South known as Hillbilly Central. It was Hazel who first called the members of the movement the Outlaws.

"It was the most exciting time," Hazel said. "I would be walking around there barefooted and Chet Atkins would walk in or Shel Silverstein. The president of RCA came in often because they were making him big-time money. If Elvis had walked in, I woulda said, 'Hi, Elvis, I'm Hazel. What could I do for you?' " For a country girl, the scene was eye-popping. "It was twenty-four hours a day people there," she recalled. "I'd come in and someone would be asleep on my desk with a bottle of wine in their hand. Once Bobby Bare left the bathroom door open while he urinated and I almost fainted." Also, since most of the men were high on drugs, it was often tense. "I remember this old Cadillac Waylon had that was the color of the sun setting down on Anna Maria Island. It looked like some great big fat woman's dress. One day Tompall came in and said, 'Waylon, who's your best friend?' 'You are, Tompall.' 'Waylon, when everybody else turns your back on you, who do you call on?' 'You, Tompall.' 'Waylon, who just give you a brand-new guitar?' 'You, Tompall.' 'Waylon, guess who just ruined your new Cadillac?' 'You son of a bitch.' "

"So with all those wild men hanging around," I asked her, "did you ever date any of them?"

"I did not," she said matter-of-factly. "I had somebody I cared about. And they all knew that, and they respected that. At one time Dave Hickey, who was a marvelous, marvelous writer, said, 'Hazel can't love nobody but a legend.' That was all that was ever said."

"So who was that legend?" I said, sitting up in my chair.

Hazel thought for a moment, then grinned. "Baby, you haven't learned that by now? Stick around here long enough, you will."

After about fifteen minutes, the telephone rang again and Hazel hurried into her paneled office. "Hey, baby," she answered. "What's up?" The sun was just beginning to reach her backyard. A lone locust tree lorded over a rusty swing set, and a few red squirrels pranced around the grass.

"Tell me what the plan is," she said. "How much time do we have this morning?"

For the past eighteen years, Hazel has lived in a modest redbrick house in the barren bedroom community of Madison, a dozen miles north of Nashville. From her office window she could clearly see the woods, where deer strolled in the afternoon. Across the yard her neighbors' homes were decorated with iron animals and JESUS CARES placards. "People call Madison the butthole of Nashville," Hazel said, "but I like it." A few miles farther north was Goodlettsville, the town where Bill Monroe had a farm and where Garth had purchased a 400-acre piece of land, fittingly, from the previous mayor of Nashville. Farther to the north was Hendersonville, home of Johnny Cash, George Jones, and Trisha Yearwood. "In the old days, all the country stars had to live out here because the folks in Belle Meade wouldn't let 'em live around the country club," she explained, referring to what she called the "snootiest" neighborhood in town. "Loretta Lynn used to live in Madison. And Kitty Wells. Patsy Cline got in a bad wreck over yonder between here and Kroger. Every time I go through there, I think of her. Nowadays the country people got more money, and those rich folks got no choice but to respect them. They even let them play golf."

That story of country music's rise to prominence parallels the story of Hazel Smith's life and, in part, is responsible for how she grew from part-time tobacco hocker to national radio personality. After Hazel had spent several years answering the telephones and dodging drug busts at 916 19th Avenue South, the editor of *Country Music* magazine, the most influential of the several fan magazines devoted exclusively to covering Nashville, asked her if she would write a column. Called "Hillbilly Central," it contained homespun reports on the various parties, seen-about-towns, and heard-on-the-streets of Music Row. Hazel has continued writing the column through various jobs over the years, including her latest, as a compiler of K-tel greatest hits packages sold in supermarkets and on late-night television. "The qualifications were the best schmoozer in Nashville that goes to all the parties and knows everybody. My friend Bill Isaacs told them that was me."

More than just a presence at every Music Row shindig—often as many as two or three a night—Hazel soon developed a reputation as the conscience of Nashville, the person who loved the music above all else. "I know a lot of people think that country music is all about Mama, dead babies, apple pie, funerals, the flag, God, the Bible, and so forth, but it's more than that," Hazel told me. "It's the truth. Country music tells the story of the truth. Listen at Tom T. Hall. Listen at Hank Williams. 'Your cheating heart / will make you weep. I cry and cry /

And try to sleep.' I remember the first time I saw Elvis. I was sitting on the fourth row. When Elvis walked out, I could have touched him had I wanted to. He was bigger than life, and when I came to my good senses I was standing there with my hands up, like I was at an altar. I totally lost it for a little while. That's the kind of effect he had on you. It was a worshipful situation. But you've got to know why. I love the music."

In that role she voluntarily takes on any outsiders whom she considers to be detrimental to the industry. At one time that included me. I had noticed Hazel during one of my early visits to Nashville. I attended the CMA's annual awards in October and received a pass to sit backstage where various reporters, managers, and publicists gathered to watch the show. Hazel was clearly the center of attention in that room. Like a quiet power broker, she cheered her favorite artists, cried at the speeches, and scowled at any subpar performance. Those around her mimicked her every emotion. With her Mrs. Claus appearance and "Beverly Hillbillies" accent, I thought her a cartoon, an example of the lack of sophistication I had already witnessed in the music industry press. Finally, after several such encounters, I decided to introduce myself. I approached her over a platter of cubed cheese one night at the Opryland Hotel and told her of my plan to write about the music community in Nashville. Without batting an eye, she asked me a question: "Are you a lifelong country music fan?" Somewhat taken aback, I said, "No, I'm a recent convert." "Well, then," she said, "you're no good. With you it's all head and no heart."

Stung, I slithered away.

At times Hazel carried this role as protector of Nashville into public battle. Several weeks after I had my encounter with Hazel, Dick Clark, the renowned host of "American Bandstand" and a four-decade force in the music industry, called a press conference in town to announce that he would be the new executive producer of the nightly talk show on TNN and that Tom Wopat, late of "The Dukes of Hazzard," would be the host. Hazel was not pleased. At the press conference, held in the Country Music Hall of Fame, Hazel deposited herself in the front row. After Dick Clark and the various dignitaries made their pronouncements, they asked if anyone had any questions. Hazel immediately raised her hand. As she did, the head of the Country Music Association, whom Hazel describes as "the very staid Mr. Ed Benson," muttered, "Uh-oh."

"I don't have a question," Hazel said. "I have a request."

"Yes, what is it?" Dick Clark said.

"I just want to ask you not to look down your long nose at us and make us look stupid, like we've got hay in our hair, because we ain't."

Dick Clark was aghast. "I would never do anything like that," he said.

"Dick," she came back, "you know and I know and everybody in this room knows that all Hollywood has ever done is thought we was plum stupid. You make movies like that. You make TV shows like that. You look down on us. You giggle about us—way we talk, way we go to church, way we walk, everything about us. And I'm just asking you not to do that." Then she crossed her arms with an audible harumph.

As soon as the meeting was over, he approached Hazel in the crowd. "Why did you say that hay thing?" he asked her.

"Dick, if you don't ask, you don't know," she said.

"Well, what made you say it?"

"Past experience."

"For example."

"Let's start with Elvis and Steve Allen," she said, taking what was for her a familiar posture, linking Elvis with Nashville, where he recorded. "Steve Allen had Elvis sing to a dog as he sung 'Hound Dog.' I wept. What a sorry thing he did to make Elvis Presley sing to a damn dog. The only good dog is a hot dog or a dead dog anyway."

Dick Clark took a step back. "I see what you're talking about, Hazel. Now, don't you worry. We won't do anything like that."

"Dick," she said, "the worst thing I've ever seen on television in my life was the great Garth Brooks singing 'Friends in Low Places' a couple of years ago on the Academy of Country Music Awards, which you produce. And you had those 'Solid Gold' dancers parading around behind him in bodysuits that looked like they were perfectly nude. I've never been so embarrassed in my life. I almost wept then."

Dick Clark had no response to that. Four months later, when support on Music Row failed to materialize for the new host, Tom Wopat was fired.

"Goooood mahnin', country music fans! This is Hazel, down here in Music Town." In a few minutes, Hazel was introduced by the host in Indianapolis, Jim.

"The first item I have this morning is called 'Leap of Faith.' It's about Faith and Tim going hand in hand to an ob-gyn out there on Murphy Avenue here in Music Town, and then going back several weeks later for an ultrasound. It's a leap of faith, and 'Faithless' has done it again.

"And guess what happened Saturday? Dolly Parton sang 'I Will Always Love You' at her nephew's wedding. Isn't that sweet. That Dolly is a doll. She ain't gonna cure cancer, but she's as nice as they come.

"And guess who was out shopping at Sam's Club, you know the place where you can get things cheaper? The biggest money-making country star of all time, Garth Brooks, along with his wife Sandy and their daughter August." She cackled. "Garth and Sandy are expecting their third child. 'Every time I get my weight down,' Sandy told me, 'I end up pregnant again. He just won't leave me alone.' I send my best to country music's crown prince and princess."

Hazel listed a few more tidbits, but she was clearly building to her one big coup. Wynonna Judd had been, for a decade now, one of Nashville's biggest stars. Though her music helped, it was clearly augmented by her unusual life. "I was totally surprised that a mother and daughter could travel in a bus and not kill each other," Hazel told me. "Remember, Wynonna was a sloppy teenager who ate candy and threw the paper on the floor, while Naomi was a perfect mother. Naomi was driven. She'd be over there at Maude's Restaurant all the time driving that red car. I had no idea she had a grown child." Later, when Naomi bought Maude's and turned it into Trilogy, Hazel had a story. "Naomi Judd has opened a new eatery called Trilogy near Music Row. Lord, I wish I was Wynonna. Now she's got a place to eat and doesn't have to pay."

This morning the news was bigger.

"Now, before I go any further, I got a scoop for all you Wynonna fans out there. By this time Monday, Wynonna Judd will be Mrs. Arch Kelley III. 'No,' you say. How do I know this? On Monday I was walking at Rivergate Mall and I ran into the guitar player for the Nashville Bluegrass Band. He said Ashley Judd had called him over the weekend and said Wynonna's getting *married* on Sunday and wants us to play at the reception. Ashley said she and Wynonna had gotten custom-made dresses, but that Naomi had gotten her dress over at McClures. Off the rack, ladies! Don't you know that Naomi could have just killed her for saying that. Anyway, don't you worry, I'll be back next week with all the details."

Hazel slunk back in total satisfaction, almost as if she had just had sex.

"Well, that's all I've got for today. I have my poem ready, though. With all this wedding excitement down here, I thought it was my Christian duty to help you women out there in Radioland save your marriages, so here we go. Listen closely for the name of my sponsor, the Little Bit of Texas dance club: 'Girl, here is the way to keep your man going right / Keep him well fed by day, and well loved by night / After early romance show him a Little Bit of Texas floor / Make that man dance and dance, till he wants nothing more.'

"Good weekend, America. Love ya, Jim. Bye-bye!"

At the sight of the first helicopter in the sky above Trilogy, the group of paparazzi huddled on the curb seemed, for a second, to be truly impressed. At the sight of the second, they chuckled. And by the time the fifteenth police car with flashing blue lights passed the pink stucco building, the photographers had managed to remuster their disdain.

"Who the hell cares about Wynonna anyway?" said Fred from the evening *Banner*. He had long hair, a backward baseball cap, and a distinctly unnuptial attitude. "People in this town make such a big deal out of country music. I remember when rock 'n' roll ruled, even here. I used to manage the Exit-In back in the seventies when Robert Altman was in town shooting *Nashville*. The Police came in, before they were big, Emmylou Harris, even Dylan. One night Jimmy Buffett sat at my bar and threw up on someone next to him. That's what it used to be about. Now we've got helicopters, consultants, and *publicists*." Just then a white limousine drove by. "It's them, it's them!" someone shouted, and Fred went sprinting for the corner. When the limousine didn't stop, he returned to the curb.

"Why am I even bothering?" he said. "Somebody inside will get a picture," he said. "A butler. A maid. A driver. And they'll sell it to the tabloids for a fortune."

He was right. At exactly the same time Fred was camping outside Trilogy, hoping to get a shot of Wynonna he could sell to the *National Enquirer*, Becky Goode, a stringer for the *Globe*, was about to make the most daring move of her fledgling career as a tabloid reporter by attempting to sneak into the reception itself. Her move was so bold that the handful of other tabloid reporters outside the restaurant watched in stunned silence as the youngest member of their underground pack stepped out of her car in a killer black dress with white polka dots and proceeded to walk up to the door. In her bag she was carrying three portable cameras, a tiny pad, and $10,000 in cash.

Once completely unconnected to the music business, Becky's life was totally redirected by the boom of the 1990s. She was born in Hendersonville, just north of Madison, Hazel's beloved "butthole of Nashville." Her father sold cars. Her mother worked in the home. After graduating from the University of Tennessee at Knoxville, Becky sold credit reports for a few years, then advertising. Eventually she trickled back to Nashville and settled in a spacious apartment not far from Belle Meade, in the heart of 37205, the poshest zip code in town. There she went to work selling credit information to private investigators, reading *The New Yorker* every week, and nursing her ambitions to be a writer. When the *National Enquirer* placed an ad in the local paper saying it was looking for sources, two friends insisted she call.

"I didn't hardly know any music people," Becky said. She was bobby-sock blonde with freckles, a penchant for brightly colored clothes, and an accent that could swing from Southern cheerleader to New York cynic in the flash of her charming grin. "But I am from here, and therefore I know all the people *around* the industry—the doormen, the chauffeurs, the nannies."

The first thing the *Enquirer* asked her to do was find the address of one of the members of Alabama. "I felt it was a test," Becky said. "I hung up the phone, made two phone calls, and had it in ten minutes. They were real excited. They paid me seventy-five dollars." Eventually the *Enquirer* flew her down to Florida for a two-week stringer orientation. "I think it just flatters the stringers, makes them feel needed," she explained in the uncanny way she had of getting to the heart of people's motivations. "But it was a really neat experience. I interviewed a Miss America and got my name on a story. My mother was thrilled."

Back in Nashville, Becky went to work on the types of stories the *Enquirer* craved—Dolly Parton visits a chiropractor, Vince Gill goes on a mystery diet, Conway Twitty's children try to exhume his body. First purchased in the 1950s by Generoso Pope, the *National Enquirer* (it had been the *New York Enquirer)* was a source of close-to-the-street news. In the 1960s, Pope observed that the two subjects Americans most like to talk about were health and celebrities. He changed the focus of the paper and circulation soared, reaching 5 million readers a week at its peak. In time the *Enquirer* began paying sources for information, most famously showing a picture of Elvis on its front page taken after his death. Though cameras were not allowed at his viewing, an *Enquirer* reporter learned that after a corpse is embalmed its hair grows for twelve hours and that a barber would have to come in to shave Elvis before the funeral. According to a story in *The New York Times* about the new respectability of tabloids in the 1990s, the reporter paid the barber to smuggle in a tiny camera and take a picture of the King. That issue sold 6 million copies.

Though Elvis might have provided the paper's biggest coup (its biggest disaster had occurred the previous year when Carol Burnett successfully sued the paper for falsely reporting that she had been drunk at a restaurant), the editors had little interest in country music. With content driven by box office grosses, television ratings, and album sales, country hardly registered. When country sales started soaring in the 1990s, though, and its stars started landing on the front pages of mainstream publications, the tabloids quickly followed. Within no time, 5 to 10 percent of the paper was devoted to covering Nashville. "It's a formula," Becky said. "They'd want something every week if they could get it." Because of the tabloids' enormous circulation, many stars, particularly newer ones, actually try to place stories in the papers. Before Shania Twain became big, her publicist routinely pitched her sexy body and heart-wrenching history to the tabloids, emphasizing her parents' tragic death in an automobile accident; after she became big, when Shania's grandmother accused her of fabricating her Native American heritage, the tabloids covered that, too. In the tabloid universe, Reba was always good for a story; Vince, if he did anything wrong; or Garth. Dolly is almost a guarantor of space. Becky once rode up and down a hospital elevator fifty-four times trying to find out if Dolly's brother had cancer. He did.

But for sex appeal, star power, and downright kookiness, nobody beat the Judds. Almost blue-plate royalty in Nashville, the Judds had mythic reputations in town. Becky remembered watching them as a child on "The Ralph Emery Show" on Channel 4. "They would be there every Monday and Thursday," she said. "Naomi would give out these lye soap recipes made from boiling down pig fat. And Channel 4 would have the graphics to go with it: 'Half a cup of animal fat . . .' I said to my mother, 'Do they really use lye soap?' And my mother said, 'No, it's a marketing thing.'" Even at that early age, Becky felt sympathy for Wynonna. "She was like me," Becky said. "I was twelve years old, looking at this awkward girl with Bozo hair sitting next to her mama. Naomi would sit there and talk, talk, talk. And Wynonna never said a word. I've always wanted to know what she thought."

Increasingly, I was developing that urge as well. Though I had been largely received on Music Row as what publicists call a "cred" reporter, someone who can get publicity-craving country stars into national publications, I was told I would have little chance of penetrating the cocoon around the Judds. Wynonna rarely does interviews and was famous for almost never talking in public. This is why the tabloids loved her: She was mysterious. Becky, for one, had been trailing her for years. Wynonna goes on a date. Wynonna goes on a binge. Wynonna goes on a diet. "I had rung her gate so many times it wasn't even funny," Becky said. "But nobody ever answered. The paper kept saying, 'Go out there. Try it again.' They're very persistent. That's why they're good."

The biggest story surrounding Wynonna had been the birth of her son. "I was in the hospital the whole time," Becky told me. "It was Christmas. They couldn't find a photographer to work here, so they flew in a photographer from New York. She brought her parakeet, and it ended up having a brain tumor while in Nashville. She was supposed to be in the hospital pretty much twenty-four, seven, to shoot anybody coming or going. But every time I needed her, she was either in the hospital gift shop crying over her horoscope or running back and forth to the hotel trying to comfort her blasted parakeet. We didn't get anything."

A year later, Wynonna was pregnant again and word had started spreading around Nashville that this time she was going to get married. For months it was the leading topic of gossip on Music Row. The first date the rumors settled on was Christmas Eve in the Ryman Auditorium. Becky and a photographer huddled freezing in the alley all day, only to be proven wrong. Finally, three weeks later, Becky got the tip she was looking for.

The first thing Becky did when she learned the date of the wedding was call everyone she knew and try to get an invitation to the reception, which was being held at Trilogy. (The ceremony was being held at Christ Church in Brentwood, Wynonna's Pentecostal congregation.) "I can't find *anybody*," she complained to me just days before the big event. Since meeting Becky when I first came to town (by that time, she was working for the *Globe*, who had given her a better offer), she had stunned me with her access to information. A rumor had been going around that Shania Twain had made a life-sized nude poster of herself to give to her husband, Mutt Lange. "I don't believe that," I said to her dismissively. "I talked to the person who developed it," Becky responded. One Friday night Travis Tritt's publicist called me from her car phone to tell me that Travis had just told her he was getting married. "I was so surprised I was hyperventilating," his publicist said. The following day an official press release was put out. "Did you see that Travis Tritt is getting married?" I said to Becky. "Oh, that," she said blithely. "I sent that in two months ago." Through all my time in Nashville, I never heard anything on Music Row that Becky didn't know first.

This time, though, unable to turn up an invitation (or even the exact hour), Becky resorted to money. She had lunch for three days in a row at Trilogy and hinted to the waitresses that they could make $500—or even more—for a photograph of Wynonna cutting the cake. They refused. She went to see a neighbor who worked there as a busboy. "I've already been warned about you," he said. "Please leave." Finally, the night before the wedding, Becky learned that the rehearsal dinner was being held in a back room at Choices Restaurant in Franklin. With her editor from the *Globe*, she went to eat at the restaurant. Once inside, Becky wandered back and forth between her table and the ladies' room, hoping to bump into someone who had been at the party. It was on one such trip to the ladies' room that Becky caught her first big break. Ashley Judd was standing at the sink.

"Oh, hi!" Becky stammered, resorting quickly to her best cheerleader accent. "I'm a big fan. I loved you in *Heat*!"

Ashley smiled and thanked her.

"I saw that interview you did on TV," Becky continued. "I love how you promote Franklin."

"Yes, we have such a charming little town here."

Becky knew she wouldn't have much time with Ashley. What was the most important thing to learn? She took a breath. "So," she said, "what time is Wynonna getting *married*?"

Ashley started to answer—"Sister's getting married . . ."—when suddenly a voluminous flush came from a nearby stall and out walked Naomi, flaming red hair, pleasant smile, and deep, frightening glare. Ashley went dumb. The two Judds washed their hands side by side and quietly left the room.

By half past six the following evening, the mood in the bushes around Trilogy was growing surly. The paparazzi had chased so many limousines they were becoming impatient. The two dozen police officers kept shoving the onlookers back into the gutter. The various television film crews were jabbering into their walkie-talkies. One cameraman told me he had followed the couple since early in the morning when a limousine left Wynonna's farm and headed to a nearby apartment complex, at which point four women in bright red wigs came bounding from the car and ducked into four separate minivans. Confused, the entourage of reporters that had been tailing the limousine was left in a funk.

Meanwhile, at the door to Trilogy, Becky Goode was about to face her first big challenge. A giant sign taped to the front door said PLEASE

HAVE INVITATIONS READY. Becky didn't have an invitation. She didn't even have a date. As she approached the guards though, the people waiting in line were so anxious to find their names on the RSVP list that they swarmed the security table, overwhelming the guards and thereby freeing Becky to slip by undetected. Once inside, she quickly disappeared into the horde. "It was packed," she said. "Wall-to-wall people. Peter Frampton was there. Kenny Rogers. Steve Winwood. Donna Summer. I saw this strange guy with long hair and somebody said it was Pee-Wee Herman. His sister lives in Nashville." More surprising was the number of Becky's friends who were there. "It was like going to a big party in Nashville," she said. "I kept saying, 'Are you friends with Arch or Wynonna?' And they would say, 'Well, we knew Arch when he was fifteen . . .' or 'We bought a boat from Arch ten years ago.' He must have invited anybody he ever met. I think he was showing off."

No sooner had Becky gotten in than a set of blinking police lights appeared atop the hill. A line of vehicles approached. Three police motorcycles led the way, followed by two squad cars, a minivan with security, a white limousine containing Naomi, her ex-husband, her current husband, and Ashley, and, at the tail end, a 1952 black Cadillac, driven by Arch and containing his bride. It was as though the Emperor of Japan was making a state visit to Trilogy. As soon as the cars came to a stop in front of the valet parking stand, the horde of photographers rushed forward with such hysteria—cursing, screaming, flashing, and shooting— that they forced themselves to tumble like a band of Keystone Kops directly into the thicket of shrubs. Ashley emerged first and glared at the comic assembly. Naomi and her husbands followed, hurrying the few steps to the door. Finally Arch, in a white Nehru suit, scurried around to retrieve Wynonna. She wore a cream satin dress with bouffant sleeves and a long satin train. A garland of wildflowers and buds was woven into her hair. She was, as Hazel would say, "in the family way"— four months and counting.

As soon as the family stepped into the restaurant, Becky Goode faced a new challenge. Larry Strickland, Naomi's current husband, was tipped off that she might be inside and went looking for her. Soon enough he found her, hovering behind the shrimp. Larry lunged at her, but Becky stepped aside and quickly fled to the ladies' room. Minutes later, two male security guards hurried into the bathroom and, feeling out of place, dragged out the first blonde they saw. It wasn't Becky, but a friend of hers. The woman started crying. "What do you *want*? I didn't do *any-*

thing! It's *her* that you want." Becky considered fleeing again, but knew she had little chance. Soon two female security guards entered the ladies' room and apprehended their suspect. "They went through my purse," Becky recalled. "They patted me down. They grabbed my portable cameras. I was just so delighted I had made it that far." Outside the paparazzi, having righted themselves, applauded when Becky was escorted from the party. She went across the street to a restaurant to meet her editor. A few minutes later, her cellular telephone rang.

It was George Alonzo, calling from inside Trilogy. Some friends had told him that Becky would offer money for pictures. George used to work for Wynonna and Arch. He was in his early fifties, balding, and overextended—ample motivation, Becky figured, for him to help. On the phone, he told Becky that he had taken two photographs of Wynonna and Arch cutting the cake.

"Where are you?" she said. "I'll be there in a minute." She told him the pictures could be worth $500 to a $1,000, maybe more.

"Uh, well," he said, "I'm taking this girl home."

"Okay," she said, "I'll wait at your place."

"Uh, well," he said, "I'm going to spend the night with her."

"All right," Becky said, "I'll come to her place."

"I can't do that," he said. "It wouldn't be cool. I'll call you at eight in the morning."

At seven, Becky started calling his house. There was no answer. She kept calling. At nine, he called her. "I'm running late," he said. "I can't meet you until ten." At ten, she and her editor, Jim, went to his apartment. Nobody was home. Becky checked the front, the back, the garage. He wasn't there. At eleven-thirty he called. "Meet me across the street at Woodmont Baptist Church," he said. They raced across the street. He wasn't there. They waited, but he never came. They waited some more. After two minutes Jim said, "He isn't coming. He wanted us to leave so he could pull into his garage." "Shit, I'm sure that's what it is," Becky said. She leaped from the car and sprinted across the street. She rang the doorbell. Nobody answered. She looked in the window. It was dark. She ran around to the back and looked in the garage. There was his Lexus, parked at an angle, and behind it a red rental car. In the backseat was a copy of the *National Enquirer*. Becky pounded the hood and burst into tears.

When Jim arrived, Becky got on the mobile phone and called his house. "All right, we know you're in there," she said. "We know the

National Enquirer is there. We'll offer you more money: five thousand dollars."

"But I don't have any pictures," he said.

"Ten thousand."

He didn't answer.

"Fifteen."

He refused to come to the door.

"Twenty."

He was silent.

"Thirty."

Finally he hung up.

"He's a first-class jerk, if you ask me," Becky said. "First he sold out Arch and Wynonna. Then he sold out me."

Two days later, the *National Enquirer* ran five photos of Wynonna (including one cutting the cake), plugged on its front page as a WORLD EXCLUSIVE: NAOMI ENDS FAMILY FEUD AS WYNONNA WEDS. Pleased with their coup, their story was positive. "Naomi toasts her daughter's new hubby—the man she once called a 'gold digger.' 'Arch, I know you are going to take good care of my daughter. I thank you for that.'" Two days later, the *Globe* ran a photo of Wynonna on its front page that clearly appeared doctored. Scooped, they spun the story negative. BRIDE WYNONNA'S TWO-RING CIRCUS. "200-pound pregnant singer tripped as she walked down the aisle. Groom and his mother-in-law fought like cats and dogs. Hungry guests fled Naomi's restaurant to eat across the street."

The following week I received a call from Wynonna's publicist. Wynonna was fed up with all the tabloid coverage of her; she was finally ready to talk.

FOUR

THE TOWN

It was the funeral to which they all came. They gathered in a line of town cars and pickups in the sheeting rain just south of Nashville. They nodded solemnly to the waiting cameras as they straightened their black suits, winter hats, and dark glasses and hurried between the stark white columns into the Brentwood United Methodist Church. They paused in the marble foyer to shake hands with the man stooped under the doorway who just that morning had pulled down a few limbs of magnolia from his yard and placed them alongside the plain white lilies strewn abreast his wife's cherry coffin. They shuffled through the doors and into their seats, spilling from the pews and into the aisles until they finally had to prop open the doors. And still they came, in the rain, reaching two thousand strong. Some had driven from half a lifetime ago, from the days when radio was their best friend and the Grand Ole Opry had given life to them all. Grandpa Jones, Porter Wagoner, Jan Howard, Billy Walker, Bill Anderson. Some had come from just up the road, from a newer time, when colors were brighter and lives flickered faster across the screen. Barbara Mandrell, Reba McEntire, Vince Gill, Wynonna, and Garth Brooks, his wife, and their elder daughter. But all of them had come to honor the woman who better than anyone else had bridged the gap between then and now and who alone among them had also bridged the gulf between the country where they had come from and the city where they now lived.

Minnie Pearl was unique in Nashville, a remarkable time capsule who embodied the ongoing clash of classes that has always defined the relationship between country music and Music City. Born Sarah Ophelia Colley, she came from a line of silk stockings and privilege, but gave it up for a life as a traveling thespian. As Cousin Minnie Pearl, the cotton-legginged good ol' girl who grew out of those years performing across the underbelly of the South, she earned back the money she had shunned as a girl and gained the one thing that had always eluded her colleagues: genuine social status. Her death, coming just days after Wynonna's wedding, marked a rare moment in the life of the community. With both the founding and the current heartbeats of the Opry there, along with the political and social backbone of the city (the sitting governor came, along with three of his predecessors, all of whom had lived next to her 6-acre estate in the fashionable enclave of Oak Hill, not far from tony Belle Meade), it could have been a moment in which the two dominant sides of the city—Belle Meade and Music Row—finally achieved the union they had been inching toward for decades.

Sarah Ophelia Colley was born at a time before Music Row, or Belle Meade, even existed. In 1912, Nashville was a provincial, soot-covered city of roughly 100,000 people about to become the centerpiece of a new urban migration across the South. Laid out on the gentle, if rocky terrain that dominates middle Tennessee, the city was perched on a serpentine stretch of the Cumberland River that runs five hundred miles from the Appalachian Mountains to the Mississippi River. The state was so long and thin, in fact, and its cultural heritage so diverse, that one saying about Tennessee would later proclaim: "There's blues to the west, bluegrass to the east, and a whole lot of country in between."

The state capital, Nashville, is indeed halfway between Memphis to the west and Knoxville to the east, as well as Louisville to the north and Birmingham to the south. Founded in 1780 by a band of Carolinians who trekked three hundred miles on the frozen Cumberland, then later exploited the unfrozen river to transport cotton and other products to New Orleans, Nashville (named after Revolutionary War General Francis Nash) has always been something of a benchmark town. In the eighteenth century, it served as a hub for a series of Indian trade routes, including the Natchez Trace, that early settlers used to move goods from the Ohio Valley to the Mississippi. In the nineteenth century, three early presidents, Andrew Jackson ("Old Hickory," number seven), James Polk ("Young Hickory," number eleven), and later Andrew Johnson (number

seventeen, the only one impeached), came from Nashville. And, because
of its network of railroads, the city was a pivotal resupply town during
the Civil War, first for the Confederacy, then, following a brief battle in
1862, for the Union.

In the century that followed, Nashville fluctuated between two conflicting poles. One was as a center of education and enlightenment, the
"Athens of the South." Vanderbilt University, founded in 1873 by New
York–born industrialist Cornelius Vanderbilt, was a leading Methodist
institution aimed at "strengthening the ties which should exist between
all geographical sections of our common country." Fisk University,
founded six years earlier for similar missionary purposes, was the first
African American university in the country. In 1897, on the commemora-
tion of the state's one hundredth birthday, the city of Nashville decided
to promote its enlightened image by constructing a 228-foot-long, 65-
foot-high full-scale concrete replica of the Parthenon, complete with a
frieze depicting Zeus lying ill on the ground, his son Hephaestus acciden-
tally striking him on the head with an ax, and Athena springing from
the wound. Today the world's only fully constructed Parthenon (the
other one long since having been marauded) still stands proudly across
the street from Vanderbilt, separated only by a Wendy's. Its inside now
has a 41-foot-10-inch-tall statue of Athena; its outside provides the backdrop for a laser light show every summer.

Nashville's alternative image, though, is as a center of evangelical
Christianity and Southern spirituality, the "Buckle of the Bible Belt."
Because of its central location, Nashville has long been a Mecca for
religious institutions, with more churches per capita than any city in
America. It also became headquarters for operations by various denomi-
nations of Methodists, Presbyterians, and Baptists. The association has
not always been positive. In 1925, the state came under ridicule when
public school teacher John T. Scopes of the east Tennessee town of
Dayton was put on trial for teaching evolution, which was against state
law. H. L. Mencken, who had earlier branded the South the "Sahara of
the Bozart," reported from Dayton that it was "the bunghole of the
United States, a cesspool of Baptists, a miasma of Methodism, snake-
charmers, phony real-estate operators, and syphilitic evangelists." Gran-
diloquence aside (though with a catchy tune, Mencken's words could
have made a good country song), the image he described has never quite
disappeared. The week of Minnie Pearl's funeral, Nashville once again
found itself on the front page of national gaping when the state legisla-
ture nearly passed a new version of the old Tennessee "Monkey Law"

that would have again outlawed teaching evolution as fact. Not long after that, the state senate did, in fact, pass a bill that urged Tennesseans to follow the Ten Commandments and to post them in their homes.

From its inception, this struggle for Nashville has always, to one degree or another, come down to class. Nashville has long been dominated by a large ruling elite, made up of families in finance, banking, and insurance. The centerpiece of their world was Belle Meade, a former horse plantation turned residential community. Belle Meade had been one of the most storied names in American equestrian history. In 1881, Iroquois, racing under the plantation's maroon silks, became the first American horse to win the English Derby. In the early twentieth century, suffering financially, Belle Meade was sold to a group of investors, who auctioned the horses, freed the elk, and converted the deer park into tracts of land. Soon socialites (and socialite wannabes) were flocking to this new enclave of magnolias and exclusion, where they shielded themselves from their backwoods neighbors. Gentrified Nashvillians were particularly concerned that the Scopes Monkey Trial had darkened their reputation. The image-conscious leaders of Vanderbilt even announced a so-called "answer to Dayton" in which they would "advance the South" by building more laboratories and teaching more science. But just as blue-blooded Nashville had beaten back the stench of Dayton, another cultural hurricane arrived from rural Tennessee to muddy their dinner party.

In a sense, the early history of country music can be told as a simple story: Country comes to town. It's the story of rural Americans who left their agrarian roots for greater opportunities in the city, then used music as a way to preserve and to reclaim what they'd left behind. The music that came to be called "country" first came together about a century ago from an eclectic array of sources, including Irish and Scottish string music, Mississippi blues, Christian hymns, and, later, jazz. At the time few people considered any of this music more valuable than, say, nursery rhymes. By the 1920s, when the phonograph was becoming a common household item, big-city businessmen eventually realized the financial potential for selling this kind of music. The "Big Bang" of country music, known then as "old-time music" or "hillbilly music," occurred in August 1927 when Ralph Peer, an admittedly condescending carpetbagger from New York, recorded songs by the Carter Family and Mississippi crooner Jimmie Rodgers within days of each other in Bristol, Tennessee. Eventually this music found a home on burgeoning commercial broadcasting outlets, especially after rural listeners flooded myopic big-city programmers with letters of appreciation.

As contradictory as it seems, one principal reason for the success of country music at this time was that vast numbers of people were actually *leaving* the countryside. The 1920s brought the beginning of a wholesale migration of rural Southerners from the country to the city. They came for jobs, education, and fun. Nashville was particularly changed by this migration. Thirty thousand migrants moved to Nashville in the 1920s— that in a city that began the decade with only 150,000 people. Country music, with its themes of religion, family, and home, became a link to the places these people had left. "It evokes a warm image of the culture of the common folk in the South and a plea to preserve it against incursions of the modern world," wrote Andrew Lytle, one of a group of radical thinkers at Vanderbilt in the twenties known as the Fugitives, who advocated a return to Southern Agrarianism. "Throw out the radio and take down the fiddle from the wall. Forsake the movies for the play parties and the square dance." The performers who joined the Opry in those early decades, the Delmore Brothers from northern Alabama (1932), Roy Acuff from eastern Tennessee (1938), and Bill Monroe from western Kentucky (1939), embodied this sense of nostalgia. Many of Monroe's classic songs, like "Uncle Pen," were in deference to the family he left in Kentucky after he moved to Illinois during the Depression.

Few people understood the class tension that many of these agrarians faced when they got to town better than Sarah Colley. Sarah was born to money. Her mother, Fannie, the "epitome of a Southern lady," grew up in a home with Oriental rugs, Shakespeare, and white-coated servants. Fannie horrified her parents by marrying a lumber company operator, Thomas Colley, and moving with him to Centerville, fifty miles to the south. The couple thrived, though, eventually having five children (Sarah Ophelia was the youngest). Like many such families in the South, musical taste was tied to social status. Sarah's mother disliked hillbilly music, preferring symphonies and show tunes. When traveling theater troupes passed through town, she would accompany them on the piano at the local opera house. Sarah's father, by contrast, loved rural music, particularly the Grand Ole Opry.

Sarah, spoiled, a clown from the beginning, fancied herself on stage, but it was clearly the stage of her mother. During the Depression, she bypassed college for two years at Nashville's Ward-Belmont finishing school (cost: $1,200 a year), known for its training in dramaturgy, where she "worshipped at the feet of Katharine Hepburn, Lynn Fontanne, and Bea Lillie," and dreamed of Broadway. Located at the top of what is now Music Row, Ward-Belmont (started by two women from Belle Meade) was headquartered in an 1850s Venetian-style mansion. "While America was in the throes of a major financial depression," Minnie remembered, "Ward-Belmont remained aloof, untouched, an oasis of Bourbon opulence. My first night in the dining room, where round tables had been set with lovely linens, beautiful china, sparkling crystal, and gleaming silverware, I sat with girls who talked of debutante balls, vacations in Europe, winter homes in Palm Beach, English nannies, and upstairs maids."

After graduating, Sarah Colley shunned that gilded legacy and set out in pursuit of a career in the theater. Over the objections of her parents ("Show people were still considered wild and bohemian," she recalled), she joined a theater troupe her mother had accompanied on the piano, the Wayne P. Sewell Company of Atlanta. For ten dollars a week, she traveled the Southeast performing in stock productions. It was during one such trip that Sarah became stranded in a snowstorm on Sand Mountain in northern Alabama, then, as now, one of the poorest areas in the country. The school principal asked an elderly woman who lived in a nearby cabin if she would house their visitor. "I've had sixteen young'uns and never failed to make a crop," the woman told her. Ten days later, she had found what she considered a seventeenth. "Lord a'mercy, child, I hate to see you go," she told Sarah. "You're just like one of us." For weeks afterward, Sarah mimicked the woman, slowly developing her into a character of sorts. Looking for a suitable name, the actress chose two familiar Southern names: Minnie and Pearl. "She didn't mean to be funny, she just was," Sarah Colley said later in the habit she had of speaking of Minnie in the third person. "I found it very interesting that I carried so much of her persona and character. That's when I said good-bye to Katharine Hepburn."

In 1940, out of work now and penniless, Sarah was encouraged to audition her character for a guest spot on the Grand Ole Opry. She was accepted, but officials were so worried that her privileged background would make her act seem like a putdown that they scheduled her first appearance for 11:05 P.M., after most listeners would have turned off their radios and most guests at the War Memorial Auditorium would have gone home. Following a classic Opry commercial for Crazy Water Crystals Company (a white concentrate of mineral water taken from the Crazy Well in Texas that claimed superhuman laxative powers), Minnie told jokes for three minutes. Asked for her appraisal afterward, Sarah's mother said, "Several people woke up." In the days that followed, though, three hundred people mailed letters of support to the station. The next weekend Minnie Pearl was offered a permanent job as the only female act in the Opry, "the first woman to scramble with my fingernails up the side of a wall to try to get some recognition in a man's world."

From the beginning, though, it was Sarah's class, more than her gender, that was an issue. Minnie was clearly a rube. She wore a pair of one-strap Mary Jane shoes, cotton lisle stockings ("You never saw country girls wearing silk stockings. They couldn't afford them . . ."), gingham dresses, and what she called a "tacky straw hat" with the plastic flowers and the price tag still dangling from its brim: $1.98. In later years, the hat and price tag would become so famous that a letter addressed only with a drawing of her hat eventually made its way to the Grand Ole Opry. The act itself was also based around rural humor, specifically Minnie's "family" in fictitious Grinder's Switch: "Uncle Nabob takes a drink every now and then to steady his nerves. He gets pretty steady. Sometimes he doesn't move at all." "I remember the first time we bought Uncle Nabob a store-boughten suit. It had two pairs of pants! That was nice for the winter, but wearin' both pairs got awfully hot come summer." Minnie also made fun of herself, particularly her misfortune with boys. "The robber said, 'Gimme your money.' I said, 'But I haven't got any money,' so he frisked me and said, 'Are you sure you ain't got any money?' I said, 'Nossir, but if you'll do that again, I'll write you a check.'"

Sarah, however, grew uncomfortable with this pose. "I was embarrassed about what I was doing," she said. "I was twenty-nine years old, a mature young woman, and I just couldn't see my way clear to cut loose and act a fool." Her embarrassment was made more acute by the fact that her friends from Ward-Belmont looked down on the Opry as a haven for country bumpkins. "When I first came to Nashville, country music was not welcome in this town," she said. "Many of the local citizens wanted to sweep us under the rug. They felt that the Opry was a demeaning image for their city, which they were promoting as the 'Athens of the South.'" They looked down on women even more. "Would a nice girl be traveling all over the country with a bunch of hillbilly musicians? Certainly not!" The musicians, she confessed, harbored a similar prejudice. "In all honesty, we didn't care that much about Nashville either," she said. "We only came here because of the Opry. But when they started talking down on us, we developed a saying:

'Nobody likes us but the people.' It made us stronger and more determined than ever to prove them wrong. What finally changed their minds was that we brought all that money back and started spending it *here*. That they didn't mind."

That didn't happen until after World War II.

As it was for the entire South, World War II was a boon for country music. First, it only heightened the rural migration of the previous two decades. The war brought millions of Northerners to the South for military training and brought millions of Southerners into service with Northerners. Country music was uniquely suited to telling the stories of these soldiers. "I'll Be Back in a Year, Little Darlin' " became "Have I Stayed Away Too Long," giving way to "Stars and Stripes on Iwo Jima," then "The Soldier's Last Letter." This exposure, coupled with huge population shifts, led to unprecedented growth when the war was over and to a golden era of traditional country music. In 1944, *Billboard* magazine estimated that there were six hundred regular country radio shows in the United States and that they played to a combined audience of around 40 million people—one third of the population at the time.

As would be the case throughout its history, country grew by adapting to the times. While early country music had been based around fiddles, harmonicas, and assorted backyard instruments, the new sound (called "honky-tonk" after the new breed of beer joints located on the outskirts of Southern towns) included brasher instrumentation—string bass, more rhythmic guitar, even electric guitar—and edgier, more modern themes. There were fewer songs about pastoral dislocation and poor Grandma back on the farm and more talk of sex, drinking, and families falling apart. With new stars—Hank Snow, Hank Thompson, and the prince of urban dislocation, Hank Williams—the Opry, in particular, reached more listeners than ever. Country was also popular in the Far West, with Okies who had come to California during the Dust Bowl, and in the Northeast. A group of Opry stars even played Carnegie Hall in October 1947 and sold out. "The barriers were coming down," boasted Minnie Pearl, one of the performers on that show. "I think that was the first time I realized how far-flung country music had become."

Meanwhile, at exactly the same time country music was spreading across the nation, the industry was consolidating in the one city perhaps most hostile to it: Nashville. Country is centered in Nashville today for three basic reasons, all of which came to a head in the 1940s. The first was money. Before the war, ASCAP, the American Society of Composers, Authors, and Publishers, the performing rights organization that collects royalties for music publishers, mostly ignored Nashville songwriters. When ASCAP raised its rates in 1941, though, a rival organization, Broadcast Music Incorporated (BMI), rose up to challenge it. BMI not only broke the monopoly, but also welcomed country writers into the fold, thereby allowing more money to stay in Nashville. The second reason was geography. In the mid-1940s, several producers began cutting records in Nashville to take advantage of the growing number of musicians in the area who were drawn by the Opry. Musicians preferred Nashville, as opposed to, say, Atlanta or Dallas (which also had Oprylike programs), because of its central location. Thirty states are within six hundred miles of Nashville, a number that's significant because it's the amount of territory a bus can cover overnight. In the early fifties, Owen Bradley, a former WSM engineer, set up a studio in a Quonset hut in a rundown neighborhood not far from Vanderbilt and around the corner from the studio where Wade Hayes would record his albums. The Quonset Hut, as it was known, became the magnet for a ramshackle collection of recording studios, labels, and management offices that opened up shop in converted Victorian homes and consistently unattractive stucco low-rises in what came to be called Music Row.

The third reason the industry gathered in Nashville was music publishing (the term comes from the era when music was "published" in sheet form). As with other forms of music, publishing has always been the linchpin to country music—as opposed to, say, records—because songs have the ability to make money years after they are first recorded. They can be cut—or "covered," as the term goes—by other artists; or they may be used in television shows, movies, or commercials. In 1942, Fred Rose, a songwriter with a Tin Pan Alley background, moved to Nashville from Los Angeles because his wife, a Nashville native, was homesick. With Roy Acuff he founded a publishing company that coddled Nashville artists precisely because New York companies had so ignored them. One Acuff-Rose song in particular, "The Tennessee Waltz," first recorded in 1948 and recut by Patti Page in 1950, sold 4.8 million records in a single year, earned them $330,000, and single-handedly paved the way for Nashville's mainstream success. By 1960, *Broadcasting* magazine reported that one half of all American recordings came from Nashville. The city that liked to portray itself as the "Athens of the South" was now home to 1,100 musicians, 350 songwriters, 110 publishing houses, and 35 recording studios. The industry that liked to think of itself as a hillbilly family business now brought in over $40 million a year.

The presence of all that infrastructure dramatically changed the industry. Suddenly making money—feeding the beast, as it were—became central to Music Row. This was particularly noticeable during the first big bust that hit country music: rock 'n' roll. When Elvis Presley burst onto the music scene in the mid-1950s he immediately sucked fans away from country. As Bob Luman, later a Nashville recording artist himself, recalled of first seeing the King:

> This cat came out in red pants and a green coat and a pink shirt and socks, and he had this sneer on his face, and he stood behind the mike for five minutes, I'll bet, before he made a move. Then he hit his guitar a lick and broke two strings. I'd been playing ten years and hadn't broken a total of two strings. So there he was, these two strings dangling, and he hadn't done anything yet, and these high school girls were screaming and fainting and running up to the stage, and then he began to move his hips real slow like he had a thing for his guitar . . . That's the last time I tried to sing like Webb Pierce and Lefty Frizzell.

For many more, it was the last time they wanted to listen to Pierce and Frizzell. Audiences, particularly young people, abandoned country in droves. Radio dropped country for rock. Even the Grand Ole Opry lost half its audience. This left Nashville with a choice: Adapt or close up shop. The result was the Nashville Sound, the first significant attempt to tinker with country music to make it more appealing to a mainstream audience.

Stretching throughout the 1960s, the era of the Nashville Sound (the term refers to the new sweet sound of country records; country without the twang) is still viewed with deep ambivalence. Some see it as a brilliant adjustment, or "Chet's Compromise," after Chet Atkins, the virtuoso guitarist and RCA executive who added strings, drums, and creamy background "ooohs" to the records of Jim Reeves and Don Williams. Others viewed it as abandoning tradition. Both Reeves and Patsy Cline, for example, began singing honky-tonk and wearing Western clothes, but soon drifted into evening wear and cocktail-party stylings. What was clear is that country music had (at least temporarily) abandoned its dusty boots for a new uptown image. Commercially, it worked. The number of radio stations playing the music soared from 81 in 1961 to 606 in

1969. By 1970, country music was earning $200 million a year, five times the amount of a decade earlier. Even the stuffed shirts of Belle Meade could no longer ignore the reality of the new money machine of Music Row: Country had not only come to town, it was reshaping it year by year—one guitar-shaped swimming pool at a time.

In the winter of 1970, a group of prominent Belle Meade women set up a meeting with Buddy Killen, a former Opry bass player turned music publisher who was one of the first Music Row parvenus to attempt to penetrate Nashville's standoffish elite. "When I first came to Nashville, there was an attitude toward country music that if you don't feed it, it will go away," Killen remembered. A big man with a deep voice and large rings on his fingers, he sat behind his white wooden desk, his cowboy boots splayed out on his white plush carpet. "People were ashamed of country music. They just made excuses for us." Killen, though, craved the acceptability and made a point to go out and cultivate contacts in Belle Meade. "When I built an eighteen-thousand-square-foot home in Franklin, suddenly I became very visible," he said. When he started spreading his money around to local charities, he also became socially desirable.

On this day, though, Killen's visitors wanted him for another reason: his access to Hollywood. The women asked him if he would help them get some entertainment for the Swan Ball, one of the most prestigious events in the South, a white-tie extravaganza held every spring in Belle Meade. "You know, Bob Hope or somebody like that?" the women said. "Bob Hope? Are you kidding?" Killen said. "You're sitting here in Nashville, surrounded by the hottest names in the country, in a city the rest of the world is begging to get into, and you want me to bring in Bob *Hope?*" Chastened, the women asked what he would do. "I'll get you a country act," he said. "Oh, no, my mother would kill me," one of the ladies said. "Oh, no," Killen responded, "your mother will love you." "Do you really think so?" she asked. "Yes," he said, "and so will everyone else." The following June, under twinkling white lights and flowing canapés, Johnny Cash, the man in black, a relentless yokel from Kingsland, Arkansas, made his first-ever trip to Belle Meade. "And the people went absolutely nuts," Killen recalled. "From then on, a country music performer played the Swan Ball in alternating years."

As comical as it must have been, the Belle Meade debut of Johnny Cash, deep-voiced, darkly sexual, and by his own account high on drugs for most of that decade, was only the most symbolic of gestures that were begrudgingly uniting Nashville. Though many socialites still feared a takeover—residents in Sarah Colley's enclave of Oak Hill, for example, rose up in arms when Webb Pierce charged tour buses admission fees to see *his* guitar-shaped pool—town elders eventually realized that expanded tourism meant expanded coffers for everyone. In 1967, the city donated land on Music Row for a Hall of Fame. In 1969, Nashville's morning paper, *The Tennessean*, hired its first reporter to cover the music industry. And in 1972, National Life (whose owners were Belle Meade socialites) spent $66 million to build the Opryland theme park. The big breakthrough came in 1975 when Robert Altman released his satiric film *Nashville*. Some hated it. "When you show the anatomy of a man," said producer Billy Sherrill, who walked out, "you should try to show something beside his tail." Others adored it. "I thought it had great depth," Buddy Killen said. Minnie, as usual, split the difference. "Part of it made me very sad, but sometimes I laughed so hard it hurt." But in the end it did make money, which always impressed Nashville, and it did generate attention. By the mid-seventies, 7 million people a year were visiting what Altman called America's "new Hollywood."

Nashville, indeed, was increasingly given over to Hollywood theatrics. In the 1970s, the rest of the country finally realized that country artists, and their fans, were no longer toothless, penniless hillbillies. If anything, country had now penetrated the leading edge of American pop culture: middle-class youth. Whether it was young rockers drawn in by Waylon Jennings, Willie Nelson, and Kris Kristofferson (a Rhodes scholar), or folkies attracted by Emmylou Harris, Gram Parsons, and even Bill Monroe, mainstream Americans were beginning to embrace this once déclassé form. This new love affair with country reached its peak in 1980 with the release of a spate of Hollywood cornbread: Willie Nelson's *Honeysuckle Rose*, Dolly Parton's *Nine to Five*, and Loretta Lynn's *Coal Miner's Daughter*. The most important of all was *Urban Cowboy*, a brilliant honky-tonk travel poster starring Debra Winger and John Travolta, at the time the undisputed icon of American pop chic, who in three years had gone from the lighted dance floor of *Saturday Night Fever* to the mechanical bull of Gilley's nightclub. If Vinnie Barbarino could wear a cowboy hat with a feather in it, could America be far behind? The answer was a deafening "Yee-ha!" The number of country radio stations doubled between 1978 and 1982 to 2,100. Record sales, which had hovered around 10 percent of the market, soared to 15 percent, or $400 million a year. *The New York Times*'s Stephen Holden went so far as to claim that country had "supplanted rock for the time being as the dominant commercial mode of popular music."

Then came the megabust. Predictably, the national fascination with all things Texan proved to be a fad. When the public went searching for new sounds and found none, they quickly abandoned country. Audiences dropped, and in 1986 album sales shrunk to around 9 percent of the market, their lowest level in twenty years. A front-page story in *The New York Times*, this time by reporter-critic Robert Palmer, declared country music was dead. Ironically, this happened at the exact moment that Nashville and Music Row were, at last, achieving a kind of social detente they had been inching toward for years. In the wake of *Urban Cowboy*, Nashville gave up all pretensions of calling itself the "Athens of the South" and fully embraced the nickname that Music Row had been using for decades, "Music City, U.S.A." (ironically, the term comes not from country music but from the Fisk University Jubilee Singers, who were so successful touring Europe in the 1870s that Nashvillians who visited the Continent after them were asked, "Are you from the 'Music City'?"). The chamber of commerce even replaced the Parthenon as its public manifestation with a much more fulsome image: Dolly Parton's profile.

If Dolly was the public face of country music in the eighties, though, Minnie was still its heart (as for other similarities between the two, Minnie once said of Dolly: "I wear a hat so folks can tell us apart . . ."). If anything, Minnie's stature had only been growing since she began devoting herself to charity work and acting as an adopted auntie to many younger stars. "I'm living the life now of a suburban matron," she said not long before bouts with breast cancer and a stroke that would eventually claim her life. "I learned to play bridge. I've gone back to tennis. And I have ladies over for lunch and make homemade mayonnaise." That transformation, like so many she had made over the years, was in perfect tune with the times. In her latter years, Minnie Pearl became a symbol not for the rigidity of class differences—as she once had been— but for the gradual blending of such distinctions into a broad middle class with a generic set of ideals. As her chaplain, speaking at her funeral, quoted Sarah as saying late in her life of the fictitious town where Minnie Pearl lived: " 'Grinder's Switch is a state of mind—a place where there is no illness, no war, no unhappiness, no political unrest, no tears. It's a place where there's only happiness—where all you worry about is what you're going to wear to the town social, and if your feller is going to kiss you in the moonlight on the way home. I wish for all of you a Grinder's Switch.' "

That redefinition of the idealized country locale from a place full of rural rubes, looked down on by socialites like Sarah Colley, to a place full of all-American wholesomeness, inspirational even to middle-class suburban "matrons," is what paved the way for country's expansion. That change was reflected in country's hometown. In the 1980s, not just Music Row but Nashville as a whole would plummet into a recession that would reshape the entire city. The town's insurance and banking companies were snatched up by larger concerns, crippling the elite. At the same time, a new influx of middle-class musicians from across America would all but pave over country's rural roots ("Vince Gill's different from Hank Williams," one Belle Meade socialite said to me. "He knows how to use a fork . . ."). In time both the city and country music would be reborn in the 1990s in a reconstituted, tube-sock form that, a generation earlier, neither Sarah Colley, with her silk stockings, nor Cousin Minnie, with her cotton leggings, would have recognized. Social standing, once the defining distinction of Nashville, had all but been eliminated. In its place, the city was becoming a living experiment in the new American reality—one in which roots were replaced by root- lessness, class background by social mobility, and being Southern by being, well, American.

Indeed, it was only fitting that the emblem for this Nashville—the *New* Nashville as it would be called—would be a man from Oklahoma who had little in common with the social pedigree of Sarah or the rural charm of Minnie, yet whose suburban, all-American background would be- come the perfect symbol for the new middle-class ascendancy in the South, as well as America. He was a man who wasn't particularly country at all, but who was enough of a good ol' boy to name his first daughter Taylor Mayne Pearl Brooks, after the woman who had first reached out to embrace him when he began to redefine what would forever be her town.

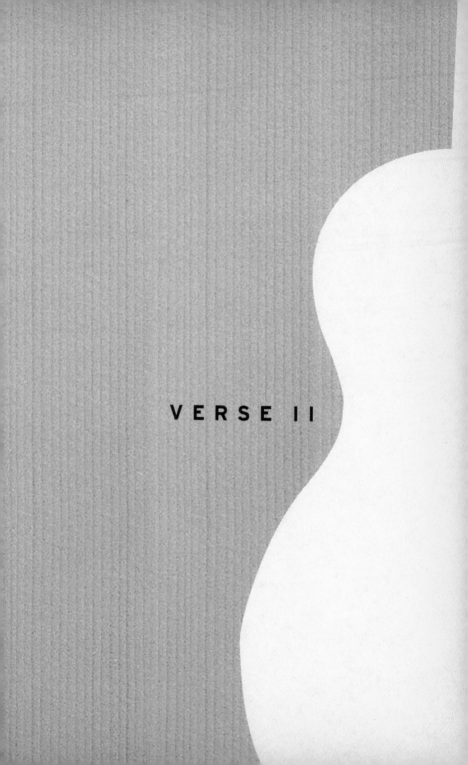

VERSE II

FIVE

THE HAT

He wasn't wearing his hat. He wasn't wearing boots either. He didn't have on jeans. Instead, he was wearing sweatpants—dusty red and fading—and a boyish black parka. His hightops were sloppily tied. Though it was just before dinner when Garth Brooks strolled up to Gate C-5 at Nashville International Airport, he looked as though he had just gotten out of bed. His lips were puffy, his eyes distracted. He hadn't shaved in several days.

"Hey, look! Everything I have on is free!" he boasted, boylike, when asked about his baseball cap.

"You mean Nike gives you free stuff?" His fresh-from-the-box sneakers were neon yellow and white.

"I've been courting them," he said, suddenly grown-up again. "They're very friendly, but all they sponsor is athletes now. I want them to do a campaign: 'Music Is a Sport.' "

That campaign, of course, would star himself.

We headed for the plane. His publicist and road manager walked several paces ahead. It was midwinter by now. Garth had fully re-emerged into view—battered by some of his negative press, but still pushing forward to promote his new album. In a breathless span of several weeks, he had traveled to London, Dublin, Los Angeles, New York, Washington, back to Los Angeles, and now back to Washington, this time for an appearance on "Larry King Live!" Though he cultivated his image in these appearances with military precision (he would take several shirts from talk show to talk show so he would always appear fresh), away from the lights he was surprisingly easygoing. There was no groveling entourage here. No pretense. He had a blue collegiate back-pack over his shoulder and was carrying his publicist's flowered garment bag in his hand. The one hint of his John Wayne-like power: Under his arm he balanced a black plastic version of a nineteenth-century ladies' hatbox.

"A friend gave it to me," he explained. "I use it on trips like this."

"What if you didn't have it?" I asked.

"I suppose I'd just wear the hat," he said. "But the problem is, people are attracted to it. It's almost like a magnet. People come up and want to talk to it."

"To 'it'?" I asked.

"Yeah, to it," he repeated. "To that person." He raised the box in the air. "'GB.' I don't mind, really. I love it. It's my chance to meet one-on-one with the fans. But they have to get where they're going." He gestured toward Karen, his publicist, and Mick, his road manager. "And this attracts too much attention."

This was a black beaver Stetson, size 7⅝, which in a manner more reminiscent of Greek mythology than contemporary American cynicism had come to embody the larger-than-life figure that Garth had managed to turn himself into. In the span of half a decade, Garth Brooks had completely rewritten the rules of Music Row and taken country music to its highest point in history. Riding his coattails, country's sales now topped $2 billion a year, three times what they were in 1990 and five times what they were in 1980. With ticket sales and merchandise, Garth Brooks made more money in *each* of the early years of his career than the *entire* industry made in 1970. He did this in large part by creating albums, designing concerts, and devising a persona that all reinforced a single image: Garth was a contemporary cowboy. He was humble, cour-teous, hardworking, and fundamentally all-American. He sang about cowboys. He dressed like Gary Cooper. He wore a hat.

In time his round-'em-up, let-'em-rip persona became so all-consum-ing that even Garth went so far as to give it a nickname, "GB," in an effort to distinguish it from himself. But just by creating that nickname, then by referring to it as if it were a prop, Garth risked sounding a little, well, odd. "Garth is not difficult to understand if you look at him as two different people," Garth—or was it GB?—wrote in his tour book at the start of his career. "There's GB the artist and Garth the lazy guy just hanging around the house. Here's how the two differ: GB likes the view from the edge; Garth hates heights. GB loves to try new things; Garth is a meat-and-potatoes kind of guy . . . GB loves the control, the responsibilities, and the duties that come with the road. Garth enjoys being lazy, dreaming, and other senseless things that people call foolishness."

Some people might have called that confession foolishness. By sharing his own mixed identity with the world, Garth all but guaranteed that his duality—his managed schizophrenia, if you will—would become the source of controversy. Garth Brooks, on paper, can make no claim to being a cowboy. He's a product of the cul-de-sac, not the farm. He didn't ride a horse until college. So where did he get off creating such a buckaroo image? Was "Little Boy Garth," the authentic cowboy, just some super marketing creation by "Grown-Up Garth," the ten-gallon tycoon? Ultimately, this tension boiled down to perhaps the most frustrating issue in contemporary country music: authenticity. Can a child of modern America—the frontier of plastic—actually transform himself into something as steeped in integrity as the cowboy? Or must such a persona, coming as it does from an ambitious entertainer steeped in the American tradition of packaging, necessarily be fake? Even among those who knew him best, this was the central question about Garth Brooks: Was he for real?

We arrived at the plane. Garth was talking about American Air Lines. He had thanked them in the liner notes of his last album, and I asked if they sponsored his tour. "No," he said, "they just take care of me— let me put things on planes without a ticket. So we take care of them." Once on board, Karen and I retreated to our seats as Mick began storing his briefcase. Garth, though, didn't have his ticket, and the flight attendant refused to let him board. When told that Mick had the ticket, she turned toward the cabin. "Excuse me," she called. Mick didn't hear her. "Excuse me," she repeated, louder. "Do you have a *ticket* for this person?" By this time, everyone in the plane had turned to see what was happening. And for a second I wondered: Did she not recognize him? Did she recognize him and not care? And then I thought: For what other person of his stature would I even ask?

The most remarkable thing about spending time with Garth Brooks is how *un*remarkable he is most of the time. He's casual, talkative, and, above all, very guylike: he loves sports and statistics; he loves stuff (T-shirts, trinkets), particularly free stuff; he loves technology—bangs and explosions. When he heard I had been in the circus, he was fascinated with the human cannonball. Also, he loves to exchange locker room banter. "Did you have that goatee last time we were together?" he asked me that afternoon. "Yes, but I'm going home in a few weeks and I'm not sure it's going to survive my mother." He loved that. "The question is not whether *it's* going to survive," he said. "The question is whether *you're* going to survive." Above all, like an athlete, he's extremely competitive. Later, when the subject turned to football and the abysmal Washington Redskins' recent victory over the Dallas Cowboys, Garth was intrigued that the underdog had beaten the champion twice in one season. "I think Norv Turner's got something on Troy Aikman," I mentioned, referring to Aikman's former quarterback tutor, now head coach of the Redskins. "He probably calls him up in the morning and says, 'I know what's in your mind.'" Garth's eyes grew wide. "Football off the field," he cooed. "I love that." If music is a sport, with Garth it's mental.

Both those traits—his casual everydayness and his keen competitiveness—have been with him since childhood. The future mayor of American music was born in what demographers would later call the most "typical" city in America: Tulsa. His father, Troyal Raymond Brooks, was a stern oil company draftsman whom Garth describes as having a "thundering, velvet hand." His mother, Colleen Carroll, was a sometime country singer (she had cut a few songs for Capitol Records in the 1950s) with an effervescent personality and salty wit, whom Troyal Brooks first saw singing on a local television show in 1957. The two met, married, and created a sort of "Brady Bunch" family. He had one child from a previous marriage; she had three; and together they had two more: Kelly, a studious boy, later to become Garth's tour manager, and Troyal Garth, the younger, born February 7, 1962.

When Garth was four, the Brookses moved to Yukon, a bedroom community of five thousand located fifteen miles northwest of Oklahoma City, which Garth once described as being "an average city in the middle of average Oklahoma in the middle of average America." The family settled into a compact, split-level house on 408 Holly Street. The house, modest by even middle-class standards—"The houses were small, but there was no shame," Garth told me in his often self-mythologizing way—was located in a flat, well-tended neighborhood. A short walk away, on what is now Garth Brooks Boulevard, was Yukon High School, home of the Millers (named after the town's dominant industry and tallest building, Yukon's Best Flour mill). A few blocks in the other direction was the town's main street: Route 66.

Life in the Brookses' house was dictated by the strong arm of Garth's father, softened by the encouraging hug of his mother. "One's a realist, my dad," Garth told me. "One's a dreamer, my mom. If the gas tank's on empty, and she's got a hundred miles to go, she believes she can make it. Dad's going, 'We've got to do something.'" From his father Garth inherited his stubbornness, his work ethic, and his sense of right and wrong. Once, Garth told me, he went to see the screenwriter, William Goldman, to talk about film projects. Though Garth was intimidated, the two hit it off famously. "Everything I said," Garth told me, "he was like, 'Man, that's exactly how I feel.' I'm hanging, man. I'm thinking, 'This is unreal.'" Then the film *Pulp Fiction* came up, which Garth hadn't seen. "And I lied," Garth said. "I told him I had seen it, but that I didn't like it, 'cause I knew that's what he was going to say." Garth guessed wrong. "No kidding," Goldman said. "I loved that movie." "And just then," Garth told me, "I heard my dad laugh. He was fifteen hundred miles away, but I heard my dad laugh. I knew what was right; I knew what was wrong; and still I chose the wrong. That wasn't the John Wayne thing to do."

From his mother Garth inherited his penchant to dream big. As a child, he dreamed of becoming various things, but always as a hero. He would be a baseball player, he said, snagging the final catch, smashing the game-winning home run. He would be an actor, dashing in to save the heroine, riding off into the sunset. He would be a forest ranger, stomping out fires, rescuing stranded campers, nursing injured bears. To his mother, it came as no surprise that later he tried to become a hero through music. "Garth says he wants to bring prayer back to the dinner table and an American flag back to the front porch," she told me. "I think that's wonderful."

Though Garth may have fantasized as a child, he did little to realize his goals. "I was always more of a talker than a doer," he said. A popular if unspectacular student, Garth eased through high school with little distinction, except for his ability to throw. "His talent was his arm," said Mick Weber, a friend since second grade. "He could throw things a million miles: football, baseball, javelin, rocks." After graduation, Garth proceeded to Oklahoma State at Stillwater, eventually earning a partial track scholarship. In his senior year at college, his athletic career fizzled, freeing him to follow a newfound passion: playing guitar. "The real big Shakespearean bell went off in Lincoln, Nebraska, in 1984," he told me.

"I'm lying in the pole vault pit, right after not qualifying for finals. It was the first time in my four-year-college career that I did not qualify for Big Eight finals. Suddenly this female trainer walks by and says, 'Now you can get on with what you're supposed to be doing.' *Wow!* I knew exactly what she meant. For me, music was the other side of the coin from sports. It was no talk, just get out there and go."

Where he went was something of a surprise. With eight different people living under one roof, Garth had grown up with all sorts of music, from George Jones, whom his father liked, to James Taylor, his brothers' favorite. Garth tended toward soft seventies rock—Elton John and Dan Fogelberg— as well as the megarock bands Journey, Boston, and Styx. At the time he had little interest in country, which he thought was "prettified and slick." That was the age of Olivia Newton-John and Kenny Rogers, after all, two pop stars who crossed into country. All that changed when Garth first heard George Strait, a true-bred Texas rancher who emerged out of Nashville in 1980 with a sparse country sound and throwback appearance that instantly distinguished itself from its *Urban Cowboy*–era competition. Suddenly a new path became available: Garth Brooks could try to become "George Strait II," a sort of cul-de-sac cousin of country, an Okie not from Muskogee, but from a Clean, Well-Lighted Place.

He started playing clubs. On weekends, he played "revved-up" country with a swing band. On weeknights, he played solo. "It was everything from Elton to Merle Haggard," he said. "My repertoire got up to over three hundred songs. You knew that the more songs you learned, the more money you were going to make. And that helped because the more I knew, the more diverse my music became." All the while he fretted about that next move: Nashville did have George Strait, but still only a niche share of the marketplace; New York and L.A. had crossover possibilities, but for a mostly coddled Oklahoman, frightening reputations. "New York was definitely out," he told me, "because, to me, the only time I'd seen New York was on a show called *Escape from New York* and I thought that's what it looked like. L.A. was going to be an island someday. The biggest city I had gone to before Nashville was Des Moines, Iowa, for a track meet."

In 1985, Garth made his first trip to Nashville (driving, appropriately enough, the country's most popular car, a Honda Accord). Through a friend he obtained a meeting with Merlin Littlefield, a vice president at ASCAP. Unimpressed with Garth's demo tape, Littlefield discouraged him. Thousands come to Nashville every year, he said, only a handful manage to make a living. During the meeting, a songwriter stuck his head in Littlefield's office and asked for a $500 loan to fend off foreclosure of his home; Littlefield turned him down, stunning his visitor. "See, you've got your choice," he told Garth, "you either starve as a songwriter or get five people and starve as a band." Humiliated and now insulted ("I hated his guts," Garth said), he fled back to Oklahoma and hid in his parents' home for weeks. "I thought Nashville would be like Oz," he said later. "You came here and all your prayers were answered. I thought you'd flip open your guitar case, play a song, and someone would hand you a million bucks, tell you, 'Come into the studio right quick, son, we got ten songs we want you to cut.' You cut them that day, go back home, and people would be asking you for your autograph that night . . . The people in Stillwater had passed the hat and got the money for me to come to Nashville. I was going to be their hero. Now I had to go back."

Once back, he married his college girlfriend, Sandy Mahr, whom he had just humiliated by not even telling her he was going to Nashville. (Sandy, a soft-spoken but steely blonde from Owasso, Oklahoma, had met Garth in Stillwater after she put her hand through a bathroom wall at a bar where he worked as a bouncer. He walked her home, but refused her request to kiss him good night. "He was the most mannered man I had ever met," she told me.) Then he put together a band, Santa Fe, with whom he made a pact ("Actually, I made the pact for us," Garth said): When they had played every place they could, they would move to Nashville. In 1987, five band members, two wives, two kids, a cat, and a dog did just that, setting up house north of town. When their attempt to get a deal floundered and the band split up, Garth and Sandy took jobs in a Western apparel store while he looked for a deal. A songwriter introduced him to Bob Doyle, also an executive at ASCAP. Doyle was so enthusiastic about Garth's work that he quit his job and formed a publishing company. He gave Garth a $300-a-month stipend and rallied jobs for him singing demos. Looking for a comanager, Doyle, a quietly passionate man, recruited Pam Lewis, a fiery publicist with New Age crunchiness who had once worked for MTV in Manhattan before moving to Nashville to work for RCA.

Despite all the good omens, all seven major labels promptly rejected Garth. He had an undistinguished voice, they said, was a mediocre songwriter, and wasn't very attractive to boot. The one catch: No one had seen him perform in public. On May 11, 1988, Lynn Shults, an executive at Capitol Records, finally did see Garth before an audience at the Bluebird Cafe, the quaint songwriters' club in a strip mall not far from

Belle Meade that was later lionized in the movie *The Thing Called Love*. After watching Garth sing, Shults offered him a handshake deal on the spot. "Being on the road with a lot of great artists, you get a frame of reference for what is exceptional," Shults said. "And that night Garth was exceptional. He didn't even have a band. This was just Garth Brooks with an acoustic guitar. But his vocal performance and the magnetism of his personality connected with the people who didn't even know who Garth Brooks was. What went through the mind was that I had just seen somebody who was as good—if not better than—anyone I had ever seen." It was less than a year after he moved to Nashville, and already Garth's earnestness, his ability to connect one-on-one with the fans, was becoming his calling card.

The next step was capturing that sincerity on tape.

From the beginning, Garth fulfilled the one expectation of country music and captured its one true secret: He sounded like the place he had come from. What made him groundbreaking was that that place was subtly different from what had come before him in Nashville—and profoundly similar to where the rest of the country had come to at that time.

Yukon, Oklahoma, is a vintage barometer of the changes in American life. As one local historian wrote in 1990, "Yukon started as a small village on a new railroad and has grown into a large suburban city during its ninety-nine years." Originally part of the Chisholm Trail, on which early ranchers moved cattle from Texas to Kansas, Yukon was founded in 1891 by squatters, who named the town after the Yukon River in Canada where gold was being discovered. Though it remained a frontier outpost for much of the next century, by the 1960s, the town had been overrun by franchises: McDonald's, Dairy Queen, Ace Hardware, Mobil. In 1970, the town experienced its first big boom when residents of nearby Oklahoma City started flocking to the mostly white community to escape mandatory busing. The population doubled in the next two years.

This transition from old-fashioned town to up-to-date suburb dominated life in Yukon, right down to its street plan. Garth's childhood home on 408 Holly Street, not far from the center of town, was almost exactly halfway between the town's two main roadways: Route 66, the legendary Western highway that ran from Chicago to Los Angeles, and Interstate 40, the new main east-west thoroughfare across the southern United States that runs from Wilmington, North Carolina, through Knoxville, Nashville, and Memphis, to Little Rock, Oklahoma City, Albuquerque, and finally to the San Fernando Valley in California. Much of the story of contemporary country music can be seen in the tension between those two roads: the mythical and the modern. And though I-40 may lack the allegorical clout of Route 66, it has become, in effect, the new yellow brick road of country music. Fully two thirds of contemporary country artists grew up within 100 miles of that road, from Alan Jackson and Travis Tritt in northern Georgia, to Wynonna in Kentucky, to Garth, Wade, Reba, and Vince in Oklahoma, to Dwight Yoakam in California. Even in the age of airplanes, most of those artists drove an old family vehicle, a borrowed truck, or a dented van into Nashville not along a dusty dirt road, as they might have us believe, but along the six-lane comfort of Interstate 40, which, when approaching Nashville from the east, passes within honking distance of the Grand Ole Opry and promptly deposits present-day Dorothys precisely where they want to go, the Shoney's at the base of Music Row—Oz with grits.

Garth's self-titled first album, released with little fanfare on Capitol Records on April 12, 1989, was a classic, if not entirely groundbreaking synthesis of these two strands in country music—the allegorical and the mundane, the brash lone wanderer of Western myth and the sensitive, angst-ridden man of modern, six-lane America. The first single off the album, which is to say the first song the label sent to the two hundred stations that reported their playlists to *Billboard* and *Radio & Records* (and thus forced the two thousand other country stations to play), was "Much Too Young (to Feel This Damn Old)." The song, written by Garth and Randy Taylor, is a fairly typical country romp that fit squarely into the back-to-roots tenor that had taken over country music after *Urban Cowboy*. It tells the story of a veteran rodeo rider hustling down the highway, complaining about his lover not answering the telephone, and fretting about his younger competition. To a warbling solo fiddle, quiet drumbeat, and an understated steel guitar whine, he laments the white lines getting longer and the saddle getting cold. Though Garth, hinting at his darker side, suggested the song might be about cocaine cowboys, it was received by critics as the debut of a promising devotee of honky-tonk tradition. "By God country to the core," wrote Bob Oermann, the flamboyant critic who is Hazel's rival for chief-flitterer-about-town. "A hurtin' vocal, chiming steel, sawing fiddle and toe-tapping hillbilly beat. Garth Brooks has my heart as 'Discovery of the Day.'"

Country radio, however, was less welcoming. The song lingered on playlists for several weeks, but eventually washed out in the forties, an unembarrassing but hardly encouraging result. Jim Fogelsong, the genteel, if old-fashioned head of Capitol Records, declared the song a success and told his radio promoters to stop urging stations to play it. Garth and his managers refused to roll over, though, and started calling radio programmers themselves, a naive if quaint reminder of the old days when Loretta Lynn could pay a flirtatious visit to a disc jockey and have her song on the air that afternoon. Miraculously, it worked. The song regained its forward momentum and eventually peaked at number ten.

The next single they released was "If Tomorrow Never Comes," a more touching song, both musically and lyrically, about a man wondering if his wife—or daughter—will know how much he loved her if he dies in the night. It quickly reached number one and, more importantly, became an early benchmark establishing Garth's persona as a bastion for male sensitivity, a warm hug set to a campfire guitar. It was followed by "Not Counting You," a nondescript Texas two-step, which also reached number one, but not before Garth criticized his label for selecting a single he felt did not make a strong statement. It was the first hint that Garth might have a larger agenda in mind (and that he didn't mind picking fights in public). "Being able to stand up for three minutes in front of the nation is quite a gift," he said. "And I think it's very important that you say something. I love this country, but the downfall of the United States is that nobody makes stands. Everybody wants to be a fence-walker." His current single, he said, was part of the problem. "I like doing ballads and the serious songs a lot more. Music is really for the heart and soul, yes, but it's also a teaching process. I know whenever I'm depressed I still go into a room and put on an album by Dan Fogelberg or James Taylor. That's what music is for, to create emotion."

Garth would soon get his emotional song.

Among music industry insiders, there are various types of songs: a song that gets cut, a song that makes an album, a song that becomes a single, a song that charts, a hit song, a number one song, an impact song, and the most coveted honor of all, a *career* song. "The Dance" was the career song for Garth Brooks and, as it turned out, the defining song for the entire modern era of country. Six years after its release, "The Dance" was voted the number one song of all time by readers of *Country Weekly*. The song was written by Tony Arata, a thin, wire-spectacled copywriter from Savannah, Georgia. In structure, as well as message, it is subtle and deeply emotional. It opens with a haunting one-handed piano climb, then expands with strings, at the time still taboo in Nashville, until it mesmerizes the listener with Garth's stricken yet reassuring voice. As the narrator, Garth—and it's hard to imagine anyone else singing this song—looks back on the memory of a relationship that's ended: By choice? By fault? By death? The participants in the relationship are not articulated, nor is its ending defined, but the pain at its collapse is deeply felt. Yet, in the hopeful way of the religious ode the song so clearly emulates, the man left behind treasures the memory of the dance and the moment it took place. He could have "missed the pain," he declares emphatically, but only by missing "the dance."

As clearly as this song has been associated with Garth Brooks, he almost didn't record it. He first heard Tony Arata sing it at the Bluebird and passed it on to his producer, Allen Reynolds. After cutting it, though, Garth lost interest. With its sentimental, Robert Frost-like themes, "The Dance" didn't seem to fit the more traditional sound of the album. Reynolds pushed, though, and Garth relented. "The Dance" was placed at the end of the album, a position of ceremony. By spring 1990, with Garth already sitting on two number ones, his label didn't want to release "The Dance" to radio. It was "too pop," said Jimmy Bowen, the craggy but brilliant executive who had come to Capitol the previous December. Why jeopardize Garth's graces with the country audience? This time Garth pushed, and Bowen relented. He was glad he did. The impact of the song was immediate and overwhelming. Audiences embraced it as they had few songs in recent memory. In the year since its release, *Garth Brooks* had sold approximately 300,000 records—a respectable sum, but far less than rival Clint Black, a handsome hunk in a hat from Houston. In the month after "The Dance" was released, however, Garth's album sales tripled. They reached 1 million by early summer and 1.5 million by Labor Day.

By that time, radio was already itching for new material. A programmer in Oklahoma City who knew Colleen Brooks talked her into giving him a copy of Garth's upcoming CD, *No Fences*. Against all rules of industry decorum (as laughable as that seems, there are such rules in country), the station started playing the first cut, a rousing, barroom singalong with the irresistible hook, "Friends in Low Places." Listeners loved it—the song captured the fun-loving honky-tonk side of Garth that made his treacly self-confessional side less threatening—and soon programmers across the country were clamoring for copies. The controversy, of course, only generated more attention. "Leaks like this happened to me when I worked in pop music," Bowen said, "but never in my career in country music have we had a situation like this." Helped by Bowen's aggressive marketing campaign, *No Fences* sold 700,000 copies in its first ten days and reached a higher position on the pop album chart (number three) than any country release in five years. By early fall, Garth was earning all the honors that being a star in Nashville had to offer. He was asked to be a member of the Grand Ole Opry, his name was painted on the water tower in his hometown, and he won the CMA's Horizon Award for newcomers. "I want to thank God," he said upon receiving the award. "He's done a hell of a lot for me."

Garth apologized for that remark later, but the unconventional nature of his ascent was already well defined. In the next year alone, his first two albums would sell 8 million copies combined, figures unheard of in Nashville history. *Garth Brooks* would become the bestselling country album of the 1980s. *No Fences*, with sales of 13 million, would become the bestselling country album of all time and the tenth-bestselling album of any genre in history. "Music has a new name and it is Garth Brooks," read the opening statement of the press bio, and for once the hyperbole seemed understated. What Garth was in the process of doing, in a manner he clearly understood at the time, was making himself into more than a musical phenomenon. He was turning himself into an icon. It was not just his music that was capturing attention. Others in Nashville had done that before him; others would do that after him. But Garth was on his way to becoming something more. He was becoming emblematic of an entire era. It was an era, like many, that would come to be embodied by a single accessory of style.

The hat.

We arrived at the airport in Washington, D.C., and headed for the car. All the way through the terminal and into the late-afternoon sun, no one seemed to notice Garth. No one stopped him for an autograph. No one asked him for a kiss. It was almost as if he was completely anonymous without his hat. In full panoply—lace-up ropers to dimpled crown—his six-one frame, strong arms, and mesmerizing stare were subtly striking, even handsome. Stripped of his costume, though, he looked numbingly normal, a gerrymandered version of American midlife afflictions. First there was his weight, which fluctuates freely in a 50-pound zone and was, at the moment, at 217, somewhere in the middle. ("If they ever opened a theme park called Garthland," he said, "the food would have to be free, and there'd have to be lots of it.") Then there were his swollen lips, his fleshy arms, and his broad neck. And finally, even with evidence of a transplant, there was his first-team-all-American receding hairline. What's more, the hair he had left was now graying as well. Elvis Presley may have been the "most beautiful person on Earth," as Carl Perkins once described him; Garth Brooks would hardly have been the most noticeable person on a used car lot.

"I've always been one of those guys that if I could change one thing about me it would be everything," Garth told me. "I'm kind of squatty. I don't have long legs. The word 'round' keeps coming to mind." (Sandy, asked about this description, was more charitable. "He's an attractive naked man," she said.)

Over the years, Garth had become quite adept at sidestepping this problem. Before photo shoots, he would starve himself for days, work out aggressively, and lose as many as ten pounds. Onstage he'd wear Wranglers that were three sizes too small. "When I was one-ninety I was wearing twenty-nine Wranglers," he told me. "Even now, my sizes will vary four or five sizes from what I wear offstage to what I wear on." His reason: "You want to look good for the people."

As a relative newcomer to country music, I found Garth's obsession with perfecting his image fascinating, but troubling. After all, what's the point of being successful if you can't wear pants that fit? What I realized, though, is that while this single-mindedness certainly helped him become successful, it also defined him in a fundamental sort of way. More than a means unto an end, becoming a cowboy was an end unto itself with Garth. Indeed, the opportunity to become a larger-than-life persona was the reason he got into the business. Garth wanted more than money, sex, or attention—all of which he got more than his share of. Garth wanted to become an icon. He wanted to become like those all-American heroes he had idolized as a child. He wanted, in essence, to leave his seat at the last picture show and step onto the screen. When, early in his career, an interviewer asked him who he wanted to play himself in a movie, Garth had his answer ready: the Duke.

Within minutes, we arrived at the car and Garth started loading his luggage—as well as that of his publicist—into the trunk. His hatbox was the last to go in.

"So when did you start wearing the hat?" I asked him.

"Mostly in college," he said. "In high school, I wore baseball caps. We didn't have any land then, we couldn't afford it, so I didn't have any occasion to wear cowboy hats and do all that ag stuff. Really, the reason you're supposed to wear them is to keep out the sun and the rain. Now that I have land, if I could go out to Colorado and work outdoors twenty-four hours a day, I still couldn't make up enough time for all the years I haven't spent outside."

We got into the car. Garth, Karen, and I got into the backseat. Mick sat up front. Garth continued talking. "I like the cowboy thing," he said. "To me, cowboys are honest. Cowboys are real. It's all the 'ities': integrity, responsibility, honesty. One of the coolest stories I ever heard was at Oklahoma State. Me and the boys were just sitting around talking. This gal comes in, sits down, and says, 'Man, this cowboy I picked up last night turned out to be the biggest asshole.' Everyone laughed, and this one kid, who was always quiet, said, 'He wasn't a cowboy then.' And I thought, 'Yeah, all right. Good deal.'"

"So are you a cowboy?" I asked him.

He thought for a second. "My goals are the same," he said. "If I reach those goals, it's someone else's place to say that I'm a cowboy."

Those goals, it turns out, are surprisingly complex and together form a kind of personal code that defined the parameters of his ambition. For Garth, a sort of idiot savant of American culture (he doesn't read books, he told me, or study movies, he just deals in myth), there are three basic elements to being a cowboy: the public face, the private values, and the lifestyle. He attacked them one at a time. First, he appropriated many of the outward symbols of the cowboy: the lace-up ropers (casual boots worn by cowboys while working); Wrangler jeans (cowboys wear Wranglers because the rear pockets are higher and the inner seams are less abrasive); and his familiar starched cowboy shirts. He even adopted the slightly bowlegged walk that riders use when they dismount from a horse.

At the heart of this image was the hat. Invented in the nineteenth century and standardized by John B. Stetson in the 1860s, cowboy hats have been part of country music from its earliest days. In the 1920s, rural Southern musicians donned cowboy garb to distinguish themselves from hillbilly imagery imposed on them by outsiders like New York executive Ralph Peer and Indiana native George Hay, who made Opry performers wear overalls and floppy hats and adopt names like Possum Hunters and Fruit Jar Drinkers. The first Southerner to appear on a record, Eck Robertson, wore cowboy clothes, and that was in 1922. Even at this pre-Hollywood stage, cowboys were equated with rugged individualism and hearty Anglo-Saxon stock, even those who met neither of those criteria. Jimmie Rodgers, the first country music superstar, wore cowboy attire, even though he was a sickly train worker from Mississippi.

In the 1930s, Hollywood cemented the cowboy, particularly the *singing* cowboy, in the American imagination. Gene Autry, known as "Oklahoma's Singing Cowboy," made ninety movies between 1934 and 1954, and suddenly millions of Americans wanted to be cowboys—not real cowboys, but imaginary ones. By contrast, few Americans wanted to be hillbillies. The Appalachian imagery so exalted at the Opry, the happy-go-lucky "musical mountaineer," as historian Bill Malone labeled it, was dying out. In its place, the wrangler was carrying the day. By the 1950s, Hank Williams, from Alabama, called himself the "Drifting Cowboy," and Hank Snow, from Nova Scotia, became the "Singing Ranger." Authenticity, the weapon later branded against Garth Brooks, had already given way entirely to invention. Unlike pop stars, country artists rarely changed their appearances to suit the times, but instead tried to appeal to a timeless ideal. Invariably, the ideal was more appealing than reality. The best example of this was Hank Williams himself, who, like Dwight Yoakam, Ricky Van Shelton, and later Garth, rarely took off his hat once he put it on because underneath he was almost entirely bald.

By the 1980s, the cowboy had undergone so many incarnations that it became little more than a gaudy token. During the *Urban Cowboy* years, ten-gallon hats suddenly sprouted feathers and brightly colored bands and became synonymous with a kind of brash tackiness. In the years that followed, more new artists, like Randy Travis and Ricky Skaggs, shunned the hat. Two notable exceptions were George Strait, a Texas traditionalist who actually was a rancher, and Dwight Yoakam, a California honky-tonker who, though he wore his out of genetic necessity, also managed to make it sexy by pulling it over his eyes and wearing it with butt-hugging leather pants. Garth, debuting in 1989, had a choice: risk pigeonholing himself as a honky-tonk singer by wearing a hat or open the way to greater crossover appeal by not wearing one. Ultimately his decision was based partly on practical reasons ("As much as I hate the label a hat puts on you," he said, "if you take it off, nobody knows who you are . . .") and partly on his desires to represent the larger-than-life creation that the cowboy hat still embodied.

This spoke to the second central element in Garth's desire to be a cowboy, the private values. "When I talk about John Wayne," Garth explained to me in the car that day, "I talk about the characters that John Wayne's played. I don't know anything about the kind of person he was. But that's what Garth Brooks's private life is striving for: to be the kind of person that John Wayne was in public." As surprising, or even quaint, as this sounds—that a grown man would hold onto this boyish fantasy—it is the key to understanding Garth's appeal. Though he quickly perfected the public part of being a cowboy, he still struggled to fulfill the private part. It was that struggle that made him recognizable. He had an eating problem. He was stubborn. And, as the public soon learned, he was a womanizer. Instead of dooming him, though, it was his reactions to these flaws, particularly the last one, that made him emerge as a living, breathing creature of the nineties, not just a cartoon of the fifties.

In the months after Garth started appearing on the radio, he stopped calling home almost entirely. Flush with success, he would trot up and down the stage, point to women randomly in the audience, and have his cronies bring them backstage. He became, in other words, just another typical guy with an ego who suddenly struck it rich. Sandy, inevitably, learned of his indiscretions and phoned him in November 1989. Her bags were packed, she said. The following night, as an opening act for Eddie Rabbitt and Kenny Rogers, Garth went on stage in Cape Gerardeau and delivered what would become a legendary performance, witnessed by Peter Kinder of the *Southeast Missourian:*

> This relative newcomer to stardom strode out and played for twenty minutes, managing to establish a rather good rapport with the audience, most of whom had come to see the other two performers. He was able to accomplish this despite a persistent hoarseness . . . [After several songs] Brooks exhibited another, far more debilitating problem as he seemed near tears. This became particularly acute as Brooks launched into that heart-tugging ballad for the lady in his life, "If Tomorrow Never Comes." Some forty-five seconds into the song, Brooks startled the crowd by stopping his band, falling silent, and then speaking of the "hell" of being on the road, and his love for the woman back home, and how some "bad things are going on." Upon regaining composure, he asked of an empathetic and appreciative crowd, "Could I try again?" and proceeded through a workman-like version that the crowd loved. After two more songs, Brooks departed to a standing ovation, telling a puzzled audience that he hoped they might see him on another evening, one that was "not so sad."

Once he arrived back home, he begged his wife to stay. She agreed, though slowly, and then the two of them proceeded to go public, in newspaper and magazine interviews, and, eventually, on TNN, confessing their difficulties and renewing their vows. It was the first of many public acts that would bind Garth Brooks ineluctably to his time: A flawed man confesses his infidelities, a strong woman agrees to take him back, and a singer-songwriter from America's heartland suddenly redefines the dispassionate cowboy for the Age of Oprah as someone who recognizes his weaknesses and opens his heart (and tearducts) to the world. "Cowboys Ain't Supposed to Cry" Doodle Owens once memorably sang; Garth Brooks, a postmodern cowboy, did so often and in front of the world.

And he got away with it. Why? Because in private he believed what he said in public: that he was not worthy of his own image. This was no more apparent than in his attitude toward his third basic tenet of being a cowboy: mastering the outdoors.

"Hey, Garth," Mick called out suddenly as we crossed the 14th Street Bridge into Washington, "look at that moon." The moon, just rising over the Capitol, was covered with clouds.

"Wait a minute, what's today?" Garth asked. "The fifth?"

"No, the sixth," Mick said.

"The sixth. Yeah, that means it's a full moon."

"Excuse me," I asked, "how do you *know* that?"

He looked at me, surprised.

"The Weather Channel."

"You watch the Weather Channel?"

"Sure."

"Why?"

"Because you've got to know when to cut your weeds. There's a certain weed that if you cut it on the day before the full moon, it's never supposed to grow back."

"What weed is that?" I said.

"I don't know. I didn't grow up with this stuff. It's that weed that makes cow's milk sour. What's it called, Mick?"

Mick supplied the answer.

"Yeah, that's right. That one." He wanted to know the name, but didn't. He took a second to commit it to memory. And for the first time I caught a glimpse of Garth's insecure private side. Garth's little-boy persona—the side of him that tried so hard to be an authentic Wild West figure—*was* real. Unlike most couch cowboys, content to sit around the house watching Westerns and deluding themselves they could ride that well if they tried, Garth actually did leave the pavement. He'd even gone so far, I learned later, to set up a toy steer behind his house so he could practice roping. In many ways, this dedication was almost endearing: Garth Brooks had sold more records than any man in American history, but he was still upset that he didn't know how to weed.

"I sing about cowboys," he explained. "I try to get as true as I can. But I ain't trying to hang out with those guys in Wyoming who make a living cowboying because I can't. I'm a guy who owns some acreage; enjoys planting his hay; enjoys horseback riding, even though he's got a horse that protects him because he's not any good; enjoys roping, though he's not worth a shit at it . . . I can talk now about the difference between a no-till drill and a cultipacker. I can do that. But can I tell the difference between a Thoroughbred, a quarter horse, and a saddle pony by looking at them? No. So I got a lot of learning to do."

And that desire to be something else proved to be central to his identity. Though he may not have been to the Ponderosa born, he had, in a way, done even better. Through his music, his manners, and his let's-all-study-the-*Farmers'-Almanac* rallying cry, he had created a new icon in American life: the suburban cowboy. In the process, he'd given meaning to millions of baby boomers who'd never seen the inside of a cattle pen. You can be born on a cul-de-sac and still have feelings; you can run around on shag carpet as a child and still claim your birthright on the frontier. Garth spoke to this group because he was so clearly one of them. Indeed, American suburbs were full of millions of middle-age, middle-class, college advertising majors like him. In one of his most successful songs, "Unanswered Prayers," a man returns to a hometown football game with his wife, where together they run into his high school flame. Just as the man begins to regret having lost the girlfriend and the song risks becoming a cliché, he realizes that the old girlfriend is not as lovely as he had remembered. "And as she walked away / I looked at my wife / And then and there I thanked the good Lord / For the gifts in my life."

By addressing aging boomers, Garth Brooks became their patron saint. He also rounded out a process that had been under way since the birth of the Grand Ole Opry. In the first half of this century, country music had given voice to the migration of rural Americans to the city. Now country captured their migration away from the cities and into the suburbs (by 1990, 50 percent of Americans were living in suburbs; by contrast, only 12 percent of Americans were living in cities of 500,000 or more). More than any other musical figure, Garth recognized this transition—from escaping rural life via Route 66 to returning via I-40— then let the rest of the world in on the secret: I too liked raising hell and listening to rock 'n' roll when I was younger, but now that I'm older, I want to ride horses and raise my family. Instead of "country comes to town," a much more limited ideal that has embodied many of

Nashville's stars of the past, Garth Brooks embodied the opposite: He was town gone back to country.

And even then he felt not quite there.

"So is your farm a working one?" I asked Garth as we drove into the city.

"No, not yet. We'd like to make it that way, but I haven't made enough money to lose *that* much."

"So what do you want to have. Cattle? Crops."

"Well . . . " He seemed suddenly anguished here. "I'd like to have cattle, but I wouldn't want to kill them or use them for food. That would bother me. But Sandy really, really wants to have calves, and that's the only way to do it."

(Later Sandy would tell me a story. They were sitting around the house one afternoon when a skunk wandered into the backyard. Knowing skunks are nocturnal, Sandy assumed it must be rabid and sent Garth to kill it. He crept closer to the skunk, getting to within ten feet, then turned away, letting the skunk escape. Sandy had been watching. "I didn't hear the gun go off," she said when he returned. "Oh, he ran into the thingy," Garth said. Sandy knew better. "The man had ten chances to shoot that stupid skunk," she told me, "but he couldn't bring himself to do it. Whereas a cowboy would just go ahead and do it, Garth couldn't. As badly as he wants to be tough, he just never will be. He's still a mama's boy.")

"What do you mean it would bother you?" I asked Garth.

"Just the whole idea of killing a cow bothers me." He launched into a story. "Just the other month, a friend of mine called me. It was a Sunday, I think. Yeah, that's right. A Sunday. Three weeks into football season." Like a good storyteller, he grew increasingly animated as he remembered each detail. He leaned forward in his seat. His eyes came alive. Suddenly I could see the little boy, the youngest of six children, trying to get attention around the dinner table or at the regular family gatherings called "Funny Nights" at which each child performed for the others. It was at those nights that Garth Brooks, the entertainer, was born.

"It was just starting to get cold," he continued. "Sandy said she was going to take the girls out to the zoo. They went, and I looked around the house and thought, 'Oh, my God. It's a Sunday. I have the place to myself, and it's the first chance I have had all year to watch football.' So I stripped down to my underwear, grabbed a bag of chips, and plopped down with the remote in front of the TV. And just then the phone rang. *Brrrrring.*" (He actually made the sound.) "It was a buddy of mine. He said he had a cow down in a ravine. Would I come help pull it out? So of course I called Mick." He chuckled. "We went down there. The cow was down in the mud and we tried and tried to get her out of there. We couldn't do it. He thought we might have to kill it."

"So how do you kill a cow?" I asked. We'd arrived at the hotel by now and had descended into the lobby. Garth looked out of place in high school locker room duds. He was already running late for his interview with Larry King.

"I don't know, shoot it between the eyes, I guess. But that's the point, I didn't know what to do. And I was really nervous. I thought, 'He's not going to ask me to kill the cow, is he?' And then I realized he didn't know how to kill her either. He'd never done it, and he'd been around ranches his whole life. That didn't make me feel so bad."

Eventually they tied a rope around her legs, attached the rope to a tractor, and pulled her out of the ravine. Though her legs eventually gave way, they didn't have to kill her.

"The thing was," he said, "I thought it would take about an hour or so. It took three or four. I still haven't had a chance to watch football all year."

With his story finished, Garth excused himself to go change. When he returned, an hour later, his face was cleanly shaven. He was wearing black jeans, black boots, and a blue-and-black shirt he had designed himself. He also wore his hat. The insecure farmer of an hour earlier had been transformed into a poised professional. If music was a sport, this was his game face. Now he had work to do. Now he had an interview to give. The change was startling. Suddenly people turned their heads. Suddenly he had more strut to his step. Suddenly he had that determined twinkle in his eye. "Showtime," he said and strode out the door.

SIX

THE COVER

The early morning temperature had barely crept above twenty when Wade Hayes pulled a pair of starched blue jeans off a hanger in an unheated cabin on a dude ranch north of Nashville. The jeans, a midnight blue, had been pressed so tightly they looked as if they could cut through ice. A layer of ice had, in fact, collected on the splintery panes of glass in the makeshift dressing room where Wade was about to crack open the jeans and scrape them onto his six-foot-two frame. Before he did that, a young woman from his record label, dressed in a parka and purple gloves, darted into the wobbly walled room and turned on a portable hair dryer in a valiant effort to heat up the space. She was several seconds into her huffing and puffing when the whirring of the dryer blew a fuse in the generator, instantly plunging the cabin into darkness, silencing the kerosene heater in the corner, and eliciting a howl from the photographer's assistant, who, at that moment, was hanging a klieg light from the ceiling and was forced to let it come crashing to the ground to keep it from smashing his head.

"Get me some light!" called the frantic assistant.

"Turn off that dryer!" shouted the bewildered wardrobe lady.

"Good God! My camera!" exclaimed the photographer.

All through the commotion, the marketing rep from the label was still staring at Wade's unflappable pants. "Honey, I hope you don't get a hard-on in there," she said.

All Wade could do was laugh.

With his new album now complete, though two months behind schedule (they had trouble finding "up-tempo" songs), Wade's attention now turned to shooting the photograph that would grace the album's cover. In addition to appearing on the front of the album, the photos from this single sixteen-hour session would also land on all his tour jackets, T-shirts, key chains, bandannas, coffee mugs, guitar picks, and, perhaps, even on the side of his bus, as they stopped his life and froze his face in time, thereby defining his image for the next several years of his career. The pressure to get it right—and fast—was an important, albeit onerous, responsibility for a young artist still so overwhelmed by his new life that he remained uncomfortable with the idea of his face on hundreds of thousands of CDs, not to mention billboards, magazines, and tabloids (not yet, but they were looking).

"There's just a lot of things about my looks that I don't like," Wade said of his handsome but deeply serious face, with its heavy eyebrows, a high school football-player-type smirk (I *am* tough despite my tender appearance!), and, yes, globby misshapen nose that looked like a piece of putty between a pair of brooding eyes. "I broke my nose twice," he explained. "Once in grade school when I was fighting, and the other time when I fell off a trampoline when I was thirteen. To this day, there's only a few pictures I can even stand to look at." So why not get plastic surgery? I asked. "I'm afraid it would change my voice."

Given his anxiety, the task of shooting Wade was even more difficult. The person charged with meeting the task was a man so seemingly different from Wade, as well as Nashville in general, that his anxious stoop in the corner that morning, smoking, hurrying, with stringy gray hair, added a degree of almost droll relief to an otherwise frigid situation. "Okay," Bill Johnson said in his raspy voice once the generator had been restarted. "I'd like to get a rainbow effect through that window. Give him a little more glow this time."

Bill Johnson grew up in Yankton, South Dakota, attended art school in Omaha, and moved to San Francisco during the Age of Aquarius to become a painter, "which I did until I went broke." Eventually ending up at the start-up publication *Rolling Stone,* he moved with the magazine to New York, then left to follow a fellow designer, Virginia Team, to Nashville in the late 1970s. Since that time, as senior creative director for Sony Music, he had worked on album covers for artists as varied as Willie Nelson and Dolly Parton and, along the way, had won two

Grammy Awards: one for his work with Rosanne Cash, the other with the O'Kanes. "The shoots are a lot more complicated now," he said of the industry's new high-gloss style, a far cry from the early days of country music when artists followed the Victorian ideal by sitting for portraits in their Sunday best and grimly staring at the camera. Even in the 1970s, when blue jeans and open collars first started appearing in country music photographs, image was an afterthought. "It used to be that George Jones would walk in in whatever he was wearing, with a bottle of whiskey in his back pocket. He wouldn't even call a break. He'd just turn around and pull out his bottle, we'd continue shooting, and there he'd be again, just smiling away. You never really knew what you were gonna get, but you did get something every time."

Now, shooting a cover for Wade Hayes required at least two months of planning with his management to ensure that Wade would have half a day off among tour dates in Tuscaloosa, Pontiac, and Walla Walla; six weeks of meetings on wardrobe, setting, lighting, and facial expression; and one month of traipsing all over the state or, in some cases, the country, looking for the place, with the light and the aura, that would make the artist feel comfortable and give him the confidence to smile into the camera for that one indescribable moment. Then, on the day that moment arrived, a team of two dozen attendants had to trek out to the location: one photographer, three assistants, a lighting expert, a wardrobe lady, a marketing executive, two secretaries, an intern, the label's publicist, the artist's publicist, the artist's manager, the manager's deputy, their assistant, a caterer, a coffee lady, the art director, the artist, his roommate, and, though she was running late on this day, the hair-and-makeup artist, Melanie, whose psychic told her not to drive too early for fear that she'd wreck her car on the ice. "My psychic is the one who told me to come to Nashville in the first place," she whispered as Wade finished tying his ropers. "She said it's like Mr. Creative down here. And you know what? She was right. I'll never go back to New York or L.A. And Paris? Who needs the prima donnas?"

Wade, though he looked handsome when he emerged in his military-crisp jeans, ocean blue cowboy shirt, and sly hint of white T-shirt, had none of the sexual predator feel that dominates fashion runways. That was no accident. "In New York, the focus is on beauty and fashion," Melanie explained. "It's got to be, like, totally forward. Totally next. *Aggressive*. In Nashville, you're dealing with the country market and with fans who want to see the real thing. What we do is make our artists look beautiful, but also keep them as close to themselves as possible. Why do you think country artists all have their pictures on their album covers, while rock artists have abstract photographs or concept art? In country, it's all about identifying with the artist."

As Wade sat down in his first setup of the day, a kitchen chair in front of a morning window, he did appear eminently identifiable, a lanky kid brother with few trappings of stardom. If anything, he could have used a little more sparkle. Tamara, the photographer, sent an assistant to adjust the light so it covered only half his face as he peered out the window. Bill, who was supervising, sent another aide outside to tinker with the umbrella that shielded the sun. Melanie, meanwhile, spritzed gel onto his hair, while Marcia, the marketing hand, stood behind the camera trying to get an early morning rise out of Wade.

"Come on, honey, livelier!" she cooed once the shooting started and Wade started moving through poses—looking out the window, down at the floor, up at the ceiling, directly into the lens. "That's right, baby. Give it to me. Look me right into the store." Wade, busy concentrating, showed little emotion. "Think of multimillion albums," Marcia continued. "Think of naked girls throwing themselves on the stage." The room chuckled a bit. Wade cracked a grin. "Don't stop!" she pleaded. "You're almost there. Just imagine: Albums. Girls. Albums. Girls—" And then, out of nowhere, she got what she wanted: a bounce of the eyebrows, a sliver of a dimple, and, finally, (could it be?) a wide-toothed smile. "He *smiles*!" she shouted. "He smiles! He smiles!" She raised her hands like an Olympian—"Yes!"—then sprinted over and gave him a kiss.

From the moment he entered the business, Wade Hayes seemed to embody many of the tensions straining contemporary country music— between youth and maturity, between sexuality and temperance, be- tween wearing a hat and not. As a result, a year into his career, the question of what direction to take his image was already testing his camp.

By the time Wade's debut single, "Old Enough to Know Better (But Still Too Young to Care)," was released in October 1994, country music had experienced a drastic revolution in the five years since Garth made his debut. The revolution had to do with the size—and the makeup—of the audience. Traditionally country had appealed to older fans. Indeed, the genre's growth in the 1990s, led by the so-called Class of '89 (including Garth, Clint Black, Alan Jackson, Mary Chapin Carpenter) was fed in no small measure by a demographic accident. By 1989, the leading edge of the baby boom was just reaching forty, and it was these people who first reached out to Nashville. "Country music is about lyric-oriented songs with adult themes," Lon Helton, the country editor of *Radio & Records* (*R&R*), told me. "You've probably got to be twenty-four or twenty-five to even understand a country song. Life has to slap you around a little bit, and then you go: 'Now I get what they're singing about.' As those people aged, they came right into the wheelhouse of country music." According to the Country Music Association, half of country music fans in the first half of the 1990s were in their thirties and forties.

As the music caught on, though, and more stars started appearing on television, younger people started taking notice. According to *R&R*, the percentage of country fans in their teens tripled in the nineties; the percentage between eighteen and twenty-four increased by half. Suddenly 40 percent of Americans in their twenties said they were country music fans. More than any other mark of the Country Era, the arrival of young people represented the biggest change. "When I first started in country radio back in 1971," Lon Helton told me, "you had to convince people that country music didn't cause cancer. Their opinion wasn't neutral; it was openly negative. Now for the first time ever we've gone from having a whole generation of people having a negative view to one with a positive view. We didn't even go through neutral—went right to positive." A principal spur in this transition was the introduction of videos. "You want to know why country music got so big?" Hazel once said to me in her Grandma-told-you-so tone. "All those sexy videos. Women fantasize as much as guys do. When they see a Vince Gill video, or a Clint Black, or a Wade Hayes—'*Oooh!*'—they are going to fantasize about that guy when 'Now I lay me down to sleep.'" Country videos received an enormous boost in 1991 when TNN launched Country Music Television. Almost overnight the appearance of artists began to matter more: gone were the receding hairlines and sagging wrinkles of older stars like Merle Haggard and Johnny Cash, replaced with skin-tight jeans, powder-keg dimples, and a sudden rampant swagger of the hips not seen since Elvis Presley.

In time the presence of all those pimply fans had an enormous impact on the music. The members of the Class of '94, among them Tim McGraw, Faith Hill, Bryan White, and Wade, routinely performed music that sounded more rocking (with heavier drums and electric guitar), but often involved softer-focused pop themes (fewer rodeo riders, more teenagers at the mall). Wade's first song, for example, "Old Enough to Know Better" could not have been more different than Garth's debut song, "Much Too Young (to Feel This Damn Old)." Though both were dominated by fiddles and steel guitars, Wade's song had more spark to it, more rebellion. It's about a working-class kid who parties so hard on the weekends he can barely ready himself for work on Monday, while Garth's is about an aging cowboy succumbing to health problems. Their next songs showed the differences even more: Garth's follow-up single, "If Tomorrow Never Comes," is a classic evocation of a young father confronting his own mortality, while Wade's second song (and second number one), "I'm Still Dancin' with You," is a sentimental ballad about a young man who can't stop thinking of his ex-girlfriend: "Even when I'm holdin' someone new / I'm still dancin' with you." Though both Wade and Garth were in their late twenties when these songs were released, the differences in their personalities—and their moments— were clear: Garth was speaking to fans as old as twice his age, while Wade was speaking to fans as young as half his.

Perhaps the most striking difference, though, came in their videos. Garth originally resisted the idea of making a video for "The Dance," his fourth single, claiming that videos distract from the song. Once persuaded, though, he steered away from the straightforward imagery that had defined the genre. Abandoning the idea that Tony Arata's song was about a romance that ended prematurely, Garth took the position that the song was about risking everything for a dream. "To a lot of people, I guess 'The Dance' is a love-gone-bad song," he said at the start of the video. "To me, it's always been a song about life—or maybe the loss of. Those people that have given the ultimate sacrifice for a dream that they believed in." The video went on to include clips of famous Americans who had died prematurely: John F. Kennedy; Martin Luther King, Jr.; the members of the space shuttle *Discovery*; country singer Keith Whitley; rodeo champion Lane Frost; even John Wayne (who actually died of cancer at age seventy-two). Though the song and video were controversial, they worked within the confines of country because they spoke to those themes—early death, faded glory, tainted love—that Nashville has long embraced.

Five years later, the pendulum in country music had swung dramatically: Death was out, life was in. Age was downplayed, youth had prevailed. And love, though ever-present, now shared the stage with sex. The video for Wade Hayes's third single, "Don't Stop" was as seductive as any ever made in Nashville. In it, Wade and a scantily clad brunette (the " 'Don't Stop' babe," as Wade and Don called her; she'd been imported from L.A.) prance around an empty house as Wade strums his Telecaster guitar, and the woman, nay temptress, tries to lure him to the couch. The guitar line itself has a jaunty, bump-and-grind feel to it. The lyrics—"I can't keep holding back / When you keep doing things like that,"—suggest that "Don't Stop" might be the first country song about oral sex. Even the video escalates from winks and nods, to shirts-off sweaty thrusts and grinds, to openly lustful fingers dribbling here and there. It climaxes with Wade's amplifier bursting into flames, and the final image is of an ice cube sizzling into vapor on a silver platter. The video, twice as hot as anything Garth, Clint, Alan, or anyone else had done up to that point, was so overtly sexual that CMT originally threatened to ban it. Three months later, fan support pushed it to number one.

Which is not to say Wade was comfortable parading his sexuality. If anything, he was the opposite.

After about forty-five minutes in front of the window, Wade took a break for a cup of coffee and a tuna sandwich still frozen from the cold. Momentarily he was ushered to a different setting, a table against a mirror. When that proved too dark, Bill opted to move to the far side of the cabin and to the most important pose so far—the one where he asked Wade to take off his hat. From day one of his career, the question of whether Wade should wear a cowboy hat had dominated discussions about his image. The issue was whether to align him with the more modern edge of his music by leaving off the hat or ground him in tradition by leaving it on. The decision was made more pressing by the almost smothering symbolism the cowboy hat had assumed in recent years. After the success of Garth, Alan, Clint, and Billy Ray, the cowboy hat made a cultural resurgence. Stetson sales soared, as did business at country bars and sales of Wrangler jeans. Jeep Cherokee drivers across the country suddenly thrust open their doors and paraded their sagebrush wreaths and flaming cowboy shirts for all the world to see. Nashville, predictably, responded by signing a rush of new acts and cramming cowboy hats onto their often citified heads. Eventually the tsunami of knock-off acts became so laughable that Marty Stuart, a former bluegrass prodigy, and Travis Tritt, a country rocker from suburban Atlanta, actually headlined a show called the No-Hats Tour. Overnight the hat had gone from a symbol of pride to a badge of banality, and the Stetson was immortalized on the Tritt and Stuart poster with a giant red REJECT label stamped across its brim. John Wayne, meet Beavis and Butt-head: Cynicism had come to country.

By the time Columbia Records got around to preparing Wade's first album cover in 1994, the hat was considered so cliché that executives urged him to keep it off. "He is so handsome, his eyes have this incredible soulful depth to them," Connie Baer, the marketing executive charged with making the decision, told me. "I knew women would find him incredibly attractive." Wade would have none of it. Though he had shunned the hat when he first came to town for fear of melting in, by the time the album was finished he had decided to keep it on. "I just missed it," he said. "It's not something I just put on to sing country music. I've always worn one. I don't want people to get confused about what I am." The label pushed back as hard as it felt it could, but in the end, pleased that it had an artist who refused to be sandbagged, relented. The cover for Wade's first album showed a boyish twenty-four-year-old with a Little Lord Fauntleroy curl of hair, a plaid collarless shirt that could have been bought at the Gap, and a beige T-shirt. On his lips is a dab of a little too much lipstick and on his head is a handsome tan hat. He looks country, friendly, and completely nonthreatening. Also, not particularly sexy.

His new album, though, showed a different side of him. The two best songs on the album, "The Room" and "Hurts, Don't It," were both dark ballads about a deeply sensitive man who had been hurt by a curdled love. Other songs, particularly "Our Time Is Coming" and "I Still Do," were more openly romantic than his past material. He was growing up, in other words, and opening up, too. Bill Johnson had a plan to capture that on film. "I'm trying to develop the look in response to the songs," he said as Tamara made the final adjustments to the faint light on the brick wall. "Just talking to Wade I get a feeling that he wants a stronger image. A little Marlboro-ish. What that suggests to me is a little more masculine. I'll warm up the colors a bit. I think all in all I can be a little more artsy. I feel now I can bring in some mood and a lot more emotion." For Bill that meant one thing, though: getting more expression from Wade's face.

When Wade arrived in place for the new shot the mood was looser than it had been earlier, but still uncomfortably serious. He was still wearing his blue jeans, with a dark blue shirt now, and a black belt with silver trimming. His light tan hat appeared almost ghostly in the smoky setting. Wade positioned himself in front of a red door in the middle of the brick wall. Lit from head to toe, he appeared quietly menacing. When Marcia stuck his Telecaster in his right hand, he looked like a dueling gunfighter preparing to wield an ax in battle. "Good God, now *I'm* getting a hard-on," Marcia said.

With everyone's confidence soaring, Bill decided to act. He went and spoke quietly with Wade, leading him through a series of poses. Wade continued posing once Bill backed away and Tamara, her blonde hair pulled back into a ponytail, started shooting again, this time more tenderly. We had moved from groping foreplay to candlelit seduction. After several minutes, Bill made his move—gentle, reassuring, "Okay, now . . ."—and Wade removed his hat. He didn't toss it aside, but held it, contemplatively, his disembodied self. The mood grew hushed. Melanie tiptoed forward and brushed his hair. She pulled a spritz of bangs over his forehead. The effect was transforming. Wade still stood tall, his broad shoulders starched, but suddenly his face invited closer inspection. As he pulled his hat just inches from his nose and ran his fingers over its crown, suddenly the music in its most mythic form seemed to come to life. A man stands in front of a wall, pausing at the door—coming in or going out?—and considers his hat, the symbol of himself. In Nashville, the music may be digitally enhanced, the images may be unabashedly manipulated, but when they come together—in one unexpected moment—they can still create something unambiguously real.

The crew broke for lunch after the no-hat setting, then moved outside before the sun disappeared. We were advancing to the more rugged, manly phase of the shoot, which would peak now, at sunset, with Wade standing in front of an abandoned toolshed with barren underbrush running out toward the horizon. The importance of this shot was underlined when Wade appeared in the blustery late-afternoon conditions dressed in a black hat, black jeans, a black collarless shirt, and a crisp bolero-cut caballero jacket made of rich black leather. "My God, my baby's all grown up," Marcia said. And he was: Wade was wearing his first Manuel.

Manuel Cuevas, known simply as Manuel, was one of the most colorful characters in Nashville history, an eccentric Mexican stylist-bon vivant with perpetual silver ponytail and brightly colored cravat who would have been at home in Robert Altman's *Nashville* twenty years ago, but who seemed anachronistic in Edward Gaylord's whitewashed version of Music City, U.S.A. Born in Coalcomán, Mexico, into a family of eleven children, Manuel emigrated to the United States in 1955, where he went to work for Hollywood tailor Sy Devore making suits for Frank Sinatra and Bob Hope. Bored with what he called "regular clothes," Manuel left to work for legendary designer Nudie, the *nom de* needle of Nudie Cohen, a former boxer from Brooklyn who parlayed an unsuccessful career making costumes for striptease acts into one of the most signature styles in American entertainment. Working out of his shop in Los Angeles, it was Nudie, a Brooklyn Jew, who defined what we now think of as traditional rhinestone cowboy imagery: dressing Gene Autry and Roy Rogers in the 1940s; Hank Williams and Tex Ritter in the 1950s. His model was the singing cowboy movies of the 1930s, which he accented with his flourishes drawn from the personalities (or even the names) of his celebrities. Nudie covered Porter Wagoner's outfits with wagon wheels and Ferlin Husky's with huskies. In 1957, Elvis Presley paid Nudie $10,000 for a twenty-four-carat gold lamé tuxedo.

Manuel, himself an outsider who shared Nudie's vision of an idealized America, stayed with his mentor for twenty years, eventually marrying his daughter. It was a fruitful apprenticeship. Through Nudie, Manuel was able to help outfit an entire generation of musicians. In the 1950s, he put Johnny Cash in black; in the 1960s, he designed suits for the Beatles; and in the 1970s, he put Elvis in those famously gaudy wide-lapel white jumpsuits. Leaving Nudie in 1974 ("I divorced his daughter," he told me, "I had no choice . . ."), Manuel moved to North Hollywood and continued on his own, always adjusting his clothes slightly to fit his times. In that way, Manuel became an ideal designer for country artists, a Zelig of cowboy *couture,* always tinkering but never abandoning the nostalgic ideal at the heart of Nashville's appeal. In the early 1970s, when Gram Parsons was looking for a way to capture his blend of Music Row and Haight-Ashbury, Manuel embroidered rhinestone marijuana leaves and poppies onto his wardrobe. In 1980, he designed the costumes for *Urban Cowboy.* And five years after that, when *Urban Cowboy* had become passé and the term New Traditionalism was just emerging, Manuel gave that awkward expression a stunning face by putting Dwight Yoakam in gaudy retro-1950s jackets and dressing Dolly, Linda Ronstadt, and Emmylou Harris in matching rhinestone roses on the cover of their *Trio* album. Finally, in 1989, in what was a dramatic statement at the time, he seemed to catch another trend when he left Hollywood after thirty-five years and moved to what he called the new "front porch" of American culture: Nashville. If Manuel had gone country, surely Hollywood would follow.

"I guess I did it at the right time," Manuel said in his chuckling, heavily accented voice when he appeared that afternoon at Wade's shoot. "I felt a little crowded in Los Angeles, a little too tight. I wanted to come to a place where I had visited for so many years and had always liked. I moved here to be silent, more trusting that I could just hold my pencil and draw without a fight breaking out in the streets." Since that time, Manuel had added a sense of flair that Nashville admires, but has little of itself. Quiet by day; by night he became his jackets. Often as many as several times a week, I would run into Manuel at nightclubs or new artist showcases around town, jolly, looking a bit like Lee Trevino in a flamboyant cowboy jacket. Invariably he would be surrounded by a harem of impossibly beautiful women and would be running the longest tab at the bar. "Tequila?" he would query anyone who walked in, before turning, petulant, to some woman at the counter and asking why she wasn't enjoying the view from his lap.

And through it all he continued to define Nashville style. "In any type of entertainment," he explained, "sixty percent is imaging, twenty percent is packaging, and twenty percent is talent. That may sound surprising, but it's true. It's imaging that makes you spot an artist a mile away. If you get on a bus and a nun gets on, you'll probably offer her your seat. You might find out later that she's a prostitute, but imaging was the first thing that made you accept her." For younger artists, like Wade, getting their first Manuel jacket (the one he was wearing cost $5,000) was a sign that they had made it, a postcard they could send home to Mama's refrigerator. "What do you want to be?" he would ask them on their first visit, usually after their first album hit the charts. "A country boy or a star?" Usually, he said, they would answer, "Both." " 'Well, you can't be both,' I tell them. 'People don't want to see themselves onstage. They want to see what they believe they would look like if they got all dressed up, like you.' " In Wade's case, Manuel told him to appear more grown-up. (He tried "Lose the hat," but that didn't fly.) He told him he looked too much like a kid at the mall on his first album and not enough like a legend (too I-40, in other words, not enough Route 66). He told him to put on a jacket. Indeed, it was the jacket he had been commissioned to make in response to that comment that was the focal point of this final setting of the day.

With the lights finally focused, Bill positioned Wade on an exposed piece of rock. Wade hardly spoke as he got into position: straightening his shoulders, thrusting out his chest, and staring almost seductively into the camera. "Look at me, look at me," Tamara urged. The fifteen or so people who had stayed through the day pushed in around Wade in an effort to block the wind. A smoldering intensity began to develop. "Imagine you're a bathing suit model having to do this at this time of year in a see-through bikini," Tamara said. "That would be embarrassing," Wade said. The crowd laughed. Wade didn't. "Look down again," Tamara said. "Look down."

Patchink, patchink, the camera snapped.

"Now move your head a little," she said. "Look up. Look up." *Patchink*. "Yes!"

She continued giving orders, steadily, firmly. Less desperate than before; quietly insistent. "I need you to move your foot backward." "I need your hand there." The attendants stayed, just watching the odd dance.

"A little lower with your chin. A little higher. Yes, I know I can get confusing."

In fact, the longer the day went on, the relationship between the artist and the photographer had become notably more sexual, from the early morning flirting in front of the window, through the candlelit romance of their midday encounter, to the increasingly erotic climax of sunset, with both sides standing in the cold, staring at each other.

"Ah, we're seeing your eyes now," Tamara said. "Sexy. Sexy. Bring your leg down. Ohhhh, that's good . . ."

The temperature increased with every command. The *patchinks* got closer together. Tamara was holding her breath.

"Ah, nice move there. Squeeze your hand. Oh, God, I love it when you bring your hand up like that. Just like that. *Yes, yes,* do it again."

Her voice was growing higher. Wade dared not move.

"Turn your leg this way. Now your shoulders. Put your hand there. No, there. That's right! Stop. Stop! Yes!! Your arm. Your face. Your hand. Oh, GOD! Yes. Your hand!! Yessss!" *Patchink*.

"Okay. Take a breath. My God. That was good."

"Yes, yes, *very* good."

Bill Johnson was skipping around his office on the second floor of Sony Music's low-rise on Music Row, headquarters to its two divisions: Epic and Columbia. It was the following Tuesday now. Bill had culled the two thousand photographs of Wade down to four hundred, which he spread out on four oversized light boards along one giant counter in his office. Alan Jackson was playing on the stereo. Framed album covers of Dolly, Mary Chapin Carpenter, and Patty Loveless hung on the scuffed beige walls. Bill's two Grammy statues were tucked snugly on his bookshelf like trophies in a high school bedroom.

"I like the ones where there's heat in his eyes," he said. From the half dozen setups they had shot that day, as well as four more they did that night in Tamara's studio (the total cost had exceeded $10,000), Bill had managed to find three or four from each that he liked. In particular, he had three or four shots that he was already considering for the cover.

One was from the first setup, with Wade leaning back in the chair; one was the Marlboro shot, taken late in the afternoon and printed in a muted sepia tone that made it look as if it belonged on a buffalo nickel; and two were from the middle setting with Wade in front of the red door. In one Wade had his hat on; in the other the hat was off. The latter shot was dark, but Bill said he could lighten it. "With the computer, we can do so many things that we couldn't do before. We can fix Vince Gill's lazy eye, for example. Or cut down on some of the baby fat in Wynonna's neck. I've taken eyes from one picture, hair from another, and cropped them onto an entirely different body."

Several days later, after a second look (and cut), Bill took a package of about 100 photos over to Wade's management company several blocks away on 17th Avenue, several doors up from Garth's. In the meeting, Wade said he didn't like the Marlboro shot and liked the hatless shot even less. He preferred several shots from the late-night session in which he sat at a table strumming his guitar. In those he wore a pea green shirt and his beige hat: He looked relaxed, but with none of the seriousness or masculinity he had originally said he wanted. Wade's manager, Mike Robertson, a forty-something Louisianan who also managed Lee Roy Parnell and Pam Tillis, preferred a different shot. And Carol Harper, Mike's deputy, preferred still a different one. Now totally confused, Bill went back to the office to make mockups of their choices, cropping the pictures to the requisite four and three quarters square inches and experimenting with assorted typestyles for Wade's name and the album's title, *On a Good Night*, chosen for the new up-tempo song added at the last minute.

Throughout, Bill couldn't shake the idea of using either the sepia Marlboro shot ("It's just so personal," he kept saying) or the hatless shot with Wade in front of the door. Finally he decided to use both. In what he knew was a risky move because of Wade's reluctance to appear without a hat, he used his computer to lay a sepia outdoor shot of Wade in the background. Then he took Wade's face from the hatless pose and superimposed it in front of the other shot. The effect was magnificent. It had one shot with a hat, one without. Also, the tension between Wade's pinkish face in the foreground and his broad-shouldered, elusive pose in the background perfectly captured the mix of darkness and light in the music. As the mock-up made its way around Sony, a collective enthusiasm rose up in the building. "This is the best album cover I've ever seen," Allen Butler, the former record promoter turned executive vice president, wrote in a memo to the art department.

Bill was thrilled, but circumspect. "It's a gamble," he said the afternoon before he sent it over to Wade. "I think the no-hat thing is a little sexier, a little more revealing. He's such a soulful singer. But with the shot in the background he's also manly and tough. This treatment may be one of my favorites of all time. If they can make an adjustment here—if they can accept the shot without the hat—we're fine."

"And how often do you get what you want?" I asked him.

He thought for a second. "Ninety percent of the time."

This would be one of the 10 percent.

The final batch of covers arrived at Mike Robertson Management on Wednesday, April 3. Mike and Carol hated the design. "We have established Wade as a hat act," they said. "We don't think we should confuse the image. He's a product. You don't change your image midstream. There's a reason Bayer Aspirin has the same package every time." They forwarded it on to Wade, who agreed. "I hear they're having trouble booking people with hats on TV," he told me, "and that's why they want my hat removed. But my feeling is, TV bookers don't buy records. Fans do." He preferred the friendlier shot with him sitting at the table, smiling into the camera. "That's my Wade," Carol said when she looked at the mock-up.

Now at a full-fledged standoff with the label, the two final cover designs were sent to Don Cook, who worked out of a separate office in Sony's publishing division. He would make the final call. "Don's an executive now," said Scott Siman, the vice president of Columbia. Don, in the first sign of a creeping conservatism that was beginning to affect Wade's career, chose the lighter of the two covers. "It just seemed like the right thing to do," he said.

The team at Sony was crushed. Bill quietly turned his attention to the rest of the package. Marcia, the marketing hand, shut her door and sobbed. Scott Siman was openly rueful. "It doesn't hurt us," he said to me on the afternoon he found out. "But it doesn't help us either."

SEVEN

THE FACE

The sight of Wynonna without makeup was startling. It was midafternoon, the Monday after her triumphant three-day return to the stage at the Universal Amphitheater in Los Angeles ("She is, without a doubt, one of the best singers of her generation," wrote Robert Hilburn in his review in the *Times*), and Wynonna was sitting in a director's chair in the second-floor dressing room at "The Tonight Show," where later that afternoon she would tape an appearance with Jay Leno, along with "talk-show host and comedian Jon Stewart, six-year-old actor Luke Tarsitano, and those wacky ambassadors of goodwill, the Harlem Globetrotters!" About to begin a preshow hair and makeup session, Wynonna was wearing black bell-bottom pants and black suede pumps. Her T-shirt was a vivid forest green, a striking color-wheel complement to her reams of near-neon pumpkin hair. Her hair (with extensions, as the tabloids had reported) wasn't draped around her shoulders, but was rolled around oversized powder blue rollers, each one about the size of a hot dog bun.

"Come in, come in," Wynonna said. Her voice was casual, with distinct Southern roots, alternating with a Hollywood, I-glide-above-it-all air and an occasional belly laugh. It was when she offered that self-deprecating laugh that Wynonna was most winning. "We just finished lunch," she said, chuckling. "My favorite time of the day."

I pushed aside a cardboard box and a few scattered jackets and settled into the couch, nearly sitting on her tiny, black-and-white rat terrier, Loretta Lynn. The room was small, with two couches and a makeup chair taking up nearly all the space. A wardrobe was filled with Wy's black coats and multicolored silk shirts, various options for the evening. Her backup singers were just across the fluorescent-lit hall, the band next to them.

"So where do you live?" Wynonna asked me. We had first met several nights before in her dressing room at the Universal Amphitheater ("I know who you are," she had said, "You and I are going to be cool . . .").

"I just moved to Nashville," I said.

"Oh, really?" She perked up. "Nashville's a really interesting place right now. There's a spiritual awakening occurring there. People are coming from all around the country. I'll have to show you around to all the cool places. Church, the farm, all the meat and threes."

"I've been wondering about that term," I said. "Where I come from, they call that kind of restaurant a 'meat and two.' "

"Well, that just goes to show you," she said without missing a beat. "In Nashville, we've got more of everything."

We started talking. She was delighted with her wedding, she said, though peeved at all the tabloid coverage ("I did *not* stumble down the aisle," she insisted, then added of the doctored photograph in the *Globe*, "and I've never worn red nail polish in my life . . ."). She was pleased that Ashley had come to her debut in Los Angeles, but baffled that she only wanted to watch the NCAA Basketball Tournament ("I'm just not like her," she said. "I'm much less intense . . ."). Above all, she was thrilled with the review in the paper. "So many people don't even bother to listen anymore," she said. "They think they know me. They think they know the life I live. They don't . . ." She interrupted herself to ask if I was married. Not yet, I said. Then she asked how old I was. We were born the same year, I told her. "So *you* understand!" she exclaimed, almost reaching to embrace me. "We've had our fun. We've had our wild times. Now it's time to settle down."

As she spoke, Kenny, her makeup man, appeared. He opened a tackle box and began fiddling with makeup containers. Base and powders on the bottom, pencils and blush on the top. In no time the room began to smell like the ground floor of a department store. Wynonna squirmed a bit, but kept her face still. Undecorated, it was a surprisingly plain-looking face, almost vulnerable. Her skin was pale with a fresh-from-the-bath tangerine tint. She had a few faint freckles on her nose and cheeks. With pigtails instead of curlers and overalls instead of silk, she could have passed for a genuine Appalachian farm girl.

"You know, it started out really cool, really special," Wynonna said of the years with her mother, "just two Kentucky girls with a guitar. Then it got really out of hand. It got weird. It was hard for me to accept that I was just another *product* in the American entertainment machine." During that time, she said, people constantly harangued her about being irresponsible, about being lazy, about being late. "It got so bad that it was all I could think about," she said. "I worried and worried and worried. I wasn't free to be myself—to be spiritual. It was like being an alcoholic, in that scent of dependency. I was performing to please my manager, my mother, everybody but me."

Her arms had started swinging slightly. Her road manager stuck his head in for the third time in thirty minutes, a sign that perhaps she needed to be alone. "It's okay," she said, resting her head on my shoulder. "It's important that he be here right now. We're *merging*." Then she pushed the door closed with her foot.

"I remember after I got pregnant," she continued, "my manager sat me down and said, 'Okay, we need to decide when you and Arch are going to get married.' I felt like, 'No, that's not the way it should work.' Arch and I should be sitting down and deciding when we're getting married and then we should be telling you. For the last year and half, I sat out at the farm feeling guilty—for Mom, for God, for my fans. Then I realized, 'I don't *have* to be perfect. Nobody else is.'"

She looked in the mirror to check on the status of her face. Her Kentucky roots were covered up now, submerged under a layer of Hollywood tan—the only perfect color in the rainbow. Kenny moved on to her eyes.

"If you really want to understand where my life's at right now," she said, "everything you need to know is in that box." I opened the box beside me on the couch to find a complete set of Stephen Covey supplies, *Living the Seven Habits*: a book, a planner, audio tapes and videotapes, a completely shrink-wrapped course in made-to-order self-esteem. "My new manager gave it to me," she said. "He wants me to use them. I'm trying. I'm trying. But if you saw the way I lived my life, you'd know how hard it's going to be." Then, for the first time, Wynonna turned and looked at me directly. "Which reminds me, how much *do* you know about my life?" she asked.

Apparently not as much as I had thought.

◊   ◊   ◊

By the mid-1990s, Wynonna Judd and her mother had been a gossipy staple in American life for almost two decades, but in the larger context of country music, only two stories are really important: first, the tale of how Wynonna, the wounded genius with the immortal voice and the fractured soul, rose to become an unlikely and, at times, unsteady symbol of women in country music; and second, the often warped, self-serving version of that story that her mother pawned off on the world.

Even after three decades, the most basic facts have been slow to emerge. The future Wynonna Judd was born on May 30, 1964, at King's Daughters Hospital in Ashland, Kentucky, an Appalachian river town of ten thousand people in the northeast corner of the state. Her father, Charlie Jordan, was a senior at Ashland High School. Her mother, a classmate with whom Chuck had a brief relationship, was Diana Judd. Diana (she changed her name to Naomi as an adult) was a flirty, popular child, the oldest of four children. Diana's father, Glen, a stoic, flinty man with few indulgences, ran a filling station in downtown Ashland. Her mother, Pauline, was a housewife and nursery superintendent at the First Baptist Church. Though Naomi would later romanticize her upbringing as quaint, quirky, and filled with country people with a "natural, endemic Appalachian wisdom," even she admitted that the chilliness between her parents left her with a driving desire to escape. As she said in her intimate, though often deceptive 1993 book, *Love Can Build a Bridge* (in which she somehow neglected to mention Wynonna's real father): "I now know why I'm such a demonstrative person . . . I never had enough communication when I was little girl. I walked around like a great big piñata, wishing someone would break me open so I could spill out my emotions, secrets, and feelings."

From her earliest days, Diana Judd invented fictitious stories of herself and her family. Because Mark, the younger of her two brothers, was so "beautiful," she said, "I told people he won lots of beauty contests." As for herself, Diana often imagined she was an Indian princess, the misplaced child of some Cherokee or Chippewa royalty. As Wynonna told me: "My mother is the kind of person that feels the need to develop her own reality to spare her from the heartache of the truth. For instance, her dad was an alcoholic. He would be sitting there at the kitchen table with the newspaper drinking a Coke and they would be walking out the door going to church. *She* thought it was because he was just so exhausted from working the night before. That was her reality." In eighth grade, a neighbor of Diana's threw a costume party. One girl came dressed in a leopard-skin leotard with a wig that covered her face.

She sat in a corner and pounded rocks together, finally revealing herself at evening's end to be Diana. As one of her friends recalled: "You know, most kids who would do something like that, they might keep it up for a few minutes before they would yell, 'Hey, it's me! Didn't I fool you?' But Diana just went on and on with it."

At age sixteen, Diana Judd's need for fantasy was amplified by two traumatic events. First, her brother Brian was diagnosed with Hodgkin's disease (in her book, she takes credit for discovering the first lump). The illness ruptured her family. "That's when I realized," she wrote, "our home was not really the cozy Walton-esque existence I had constructed in my head." That image was further undermined when Diana discovered at the start of her senior year that she was pregnant. "All the daydreams and illusions I had created were stumbling over each other and mixing things up," she said. "It was a living nightmare. I was still in high school, I wasn't married. I got $1.50 allowance per week. Brian was downstairs on the sofa throwing up . . . Mom's losing her mind and our family's falling apart. Pinch me! Hit me! Wake me up!" Instead, she tried to commit suicide—first with a knife in the bathroom, then by throwing herself off the local water tower. When she lost her nerve on both occasions, she finally resolved to tell her parents. "Do you love him?" her father asked. "No," she said. Her parents insisted she get married anyway. She did, to a different man.

Michael Ciminella was an Italian Catholic, son of a prominent businessman, and a smitten young man who had been in love with Diana Judd for years. The two had met when she was fourteen. Michael drove up in a shiny Chrysler Imperial wearing his military school uniform and smelling of Old Spice, and Diana was hooked. He was the perfect screen for her fantasies, she said, and a wonderful vehicle for her imagination. Also, he was rich. The country club was where "privileged and worldly people went about socializing on a plane that no Judd had ever reached." The two dated off and on while Michael was in college, but Diana rebuked his request for marriage. When she got pregnant by another man, he agreed to marry her. "He loved her so much he married her anyway," Wynonna told me. The two tried to elope, but, without their parents' permission, were turned down by the sheriff. Finally, on January 3, 1964, Diana Judd, age seventeen, and Michael Ciminella, age eighteen, were married in the Baptist Church in the safely distant town of Parisburg, Virginia. The couple spent their wedding night in Michael's bedroom at home. Almost exactly five months later, Christina Claire "came into the world screaming on key and searching for harmony."

The first of those statements was entirely believable; the latter is the one that seems invented from the start.

Within no time, Diana and Michael's relationship became strained. Even after they moved to Lexington so he could continue college, the two teenagers bickered—over his lack of a job and her jealousy of his fraternity lifestyle. In one of the least generous sections of her book, Naomi, who never credits Michael with marrying her, maligns him for neglecting their daughter, for being arrogant, and for feeling "saddled with a wife and a baby at the exact moment he wanted to be out chasing cheerleaders and attending keg parties." It didn't matter, the two had struck a bargain and were stuck with each other, moving first to Illinois for his first job out of college (he "barely graduated," Naomi tartly reports) and then to Southern California, where Diana gave birth to her second daughter in 1968. Ashley Taylor, she observed, was "very different looking from Chris and clearly in full possession of Ciminella genes."

Life in California started well. "That time was probably the most normal we ever were," Wynonna told me. "Dad wore a tie. My mother had a pixie cut. Sit-down dinners. Get the bike for Christmas. Very 'Brady Brunch.' " But soon enough, Diana and Michael resumed their fighting. "I saw this commercial recently," Wynonna told me. "In it a little boy was sitting on stairs, looking through the pickets, and watching Mom and Dad fight. The look on this kid's face was awful. I remember that feeling. Any divorced kid will tell you they think it's their fault. Period." The fights terrified the girls. "They're so intense," Wynonna said. "Both my parents. Plus, he really loved her and she didn't love him back. She was a martyr. They tried to stay together for the sake of the kids, but, of course, that never works." At times Wynonna would have to separate her parents. "They were brutal," she said. "I've seen Dad deck her, but I've seen her deserve it. It's one of those tricky things: Don't ever underestimate the power of a woman to be able to push buttons. She provoked him. She was in total control. She is such a mother lion. It's that real thin line between being a sweet-talking woman and an absolute bitch from hell. I don't know how he didn't maim her. She would have made me want to drive my car into a tree at eighty miles an hour."

Eventually Diana decided to leave, moving with her daughters to West Hollywood. There Diana embraced the Babylonian spirit of the age. She did drugs. "It seemed like a way to smooth things over," she said later. "I was young, eager, feeling very frisky and desperate for change." She experimented with various religions, diets, jobs, and men.

One such man, whom she described as a James Dean look-alike, battered, stalked, and raped her, she claims, until a policeman she befriended put a stop to it ("As an added bonus, the officer and I really enjoyed each other's company!"). Christina hated this time, bouncing between hotels and apartments and lurching from one of her mother's boyfriends to another. That James Dean look-alike, in particular was, in her mind, "the single most worthless scumbag on the planet." "He was one of those boyfriends from hell that you read about," she said. And in an event previously undisclosed, Wynonna, who was ten years old at the time, told me that while she and that boyfriend were alone in the apartment, he sexually molested her. "He leaned up against me on the wall a couple of times, pressed his body against mine, and ran his hands across my body," she said. "Then he walked away and left me there, shaking." Though she remembers herself as being sassy and outspoken, these events petrified her. She didn't tell her mother and kept them from Ashley. "If my mother hadn't taken us to a hotel after he raped her, I'm sure he would have tried more," she said.

Eventually even Diana realized her lifestyle was destructive. When her sister followed her to Hollywood, Diana noted disapprovingly that she fell in love with California. Diana, on the other hand, was always proud of being from Kentucky, she told her sister, and looked for ways to make her heritage work for her. She had long ago noticed that everyone in Hollywood wanted to be someone else. Armed with this realization, Diana decided to seek out more quirkiness to her life's story. In the spring of 1975, Diana enrolled in nursing school at Eastern Kentucky University and moved her family into a vacant cabin on the grounds of a friend's estate. The cabin, called Chanticleer, had no television and no telephone. "I remember it being so abrupt," Wynonna told me. "We were in shock. Here we were on top of this friggin' mountaintop with hundreds of acres around us. Now what do we do?" The answer: Entertain themselves. For Ashley, it was make-believe; for Christina, music. "Music saved my life," Wynonna told me. "I was just an emotional hurricane, very weird. If it wasn't for music, I would have probably burned down the house." Once Christina showed an interest, Diana encouraged her, buying her a recorder, guitar lessons, and stacks of secondhand LPs: Hazel and Alice, the Delmore Brothers, the Boswell Sisters. For Diana, who sang along in harmony as Christina played, a new goal was emerging. She resolved to make her daughter into the next Brenda Lee—the teenage singing sensation from the 1950s—or, even better, Brenda Lee and her mother.

Christina, of course, loved the idea. Music was the one thing that prevented her from fighting with her mother. Naturally, once she became aware of her mother's interest, Christina milked it. When she had neglected to clean her room or complete her homework, she plopped herself down in the kitchen with her guitar so her mother wouldn't scold her when she came home. "Mom would never get mad at me when we were singing." This reliance on music only heightened when Diana once again moved her children to California, this time to escape testifying in her parents' divorce proceedings. For Christina, the move proved providential. To her hillbilly mix of influences, she now poured an assortment of California flavorings: Bonnie Raitt, the Eagles, Linda Ronstadt. Diana, meanwhile, busily plotted their ascent, even rustling up a few early singing gigs.

But her identity was still not complete. To crystallize her image, Diana changed her name. "I had never forgotten the little retarded girl named Naomi who stared at me in grade school," she said. Christina, following suit, changed hers as well, taking her new name from a line in the 1940s song "Route 66": "Flagstaff, Arizona, don't forget Wynona" (the second n was added for looks). Wynona, a small town in Texas, was, of course, on I-40 by that time instead of Route 66, but that symbolism didn't seem to matter. Nor did the symbolism of her mother's new name. In the Bible, Naomi was a hardworking mother who returns to her homeland after losing her husband and sons. She asks her daughters-in-law not to return with her. Oprah agrees, but Ruth goes anyway, offering up this paean to filial attachment whose words echo hauntingly in the life of Wynonna: "For whither thou goest, I will go; and where thou lodgest, I will lodge; thy people shall be my people, and thy God my God: Where thou diest I will die, and there I will be buried." Detachment, in other words, would never occur.

After shipping Ashley back to Kentucky and taking Wynonna out of school, Naomi Judd embarked on another cross-country journey—through Los Angeles, Las Vegas, and Texas—before arriving in Nashville, Tennessee. It was May 1, 1979. Her elder daughter, the "Female Elvis," as she would be called, was about to turn fifteen.

"I just want to be big," Elvis once said, "so I can do something for my mother." That sentiment might also have been uttered by Wynonna Judd in her early years in Nashville. Certainly she wanted to be a star. She daydreamed about the accoutrements of celebrity—the accolades and cream-filled cakes, the lazy afternoons with bonbons and boys. But on the matter of how to get from the run-down, roach-infested motel she inhabited with her mother to the plush, room-service fantasy she imagined, Wynonna had no idea. "My identity was in the strings and the notes and figuring out how to make what I imagined come to life," Wynonna told me. "I would have trusted anyone, signed anything. I would have been chewed up and spit out a thousand times by the business." What distinguishes her from other stars of her caliber is that in the all-important departments of drive, ambition, and single-minded obsession, she was entirely reliant on someone else.

As it happened, though, it didn't matter: Naomi had enough for both.

From the first moment she drove down Music Row in her lipstick red '57 Chevy ("It was as if we'd entered the Pearly Gates and were riding down streets of gold," she said), Naomi Judd attacked the puzzle of how to become a country music star with a dedication and cunning rarely seen in a world where most people believed they had seen it all. She needed it: Few were buying her idea of a mother-daughter duo. Naomi, though, knew that her best asset was her heritage. She knocked on doors and handed out a typed biography. She lingered around the parking lot at Channel 4 until she badgered Ralph Emery into giving them a singing slot on his morning show. She flirted her way around Music Row nightspots, eventually wiggling her way into the good graces of some songwriters, including the drinking dean of the lot, Harlan Howard. Her reception varied. Fresh from Hollywood, dolled up to look like a Christmas ornament, driving her car called REDHOT 2, and wearing Jungle Gardenia perfume, Naomi petrified women and electrified men. But while many men wanted to sleep with her, none wanted to sign her.

Finally two men told her the hard truth: Her daughter was the one who had the talent. Jon Shulenberger and Mike Bradley were running the Soundshop Studio, the same place where Wade Hayes would later record On a Good Night, when Naomi walked in with her résumé and tape in 1979. The men agreed to produce a demo, but only if Wynonna sang alone. "At her age, she was really where Brenda Lee had been," Shulenberger recalled. "I always felt Mom was the salesman. She had more drive and determination than anyone I had ever met." But when they shopped the demo around town, the reaction was the same. No label wanted an underage singer. Also, Naomi was reluctant to give up the spotlight. "Anytime we'd go in somewhere," Shulenberger said, "the limelight would always be on Naomi, which constantly undercut my trying to pitch Wynonna as a solo artist. It's difficult when you walk in a door with an artist and her mother radiates all the charm and charisma."

The next person Naomi approached actively embraced Naomi's involvement. Brent Maher was a producer whom Naomi met while she was a nurse. She took the opportunity to give him a tape, and when he found it in his car weeks later, its mix of bluegrass, country, and blues almost drove him off the road. "Who are you?" he said to Naomi when he called. Maher stopped by their home and brought along Don Potter, the spiritual, New York–born guitarist. "The first time I sat in their living room and heard them sing 'John Deere Tractor,'" Potter said, "every hair on my body stood up on end. I looked at those two girls and said to myself there was no way they wouldn't be number one stars." Soon, Naomi met Woody Bowles, an affable young public relations expert, who agreed to manage them, along with Ken Stilts, a former insulation magnate. By the end of 1982, three and a half years after their arrival in Music City, Naomi and Wynonna Judd were poised to have their big chance at stardom. The only problem: They were no longer speaking to each other.

The relationship between Diana and Christina, always complicated, only worsened as Naomi and Wynonna inched closer to realizing their dream. "The power struggle had begun," Wynonna told me. On the surface, the tensions were easy to understand; the two are dramatically different in temperament. As Naomi once put it, "She's messy and can't ever find her stuff; I'm an organized neatnik. She loves TV and I don't. I'm punctual, while she's always running behind. Sometimes she was like a banana peel on the floor." But beneath the surface, the true nature of their relationship was even darker. "I had decided early on that I resented my mother because of how much I needed her," Wynonna told me. "I have been my mother's child since the moment she knew I was within her body."

Because Naomi had concealed Wynonna's true parentage, Wynonna believes her mother always overcompensated. "Mom always felt toward me that she had to make up for something. She had to give me more, baby-sit me more, spoon-feed me more, hold me more, love me more." Wynonna, naturally, lapped up the attention. "I was very high need," she said. "I was born into chaos. My family was always fighting. I was very sensitive. I was the kind of person who would get up into somebody's lap I didn't know. I was just hungering for love. I wanted *desperately* to be held." Naomi, though, had little time. "Mom's life was not very good during those early days in Nashville," Wynonna told me. "There was about a five-year period where she spent a lot of time having no sleep because of working two or three jobs. Then I enter the picture and give her a hard time." To make matters worse, Naomi had begun a relationship with Larry Strickland, a former background musician with Elvis. "It was awful," Wynonna said. "He'd cheat on her and she'd take him back. She was smoking a lot of pot. I didn't respect her at all. It pissed me off."

During this time, Wynonna's relationship with her mother often turned physical. "She would provoke me, like she did with Dad," Wynonna said, referring to Michael Ciminella. "The fights would often be about the three of us, with me being protective of Ashley. Mom was hard on her because Ashley was smart and popular and beautiful and everybody loved her." In a pattern Wynonna said was repeated dozens of times, the two ended up in nasty brawls. "I would just get outraged and come after her," Wynonna said, "then she'd grab me and that's how it would start. I'd get bruised. Occasionally she'd bloody me. Eventually she'd just walk out: 'See ya!' "

The two were locked in a battle that would be repeated throughout their lives: a question of who was the mother and who was the daughter. "I wanted to be in charge," Wynonna told me. "I felt like the man of the house. Sometimes I'd hurt her, sometimes she'd hurt me. She'd dump me out of the car or lock me out of the house. Then I'd turn around and throw her keys away. Then we'd cry, hug each other, and express our devoted love forever." More than normal bickering, the two needed each other—desperately needed each other—to achieve their dreams. "It was that control/power thing," Wynonna said, "where I knew she couldn't do it without me and I used that." While Naomi was out trying to launch their career, Wynonna often played hooky or threw wild parties at home. "Boy, was I awful," she snickered. After Naomi found pictures of one such incident, she kicked Wynonna out of the house and sent her to live with Michael in Florida.

At this point the Judds' saga almost ended prematurely. Wynonna, living with Michael, felt trapped, especially after he took advantage of his rare parental opportunity by forcing Wynonna to prepare for college. It was his chance to get back at Naomi by squelching her dream. But it was also Wynonna's dream he was squelching, and Wynonna sought relief. In an episode largely left out of the Judds' public story, Wynonna tried to kill herself during this period by driving her father's car into a tree. "I attempted suicide by deliberately going out and drinking and driving," she told me. "I was coming home on a long, deserted stretch of highway in Ocala. I was so wasted I don't even remember what happened, but I do remember attempting to run off the road. The car spun around three hundred and sixty degrees, at least twice, and I ended up in the middle of the road facing the other way."

And why did she do it? "Wouldn't you?" she said. "I was being forced to go to college. To me, that was prison. It's that feeling of: 'I've got to get away. I've got to leave the country. Because I'm getting ready to be shipped off to someplace I don't want to go.' I was probably the lowest I've ever been. My dad had me go and get my hair cut up to here. He wouldn't let me wear makeup. He took my guitar away from me. He was going to turn me into something else. He made me get up and read every day and listen to classical music and do stuff that was so bizarre. That was very different from what I wanted. I wanted to be Elvis and play my guitar and put on my lipstick and go out there and jam."

Several days after her abortive suicide attempt, Larry Strickland was playing nearby. Wynonna took a bus to meet him. "We sat up in his hotel room and he said to me, 'You're a singer if ever I've known one. I know this is corny, but to me you're Elvis. You have as much charisma as any person I've ever known.' " The next week Wynonna returned to Nashville. Still unable to get along with her mother, she moved in with manager Woody Bowles, his wife, and their twins. "Naomi would come over to the house," Woody told me, "and lock Wynonna in the bathroom with her. Then she would yell and scream at her for hours, 'You're a fat, lazy child. You're nothing without me. You owe everything to me. You should be grateful, you lazy slob!' " According to Woody and his wife, who listened, horrified, the shouting sessions went on night after night, often ending with bloody screeches, shower curtains being ripped, and shampoo bottles being upended. At one point Wynonna was hospitalized for throat damage because she had been yelling so much.

Just at the point the relationship seemed on the verge of crumbling, Bowles and Stilts, with the help of California-based executive Mike Curb, who signed the Judds to a production deal, arranged an unusual audition for Naomi and Wynonna in the office of Joe Galante, the aggressive young head of RCA Records. At 6 P.M. on March 2, 1983, surrounded by half a dozen executives, Naomi and Wynonna, accompanied by Don Potter, sat down in Galante's office and sang several songs. "I was mortified," Wynonna told me. "The meetings where they shine the white light in the criminal's face—that's how I felt. It was me against them." When it was over, Galante, dressed in Hawaiian shirt and tennis shoes, asked them to wait outside. Fifteen minutes later, Bowles appeared. "Congratulations," he said, "you're RCA recording artists!" "It was almost like winning the lottery," Wynonna told me. "You're floating around the room, you're numb, but there's just too much information. Everybody else was so excited, high-fiving and champagne, and I was kind of going, 'Wow, what now? I guess I've got to produce it.' And that was terrifying." At eighteen, the youngest person signed to RCA since Elvis Presley, Wynonna Judd was still wearing braces.

Back at "The Tonight Show," after nearly an hour in Wynonna's dressing room, Kenny, the makeup artist, was nearing the end of his task. On top of the layer of Hollywood tan, he had brushed large patches of rust-colored blush. Underneath her jawline, he painted deep brown for shadowing. Together these lent her face a starker, more statuesque feel. Her eyes, though, got the most attention. Around the edges, Kenny painted a pair of catlike red flames. On her upper and lower lids, he penciled in thick black lines. What followed was, for me, the most fascinating part. Instead of using Carol Channing-like fake eyelashes, Kenny opened a green compact, inside of which were tiny ridges. Using a tweezers and working with his most concentrated silence, Kenny plucked individual artificial eyelash hairs from those ridges, dipped them in a gluey substance, and pasted them one at a time along the impossibly fine line of her upper lid. Wynonna closed her eyes for this inlaying and turned her head toward the light. When she opened her eyes again, the lashes and the brown rainbow enlivened her naturally cinnamon irises. As a *pièce de résistance,* Kenny dabbed black paint on the mole alongside her right eye. Wynonna grinned, and when she turned to look at me, the change was remarkable. From the pale, vulnerable farm girl she had looked like earlier she had been transformed into a brash young star— jaunty chin, snickering smile, wicked eyes. The sinful daughter once. The sassy mother now.

Within minutes, a member of "The Tonight Show" staff appeared at the door to conduct his preshow interview. He carried a notepad and several press clippings. He was fishing for charming stories. "There was a cute story you told in *People* about Arch proposing to you . . ." he said. Wynonna wasn't pleased. "Fans like to hear stories of Arch getting down on his knees," she said, "but I think Arch is tired of hearing it." "What about a story from your wedding . . . ?" Wynonna rejected that as well. "I've got it," he said. "How about we talk about your mother . . . ?" "No, *I've* got it," she said, standing up to go to rehearsal. "How about we talk about my *music.*"

◦   ◦   ◦

Music is always what Wynonna did best and what the Judds did better than most. At the time the Judds debuted in the mid-1980s, women still played second fiddle to men. This had been the case since the dawn of hillbilly music and reflected the gender traditions of the Old South. Men went out—worked, drank, and fooled around (the fiddle was called the "devil's box" because of its association with drinking and dancing), while women stayed home. As Bob Oermann and Mary Bufwack wrote in their study of women in American music, country music "stands nearly alone as a record of the thoughts and feelings, the fantasies and experiences of this invisible and often silent group of women." The female entertainers who did perform in public tended to be in families, like the Carter Family—A.P., his wife Sara, and her sister Maybelle. Though A.P. dominated the group, the women created the musical legacy. Maybelle's revolutionary guitar style helped transform the instrument from background rhythm to dominant sound. Through songs like "Will the Circle Be Unbroken" and "Keep on the Sunny Side" the Carter Family created a folk tradition that would reach its apogee with the Mandrells and the Judds. (Another legacy of the Carters fully realized with the Judds was rewriting family history. Though Sara separated from A. P. Carter in 1933 and divorced him six years later, Victor executives asked Sara to conceal her feelings so the public would still believe they were a family.)

Female artists finally started shining following World War II. Whereas much of prewar country music had been drawn from Victorian culture—courting, praising God, mourning dead children—the genre now expanded to include more taboo subjects, like infidelity, drinking, and divorce. The two women who defined that change were Kitty Wells and Patsy Cline. Cline, born just across the mountains from Naomi, was a pop country chanteuse with lamé dresses, fur shawls, and a lusty, hard-living lifestyle. Wells, a native of Nashville, was privately a demure wife and mother, though publicly she sang of guilt, illicit romance, and broken dreams. Her breakthrough song, written by a man, was an answer song to Hank Thompson called "It Wasn't God Who Made Honky-Tonk Angels:" "Too many times married men think they're still single / That has caused many a good girl to go wrong." It was Wells, with her gingham dresses, who paved the way for traditional artists like Loretta Lynn and Tammy Wynette. Cline, more seductive, spawned a crossover legacy that ran through Crystal Gayle and, once she left her mother, Wynonna.

The Judds belonged to a different, folksier tradition. Their debut album came out at a time when Nashville was beginning to rekindle its traditional roots. Going back to Joan Baez and Judy Collins in the sixties, as well as Emmylou Harris in the seventies, acoustic music had a rich if uncommercial legacy on Music Row. In the eighties, a new breed of windswept women began to take hold: Kathy Mattea, Nanci Griffith, and Suzy Bogguss. Naomi's back-to-roots plan, in other words, had proved prescient. To drive home their retro theme, the Judds' first single was "Had a Dream (for the Heart)," an Elvis cover. The song reached only the twenties, but sufficiently set up their next two singles, both of which reached number one. "Why Not Me," written by Maher, Sonny Throckmorton, and Harlan Howard, is a resounding call to basics perfectly suited to the country's "It's Morning in America" attitude. The song was named Single of the Year by the Country Music Association. It was followed by "Mama He's Crazy," a vulnerable statement of first love, led by a voice that, as Oermann and Bufwack noted, could "trumpet, coo, shout, sigh, moan, growl, and sass." The song won a Grammy for Best Country Performance by a Duo or Group.

Over the next six years, the Judds released a total of six albums, each filled with plain-spoken pleas, "Grandpa (Tell Me 'Bout the Good Old Days)," and spunky declarations, "Girls Night Out." They also had Naomi, who was charming onstage and funny on television ("Well, slap the dog and spit in the fire!" she said after winning the Horizon Award). "Reviewers reviewed more than our music," Naomi said. "Because of my size-six figure-flattering dresses, I was called a 'Barbie Doll.' They'd compare Wy with Elvis because of her black pants, rhinestoned rock 'n' roll jackets, and matching boots." To most Americans, their roles were even more familiar. Wynonna was the randy, troubled teen who sneaks under the bleachers at the football game to cop a cigarette and a grope with her boyfriend (one early boyfriend had been Dwight Yoakam). Naomi, meanwhile, was the small-town teacher who wears hotpants and a little too much makeup and whom everyone suspects has been sleeping with the captain of the football team and teaching him about the real world. "The neurotic, dramatic Naomi Judd of life," Wynonna called her mother, the "Queen of Everything." Naomi called her daughter, the "Princess of Quite a Lot." The titles seemed fitting. If Garth hoped to be a superhero, John Wayne saving the world, Naomi wanted to be the "Queen of Nashville," meddling, prying, and ruling over all. Wynonna, by contrast, wanted only to be a princess. She had no intention of lifting a finger.

Inevitably, though, she had to. Fairy tales have a way of working themselves out: the mother must die; the hero must face the final challenge alone. Just ask Bambi. Few in Nashville, in fact, were surprised by this turn of events. As Woody Bowles told me, the Judds had always been listed with Wynonna's name first, because many around them anticipated that Wynonna would someday become a solo act. By the late eighties, that expectation had become so commonplace that a consensus had emerged in Nashville that Wynonna would benefit from separating from her mother. Even Naomi began to notice it. "I remember settling down in my hotel room to read the new copy of *Country Music* magazine," she wrote. "Our writer friend Bob Allen interviewed us separately for an article titled 'Two Down-Home Gals Ease into their Cinderella Slippers.' Wy's words prompted Bob to point out a fact perhaps obvious to everyone but me. His psychological assessment was that despite her immense talent, she often displays the gum-smacking impulsiveness of a child. Wy seemed to rail against my quiet domination, but always sought my approval." The two would have to "cut that psychological umbilical cord that was still attached." Soon enough, they got their chance.

In 1986, Naomi, now forty and still cultivating the idea that she and Wynonna looked like sisters, suffered a severe allergic reaction to an injection of collagen—ten shots in each of her cheeks. Her face ballooned, she wrote, and her strength weakened. By 1987, still tired and often unable to perform, she was diagnosed with mono. Three years later, unable to get out of bed at all, she was diagnosed, she says, with hepatitis. "You tested negative for both A and B," one of her doctors told her. "There's now a test to identify type C, and I'll get you one. You can't have D because you must have B before you can get type D, and your symptoms don't match type E." Naomi says her doctors never determined the real nature of her illness (she thinks it may have come from the collagen), but that it was chronic. Within weeks, she had decided to quit.

On October 17, 1990, in the same building where the Judds had their audition, Naomi Judd, surrounded by her new husband Larry Strickland and Wynonna, announced that she would be retiring after a year-long farewell tour. "All these men were there," Hazel Smith told me, "and they were so busy crying they didn't bother to ask any questions. Finally I had to raise my hand and ask, 'What *was* the problem? Why didn't she have any *symptoms*?'" Wynonna didn't care; she wanted to die. "I remember the tears just streaming down my face," she told me. "The first thing I said was: 'Then I quit too.'"

Naomi, though, would not let her quit. "I've always believed that I appreciated our career more than you," Naomi told her, "but the reality is you need it more than me. You need to sing like you need to breathe. I've crawled over broken glass to get us here, so don't you dare think for one instant you're going to throw it all away, young lady! You're gonna carry on the family business." But the real question, as it turned out, was whether Wynonna could ever carry on a business her mother had so controlled. In one of the more passionate sections of her book, Naomi wrote that one of the most liberating realizations in her life was that "Every successful person in a codependent relationship is still a failure!" How prophetic those words would seem when applied to the life of her daughter, who was about to go out on her own in a gesture she believed—and that everyone hoped—would finally set her free.

Downstairs in the studio, a stand-up comedian was warming up the audience. It was a few minutes before five now, and the audience members had just been let into their seats. They were squirming, snapping pictures, and hurriedly changing their Disneyland T-shirts for ones with the names of their hometowns: SPRINGFIELD TIGERS. HEAR US ROAR!!

"Now we have to make a decision," announced the comedian. For five minutes, he had been telling DJ-caliber jokes and pimping for applause by handing out free Frisbees. "Either we all stand or we all sit. It's totally up to you. But let me tell you, you have a better chance of getting on *TV* if you all stand."

The audience screamed, "Stand!!!" The comedian winked at the director.

"Okay," he said when the roaring subsided, "when Jay walks out from behind that wall, we all stand. But make it look real. Make it look American. Leaping to your feet is a *good* thing."

A few minutes after the theme music, Jay Leno finally appeared, bobbing and nodding in his small-town councilman way and weaving gracefully through a series of jokes. The audience laughed appreciably, helped by an occasional rim shot. The room itself was surprisingly cramped, about the size of a high school gymnasium. On second glance, I realized the studio was painted to look like a replica of Fenway Park in Boston, near Jay's hometown. His monologue spot was home plate. Behind the audience the walls were painted green, with yellow foul lines. And where the pitcher's mound would be, a camera stared him in the face.

Near the end of the show, Wynonna appeared. She looked heavy in her five-months pregnant state, but the camera stayed focused on her face, which was tranquil. She closed her eyes when she sang and swayed back and forth to the chant, "My Angel Is Here." These moments, as I was learning, were her only moments of peace. The audience was grateful. They stood without instruction.

When it was over, she walked the few steps to the desk and greeted Jay Leno with a kiss. She had only a few seconds of "couch time," as it's known. Few singers, though, are even invited to sit. It was a mark of her celebrity status. "Congratulations," Jay said when he sat down, "on your marriage and the baby!" Wynonna smiled. Jay was running out of time. The stage manager gave him the signal to wind down. "But I have to ask you this," he said, leaning over, touching her velvet arm, and adopting his most sympathetic expression. "How is your *mother*?"

THE LEGENDS

Waylon Jennings was feeling crotchety. On a cold winter night, twenty years after he walked into a Nashville studio, waved a Buntline Special .22 Magnum Revolver, and threatened to shoot anyone who played his instrument without feeling, Waylon was returning to a nearby studio with his pal Willie Nelson to reprise their epochal Outlaw success. The Outlaws were an informal mix of drug addicts and renegades (Waylon, Willie, Tompall Glaser, Kinky Friedman, Kris Kristofferson) who shook up Nashville in the sixties and seventies and yanked Music Row into the Age of Hippies. They overturned convention, grew gangly beards, demanded exalted royalty rates, and ultimately achieved such break-through success (*Wanted! The Outlaws*, released in 1976, was country's first certified million-copy seller) that Nashville had no choice but to hold its nose and accede to their every demand.

The Outlaws also spawned a generation of followers who tried to mimic their raw, individualistic style. "The more I thought about it over the past couple of years," Wade once told me, "and the more I listen to myself sing, the more I realized, 'I'm trying to sing like Waylon.' I always have. My whole life. It's the way his voice sounds, the way he sings: deep, wailing—hell, his name is Waylon."

As great as the Outlaws' influence was, however, and as legendary as they had become (Willie had even been invited to join the Country

Music Hall of Fame), the Outlaws were also in a bind, a dilemma shared by most of Nashville's past masters: Having laid the foundation of country's success, how should they share in its current boom? Younger stars might pay them deference ("Waylon is the reason I'm here," Wade said flatly), but those newer artists also expected their time in the spotlight. For onetime revolutionaries Willie and Waylon, who on this night were being asked to update their signature album for RCA, the predicament was particularly acute: Should they step aside in deference to the young, as they once asked their elders to do, or should they rejoin the battles of their youth?

Waylon was the first to arrive, strolling through the glass doors of the Room and Board Studio, across the corner from Reba's new multi-million-dollar palace. He was not feeling well. His diabetes had been acting up lately, furthering dimming his vision and adding a little more shuffle to his step. Though just shy of sixty (and off drugs for a decade now), he looked weathered in a sort of Mount Rushmore way. His dark eyebrows—occasionally arched, often furrowed—were still the most noticeable feature on his face. His semibeard was formidable, a goatee with wings. He looked like a small-town Texas undertaker.

Willie arrived next. He seemed gnomelike in his AMERICAN PROFESSIONAL BULLFIGHTING road jacket, black sweatpants, and New Balance jogging shoes. A red bandanna was tied around his neck. His beard was a majestic white, a sharp contrast to his blotched red nose and grayish brown pigtails, which hung, Pippi Longstocking-like, around his shoulders. A small entourage followed him into the already crowded foyer. His usual retinue included several young women in tight-fitting dresses who looked as though they could have handed out Academy Awards and a few songwriters hovering in a marijuana haze who appeared as if they were waiting for Willie to turn, clasp his hands together, and utter a Yoda-caliber musing that would send them into orbits of imagination and insight.

The two old friends greeted each other warmly, then turned to speak to the huddled mass of hangers-on, publicists, record executives, film crews, plus a few odd celebrities like singer Kim Carnes, songwriter Dan Penn, and writer Chet Flippo, formerly of *Rolling Stone*, whose breakthrough reporting helped launched the Outlaws into the counterculture pantheon. Altogether the attention, coupled with the trays of bright orange cheese and platters of pink salmon, was oddly reverential for these paragons of iconoclasm. Their grumpiness, though, always the source of their charisma, was never far away. When my turn came to shake Waylon's hand, I was struck by the convention of his slick jogging suit with three racing stripes down the side, his bronze eagle-shaped logo around his neck, and his black WAYLON baseball cap. His shoes, too, seemed to be part of the ensemble.

"So where did you get those shoes?" I asked of his black, Italian-looking hightops.

"I bought them in Seattle," he said. His voice had the scratchy whine of a recording coming through twenty-year-old speakers. "They were being discontinued. John Cash tried to get him a pair, but they don't make 'em anymore."

"Is this outfit part of your line, too?" I asked.

"Shit, you've got to be kidding," he grumbled. He took a step back and glared at me. "What the hell kind of question is that? What's that crap you got on?" I was wearing blue jeans and a purple turtleneck. "Is that your line? Don't you know who I am? Loose and easy is my style." And then, speaking to one of the many attendants crowded around him: "Do you have a gun? I'm afraid I'm going to have shoot this man for asking such a stupid question."

He turned to enter the studio.

Once in the studio, Waylon and Willie crowded in front of free-standing mikes as the half dozen musicians prepared to lay the tracks. The song, which would later be added to the eleven original cuts and nine "lost" tracks to make the commemorative package, was the dusty West Texas ode "Nowhere Road," written by Steve Earle and Reno Kling. Earle, at Waylon's suggestion, was also serving as producer. It seemed like an inspired choice. Steve Earle, like his patron, was a troubled prodigy (his 1986 album *Guitar Town* is widely considered to have launched the contemporary country-rock movement in Nashville) who had been in and out of prison, as well as on and off heroin, in a commercially stunted but critically exalted career as a modern-day Outlaw. Recently released from incarceration, he was making something of a comeback in Nashville. Though his black hair flopped uncontrollably over his forehead and his sideburns were as bushy as ever, he had less of the wild bear appearance that had marked his earlier years. He was dressed in jeans and a black T-shirt, with a FEAR NO EVIL tattoo visible on his arm.

"Okay, the song's in A-flat, the people's key," Steve said, scurrying to prepare the room.

"You got any idea how to split this up?" Waylon asked, holding the lyric sheet at arm's length to aid his straining eyes.

"You sing one verse," Willie answered, "I'll sing the other. Then we'll split the chorus." He too held the sheet at arm's length, but since he was a full head shorter than Waylon, it rested closer to his eyes. "I've got to get me some glasses," he joked. "My arms aren't as long as his." He pulled out a pair of plastic-rimmed glasses as Waylon turned his cap backward. The scene seemed straight out of *Grumpy Old Men.*

Finally Earle interrupted their primping. "It's probably best to split the last verse," he said. When they agreed, he gave the order to proceed. "Let's saddle up, everybody!" he said.

Once they started singing, Willie and Waylon were both fiercely focused, though with vastly different styles. Willie stood still, with his hands clasped together, his head nearly motionless. He moved only at the end of each line, leaning just slightly toward the microphone to let the bass in his voice be heard. Waylon was more fidgety. Whereas Willie sang as if he was telling the microphone a secret, Waylon acted as if he was trying to convert the entire assembly. He crossed his arms, rocked back and forth, and occasionally quivered, revivallike, a gesture that sent his hangers-on into head-nodding ecstasy.

At the end of the first take, the room burst into applause. Waylon, though, was unsold. "I think you and me got one better in us," he said to Willie. To which his friend answered, "My phrasing won't be the same twice." "Doesn't matter," Waylon said. "We've kicked the asses of singers and writers who've done less than that." After the second take, Waylon was still frustrated. "We could practice with this track," Steve said, hopeful. "But we don't know how to practice," Waylon said.

After the third take, he was at least somewhat placated. "You really stretched that out," he said to his partner. "You Willied it." "Well don't expect it to happen again," Willie cracked. The two men wandered into the control room to hear the playback. Once inside, Willie took off his baseball cap for the benefit of those who had been trying to read the muddled writing on the front. Was it Chinese? Waylon wondered. Grinning, Willie turned the cap sideways where the bright red letters formed the words GO FUCK YOURSELF.

Following a short break (and a few bites of cheese), Willie and Waylon returned to the studio to overdub the vocals. The television cameras had packed up and left. Most of the hangers-on had trickled out. Three takes on the vocals and Willie declared himself through. He waved good-bye to the crew. "Where're you headed?" Earle asked. "Austin," Willie said. "Good place to be headed for," Earle said. Willie stepped outside the studio and boarded his bus, Honeysuckle Rose II. On the side was a portrait of a Native American; on the back: Willie, Waylon, Kristofferson, and Cash.

Back in the studio, Waylon was still tinkering. "One thing about Willie," he said. "He's consistently out of tune." Later he would tell me a story: Mark Knopfler, the virtuoso guitarist and leader of Dire Straits, wanted to cut an album with Waylon and Willie. "I want Willie very involved in the tracks," Knopfler said. "You think you do," Waylon corrected him, "but you don't." "Trust me," Knopfler said. "No," Waylon said, "trust *me*. If you get him in here while we're doing tracks, he's going to park a bus right outside the door. He's going to smoke pot most of the day and drink the rest of the day, and every once in a while he's going to come in here with about three songwriters trailing behind him with tears rolling down their cheeks. Then he's going to say, 'I want to cut this great song.' And you know what, it ain't going to be worth a *shit*." Willie agreed. "Anything Waylon does in the studio is fine with me," he told me, chuckling. "He can help me, teach me. Because normally I have no idea what I'm doing."

Waylon, meanwhile, grew more intense. Three, four, five times he retried the harmony, never quite satisfied. "Let's hear that from the top," he said to Earle after the tenth take. "Top of the chorus or top of the song?" Earle asked the engineer. They discussed it briefly, then opted for the top of the chorus. They opted wrong. "Top of the song!" Waylon barked. "Top of the song." Then he muttered, "Jesus Christ!" Earle seemed unconcerned by Waylon's growing irritation. "My uncle's going to be real impressed that I did this," he gushed. "Hell, I think everybody's going to be impressed."

After the next run-through, Earle was pleased. "That was a really good take," he announced.

"I'd like to do it one more time," Waylon said.

"No need to do that," Earle suggested. "We've got everything we need. We can just comp it later." He was referring to the process of taking the best-sounding word from several different cuts and digitally piecing them together.

"No," Waylon said. "I don't like that comping shit. One word here, one word there. Where's the feeling in that?"

"Everyone does it," Earle replied. "Really, I think you've got it. But if you insist . . ."

He started to rewind the tape, but just as he did, Waylon ripped off his headset and threw it to the floor. He walked through the doors of the studio, steaming.

"I think you'll regret this," Waylon said. "You just don't do shit like that! I've made a lot of these myself, you know." He swiveled and stormed out of the control room, past the drying cheese and wilted salmon, and directly out the front doors. The reaction of the two men at the console was nothing short of nuclear paralysis. They sat, stunned. After a moment, Earle stepped into the foyer and lit up a cigarette.

"Man, we were down to one line. One *line*," he said, trying to explain what had happened to the few musicians waiting to touch up their parts. "He just wanted to redo it. We would have comped it, but he doesn't like comping. He hates this modern stuff. He likes doing things the old-fashioned way."

In his autobiography, Waylon Jennings offers the following description of his first trip to Nashville. It was 1965. The native of Littlefield, Texas, was twenty-seven years old. He had already apprenticed as Buddy Holly's bass player and confidant, then shucked the business in disillusionment following Holly's death. Six years later, he was lured back to music by songwriter Bobby Bare, who told RCA head Chet Atkins that Waylon Jennings was "the best thing since Elvis." It was enough for Chet to place a call to Phoenix, where Waylon was living at the time. "I started out for Nashville with a yellow Cadillac and a yellow-haired woman," Waylon wrote. "We climbed over one hill after another, Memphis aiming east toward Nashville, until finally Music City stuck its head over the horizon like a rising sun. I could hardly believe this was going to be my new home."

Almost immediately Waylon found himself in the center of a young group of insurgents—Johnny Cash, Tom T. Hall, and Kris Kristofferson—who were just congealing on lower Broadway and preparing to wreak havoc on the elders of Music Row. They felt estranged from the sweet sounds and gentle strings of the Nashville Sound. "Awash in strings, crooning and mooning and juneing," Waylon wrote, "the Nashville Sound may have been Nashville's way of broadening its pop horizons, but it was making for noncontroversial, watered-down, dull music that soothed rather than stirred the emotions." By contrast, the new wave of singer-songwriters hoped to return the music to its hardscrabble roots—fewer strings, rawer guitar. "We never said that we couldn't do something because it would sound like a pop record or it would be too rock 'n' roll. We weren't worried that country music would lose its identity because we had faith in its future and its character."

In addition to challenging Nashville's musical conventions, this group of rebels also challenged many of the social conventions of what was still one of the more conservative, uptight cities in the South. Most notably, they did drugs. "Me and John were the world champions of pill-taking," Waylon wrote of his roommate, Johnny Cash, "but we each didn't let on to the other that we knew it." The two set up an apartment together ("I was supposed to clean up and John was the one doing the cooking . . .") and went to extraordinary lengths to conceal their addictions. "I hid my stash in back of the air conditioner, while John kept his behind the television," Waylon wrote. "He'd tear the place apart if he ran out. If we had started combining supplies and sources, we probably would've bottomed out and killed ourselves, feeding each other's habits."

The drugs, like liquor to the generation before them, gave these shy men the courage to go onstage. Cash, in his autobiography, describes how the pills—mostly Dexedrine, Benzedrine, and Dexamyl—gave him confidence. "I've always loved to perform," he wrote, "but I've never gone onstage without experiencing those 'butterflies.'" With a couple of pills in him, he said, the butterflies disappeared. "My energy was multiplied. My timing was superb. I enjoyed every song in every concert and could perform with a driving, relentless intensity." Waylon felt those sensations and more. "I never hit the ground for twenty-one years," he told me. "I had incredible stamina; I prided myself on the fact that I could take more pills, stay up longer, sing more songs, and screw more women than most anybody you ever met in your life. I didn't know when to stop or see any need to."

In time Waylon transferred that confidence to the studio. Though he had some success in the late 1960s, Waylon didn't thrive until he challenged the reigning bureaucracy of Music Row and took control of his own recordings. Nashville had always been a community where producers made all the decisions: They used the same studios, hired the same musicians, and made every record on the same nickel-and-dime budget. More than custom, this was written into artists' contracts. "I'm not a brilliant person by any definition," Waylon recalled, "but I could read them sons of bitches and see one thing: I was getting screwed. They had a wonderful thing going on. They would give it to you up here and take it all away from you down here. I just told them, 'I'm going to get you. I'm not going to let you do it.' I guess I was the one who brought the system completely down."

In 1972, citing a technicality in his contract, Waylon earned the right to record in the studio of his choosing and to use his own band. Though commonplace in rock, in country it was revolutionary—all the more so when Waylon started producing records that were dramatically different. His 1973 album *Honky-Tonk Heroes*, recorded in Hillbilly Central, was a breakthrough mixture of dark cowboy ballads and sparse West Texas odes. "I liked things that weren't perfect," he told me. "I wanted my music on the edge, with some feeling in the rhythm section." Waylon took that credo seriously. He hated various techniques that other country musicians used, specifically pickup notes, the one or two notes used to introduce a phrase. "I think it's like stompin' your way into a song," he said, "announcing that you're going to knock on the door before you actually do. It irritates me to no end, and one day I decided to put a stop to it." To do so, he took a pistol into the studio. "It was one of those ones that Wyatt Earp used," he recalled. "When I got in there, I just pulled it out and said, 'The first son of a bitch that plays a pickup note or that's still looking at those goddamned charts after the third take, I'm going to blow his hands off.'"

Having drawn the line so dramatically, Waylon reveled in his freedom. In Hillbilly Central, he finally had a place where he could do whatever he pleased: Come and go at all hours of the night, do drugs, bring women, go around the corner and shoot pinball, or stay up for three days straight to write a song. More than just a band of roughnecks having a good time (imagine the characters from *Animal House* set loose in the midst of a Baptist seminary), the Outlaws were changing everything about country music, right down to its appearance. Waylon and Willie, who despite their grumbling had actually been as neatly groomed as everyone else in Nashville, even took the shocking step of abandoning their pretty-boy images for blue jeans and beards. "I had grown it just for kicks," Waylon wrote, "but when I looked in the mirror, it was like I was just starting to look like myself. We were all undergoing transformations. I mean, can you imagine Willie without a beard and those braids?" When Waylon put on a black hat and walked onstage, he was "staking my own piece of land where the buffalo roam. 'Don't fuck with me' was what we were saying."

Those strong-minded ideas, coupled with their strong-arm tactics, heralded the return of what had always been a chief tension in country music: the battle between the old and the new. "Revolution and counter-revolution may seem to be the most unlikely topics ever to be associated with country music," Chet Flippo wrote in an essay on the Outlaws, "but the tendency to rebel is ingrained in humans, and it certainly is imbedded in the persons most likely to be prominent in country music."

If anything, youthful turnover is the one constant in Nashville. A study by the Country Music Foundation found that the average age of country artists who charted in 1955 was thirty-three; in 1995, the figure was also thirty-three. The average age of first-time artists in 1955 was twenty-five; in 1995, the figure was the same. The reason for this constant influx: Country music, because it reflects the nation, naturally reflects *changes* in the nation as well. Indeed, perhaps the central misunderstanding about country music is that it's about the past. Though older performers may linger in the public's mind, each decade has given birth to a new musical form—bluegrass in the 1940s, honky-tonk in the 1950s, the Nashville Sound in the 1960s—that brought with it a new crop of stars. Nashville in the 1970s was ripe for such a change. "We didn't need Nashville," Waylon said. "Nashville needed us."

Ultimately, even the conservative denizens of Music Row realized these new artists were becoming a phenomenon and could sell far beyond the traditional country audience. In 1976, Jerry Bradley, then the head of RCA, decided to capitalize on this trend by rereleasing old tracks by Waylon, as well as Willie and Jessi Colter (Waylon's future wife), both on other labels. Though Waylon originally disapproved, Bradley insisted. "We had a meeting," Bradley told me, "and I said, 'Here's how it is. What did you make last year? Three, four, five million dollars? I have a job that pays me fifty grand. I enjoy my job. If I put this album out, I might get to keep it another year.'" Waylon agreed, even consenting to the use of the name the Outlaws, even though he, Willie, Colter, and Tompall Glaser were not a formal band.

The album, though it wasn't a creative breakthrough in any way (in fact, most of the cuts had appeared on earlier works), revolutionized country music. Like Garth Brooks a decade and a half later, Willie and Waylon had a sound—bleak desert—and a style—grunge cowboy—that perfectly captured the country's Zeitgeist. "The simple fact is, the Outlaws enabled a lot of other things to happen," Bradley said. "They brought an acceptance of a broader range of people. They escalated sales. They made it easier for Ronnie Milsap and Dolly Parton and Crystal Gayle." Chet Flippo was even more generous. "By the time it ran its course, the Outlaw movement had changed the face of country music forever," he wrote. "The producer as king—that feudal notion was shattered. Country artists gained control over their own record sessions, their own booking, their record production, everything else related to their careers, including the right to make their own mistakes."

Inevitably, of course, those mistakes would help bring the Outlaws down. In 1977, at the height of his popularity, Waylon went broke—from all his high living, he says, and from not paying attention to his own affairs. "I figured out one thing," he told me. "Country boys have to go broke once. Every one of them I've ever known has done that. You go broke, get stung real good, and then you learn." In August of that year, Waylon was arrested for possessing twenty-seven grams of cocaine after federal agents intercepted a package mailed to him from New York (when agents could not prove that Waylon ever actually possessed the package, the charges were dropped). Soon he and Tompall Glaser were suing each other over royalties, and *Rolling Stone*, once a supporter, groused that Waylon was refusing to do interviews and had started being rude to his fans. Willie, meanwhile, was lured to Hollywood.

By that time, the Zeitgeist had moved on as well. A new crop of more clean-cut artists, led by George Strait and Randy Travis, arose to supplant the bad boys of the seventies, followed by Garth, Clint, and a clan of even cleaner-shaven artists in the nineties, many of whom didn't even drink, let alone trash hotel rooms. "Joe Galante called me several years ago," Waylon said, referring to the new head of RCA. "He said, 'Clint Black loves you and wants to meet you. I'd like you to tell him some of those old Willie and Waylon stories.'" The three men went to lunch. "I told these stories," Waylon recalled, "about me getting coked up and things like that, and after a while it got a little quiet and Clint Black said, 'Well, I tell you one thing. I can see right now, I've got to get rid of this Goody Two-Shoes reputation.' And Joe burst in and said, 'No, wait. You don't understand.' Joe was scared to death that Clint would become like me."

(Wade was a perfect example of the clean-cut nineties country star. As much as he idolized Waylon, he couldn't fathom living that lifestyle. Not only did he never indulge in groupies, he told me, but when he read Waylon's book and learned of his decades-long drug use, Wade was dumbfounded. "I've been trying to quit chewing snuff forever," he said. "I can't imagine trying to quit cocaine.")

Predictably, there was a backlash when the oldtimers were overthrown. COUNTRY GRAYBEARDS GET THE BOOT bemoaned *The New York Times* in 1994. "George Jones may be the greatest country singer of his time," the *Times* said, "but he has a small problem these days. Like Waylon Jennings, Willie Nelson, Merle Haggard, Tammy Wynette, Loretta Lynn, and virtually every star from the pre-Garth-Brooks, pre-video, pre-country-hunk era, he can't get on the radio anymore." WHY

CAN'T YOU HEAR COUNTRY'S GREATEST STARS ON THE RADIO? screamed a
1996 issue of *Country Weekly*, a cover that featured both Waylon and
Willie. Most people on Music Row knew the answer to that question.
"It's like my daddy said," Jerry Bradley told me. "How much chocolate
ice cream can you eat? How much of Webb Pierce can you listen to?
How much Waylon Jennings can you listen to? How much Garth *Brooks*
can you listen to?" Moreover, as Bradley knew, that complaint had been
around for years. "It's the thing that Bill Monroe said in 1960," he
noted, "that Hank Snow said in 1970, that Loretta Lynn said in 1980.
Hey, guess what, you have that period of productivity and it doesn't
last forever."

Waylon, to his credit, claimed not to be surprised by the turn of
fortune. "I was always aware that it was going to come down," he told
me. "Their loss," echoed Willie. Far more distressing, though—to them
and to many of their admirers—was the fact that many of the freedoms
they had earned seemed to be disappearing along with them. Few artists
challenge the musical norm anymore. Even fewer bring their bands into
the studio. "I wish it was still like it was in those days," Travis Tritt, the
closest thing Nashville has to an Outlaw today, told me. "You're in a
situation now where the money is the generating factor behind everything. It's a shame because the word 'art' falls out of the word 'artist'
at that point. Creativity does, too. You're not writing from the heart,
you're doing albums based on what's commercial at the time. The cool
thing about Waylon, Johnny Cash, and Willie Nelson is they didn't pay
any attention to that shit. They just did their own thing."

"That's what's pissed me off about that night with Steve," Waylon said.
It was some months after that evening in the studio, and Waylon was
sitting in a plush leather chair in his office on Music Row. "What pissed
me off is that he wasn't listening to anybody. I would never tell an artist
not to sing something one more time. And to think I kept him from
going to prison a few years back."

Waylon was feeling feisty. He was perched underneath a wall full of
photos from his glory days—Waylon in a mushed-down cowboy hat,
Waylon and Willie sitting behind a microphone, Waylon with sweat drip-
ping off his bangs. Across the alley, Garth's management company was
visible through the trees. ("He's the most insincere person I've ever
seen," Waylon said of Garth. "I remember a few years ago an old buddy
of mine who worked with Ernest Tubb was giving him an old record.

He tried so hard to cry, but he just couldn't make himself do it. He thinks it's going to last forever. He's wrong.")

The conversation, though notably casual, had taken months to set up as Waylon went through bouts of illness, isolation, and just plain orneriness that left his handlers cowering in the face of his unpredictability. (Willie was no different. When he failed to emerge at the scheduled hour several days later for a conversation, his publicist called to report: "I've phoned him at home and at the World Headquarters in Austin. I faxed the bus and had him paged on the golf course. Be patient. Sooner or later, he'll show." Several hours later, my telephone rang. "Hi, Bruce," came the gravelly voice on the other end. "It's Willie.")

Waylon chuckled when reminded he had threatened to shoot me. "It was part of my joke," he said. "I got a lot of credit for things over the years when I was just joking. One time RCA had been giving me some shit. I said, 'You sons of bitches. If you keep messing with me, I'll get in the vault and burn the goddamn tapes.' Well, they was so scared they went and hid my tapes in Indianapolis. They didn't find them for five years. When they finally came and told me, I said, 'What a bunch of assholes. I wasn't going to burn my tapes. I might burn Elvis's tapes, but I certainly wasn't going to burn mine.'" At times, too, he seemed almost embarrassed—by his fame, by his drug abuse, and mostly by his reputation. "John Lennon one time said to me, 'You know what? I thought you were crazy.' I said, 'You're funnier than hell.' He said, 'No, I'm serious. People in England think you shoot folks.'"

But underneath, Waylon seemed most surprised that his once overwhelming power had evaporated. Even his disciples no longer seemed interested in listening to his opinions. "I had ideas I wanted to try that night," he said of his evening with Steve Earle. "I had a guitar part I was going to do. I never got it out of the case. I think Steve was overly excited. Me and Willie had talked about it; we sure didn't want to do a Steve Earle record. We wanted to do our own version. All he had to do was listen to us." That he didn't listen was what disappointed Waylon most. "This ain't something that I don't know about," he said. "I have found out that every damn time I let somebody get too much control, it ain't me."

Willie, for his part, was much less confrontational. "It takes a lot of energy to fight those kind of fights that are really not winnable," he said later. "I loved that night. I know Waylon was listening for something he wasn't hearing. I just hope he didn't hurt anybody. I think it was Waylon who said, 'It takes a bad son of a bitch to whip my ass, but it don't take him long,'" He chuckled. "Truth is, I never fought 'em as much as Waylon has. He's fought 'em pretty good, and more power to him because he was right." It was that difference in temperament, Willie believed, that has so far kept Waylon out of the Hall of Fame. "Waylon is not the greatest politician in Nashville," he said. "He shot himself in every foot he has, plus a couple of fingers."

(As for Earle, he felt the studio was too crowded. "My sense of the vibe was you guys shouldn't have been here," he said, referring to all the guests. "And I really was upset about it. It came within a hair's breath of me clearing the studio, and I should have in retrospect." As producer, he had never gotten into a fight with an artist, he said. "Of course, I'm not the first person that Waylon ever got pissed off at in a recording studio. But I feel bad that it happened. I think the atmosphere in the studio cut down on my ability to communicate with him. He was gone before I really realized he was uncomfortable.")

Either way, what Waylon realized anew that night was how removed he had become from the town he once defined. Though he continues to live in Nashville (the result, he says, of his seventeen-year-old son, Scooter, who refuses to let him leave) and records there as well, neither he nor Willie has released a record on a Nashville label in years. Both are now distributed by independent labels. As Waylon sang in 1994, in a seeming nod to this passage: "We were the wild ones / The ones they couldn't control." "We were survivors," he declared.

Though he claimed not to be bitter by his forced exile—("The good thing about it is I won," Waylon said) in truth, it was easy to detect a note of betrayal. "I don't want to be remembered for the type of music that's here now," Waylon said. "It's not that good. Have you heard anything out there that you recognize? They've got a couple of artists that it took them sixty hours to get their voices in tune." Some artists, he said, excited him: Mark Chestnutt, Trisha Yearwood, "and that kid Wade Hayes—he doesn't need his voice to be tinkered with." But most, he confessed, left him flat. "All the emphasis is on being pretty and dancing around," he said. "I would go nuts. There's no feeling in that."

Instead, the two aging rebels were left to seek new musical ground. Willie dabbled in reggae and alternative rock and drew in a new audience inspired by his ageless hippie image (his support for Farm Aid didn't hurt, nor did his multimillion-dollar battle with the IRS, which he was slowly paying off). "There's a lot of young people today that have never heard all the great country music artists," he said. "They're still interested in hearing Lefty Frizzell and Hank Williams and Bob Wills and all those guys that you don't hear on the radio every day. The difference between those guys and me, plus being alive, is that I'm out there working every night."

Waylon took up Appalachian folk music, produced a blues album, befriended and mentored a group of young rockers (including James Hetfield of Metallica), and even appeared on the Lollapalooza Tour with the Screaming Trees. "It was the best audience I've ever played to," he said. "I just knew my ass was in trouble. But the thing that made it great was when you looked out there in those faces and they were grinning."

In a way, it was the best thing that could have happened to them. Instead of taking Nashville's normal route—retiring to Branson, Missouri, or the Grand Ole Opry, the twin Jurassic Parks of country—they set out once more in pursuit of the fresh, renewing the vow of rebellion they had made three decades earlier. "You have your time," Waylon said, "then you go about your business. I put some people out when I first arrived; they had to make room for me somewhere. So now they're doing the same to me. That's just the way it works. I go out every now and then and perform a few shows. Every once in a while, I cut an album. I'm happy."

The best thing about the turnover, he allowed, is that more young people are interested in the music. And even if they don't always revere old-timers like him, well, that's the way it is with young people. Who would know that better than he? "The truth is," he said, "I never wanted to be a hero. I don't want that much pressure. If these people want a hero, they need somebody besides me. I'm a singer, that's all, and a guitar player. When I go out there, what I owe those people in the audience is a good performance. I don't owe them shit past that. That's not being ungrateful. That's being real. I've worked hard and I own my talent."

Which is perhaps his greatest achievement of all. Having once, high on drugs, slayed a generation of heroes, the Outlaws, a generation later, had managed to escape the most potent drug of all—becoming heroes themselves.

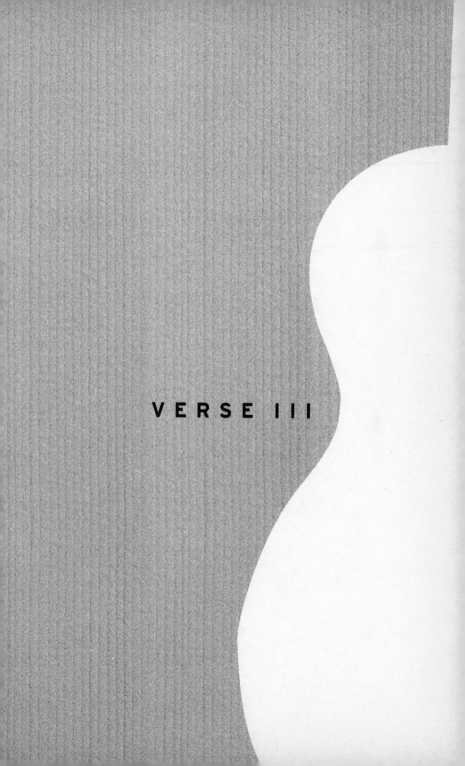

VERSE III

NINE

THE INTERVIEW

The morning snow was piled so high on 66th Street that the white limousine had to make a six-point turn in order to back into the cramped loading dock on the first floor of the ABC Studios just off Columbus Avenue on the Upper West Side of Manhattan.

"Now, you gotta be outta here by ten-fifteen!" the attendant barked at the driver, who wasn't paying him the least bit of attention because, at that moment, the front bumper of his car was stuck on a small glacier of ice. "Susan Sarandon is coming to tape a show and we can't keep her waiting. She's a very important lady, you know."

Inside the limousine, Garth Brooks nodded in bemusement and shivered from the cold as he squeezed himself toothpastelike out of the car, tipped his baseball cap at the attendant, and squiggled through the back door of the building into the refrigerated linoleum hallway. "Boy, I love New York," he said under his still-visible breath.

Inside the Green Room of "LIVE! With Regis and Kathie Lee," the pre-talk show flutter was well underway. Producers with clipboards and stagehands with cue cards scurried in and out of the cramped office space, plucked pieces of pineapple from the buffet, and hurriedly shuffled sample questions from the hosts to the guests. Presently, Peter Jennings arrived for a segment on his Bosnia special and introduced himself to Garth. "Hmmm, Capitol Records," he said of Garth's lightweight black jacket with the purple logo. "Do you wear it for love or money?" Garth looked down at the jacket. "Neither," he said. "It's free." Jennings chuckled and sat down on the sofa. "He sure seems like a nice guy," he said.

Momentarily Garth was beckoned into the studio. Though still puffy from the early hour, he shadowboxed to give himself energy, and, as he had done already several times that week—on Fox's "After Breakfast" and Lifetime's "Biggers and Summers" (out of loyalty to Jay Leno, an early supporter, Garth refused to appear on David Letterman)—strolled out into the TV glow. Once in view of the camera, he doffed his hat, brought his hands together, and performed his trademark altar-boy bow. The audience, mostly women, rose to its feet.

What followed was fairly standard morning television fare, watchable in that breezy, happy-happy way that Regis Philbin and Kathie Lee Gifford have perfected over the years. The interview consisted of some gentle teasing from the hosts (Regis: "You don't go anywhere without that hat now, do you?"); a touch of self-deprecation from the guest (Garth: "I have to cover my balding head with something . . ."); and a little family bonding along the way (Kathie Lee: "You have two kids now, yes? And another on the way. I remember you said you might retire a few years ago if your music conflicted with raising your children. I think that's lovely . . ."). There's a reason country has prospered in the era of the television talk shows: Their intimate, confessional natures are uniquely suited to each other.

But what happened next is what defines a master of the media. During the commercial ("Two minutes, two minutes to air . . ."), several of the members of the studio audience began jumping up and down, waving trinkets and carefully wrapped gifts they had brought to give to Garth. One of the pages went galloping into the bleachers and retrieved the gifts, including a small teddy bear and a single red rose from one woman near the back row. It could have been a moment that passed unnoticed. But as the stage manager started counting the time ("Thirty seconds and counting . . ."), Garth leaped out of his seat, grabbed the rose and bear from the page, and went bounding up into the seats, two steps at a time. Arriving at the last row, he gave the woman a full-body hug. Then, just as quickly, he turned, sprinted down the stairs again, and as the red light reappeared on the camera, ("In five, four, three, two . . ."), squiggled back into his seat.

"Oh, my God!" Regis gushed. "You're not going to believe what just happened! Can we get a shot of that woman?" And suddenly there she was, on national TV, shaking and smiling, with tears streaming down her face.

Later that morning, Howard Stern's office called and asked if Garth would like to make an appearance at Howard's birthday bash that evening. Mr. Brooks politely declined.

For much of its history, country music was all but ignored by the national media. Various reasons explain this lack of interest. First, a chronic regionalism in American life. Country, of course, was based in the South, which for much of the Age of Television was treated as an isolated bulwark of pigtails, incest, and preternatural racism. Second, a tradition of classism in America. Because of its association with overalls, car grease, and Jell-O molds, country has been linked with a lingering fear in America that white trash "elements" will overrun the Establishment. Blips in Nashville, like Johnny Cash or Patsy Cline, might be noted in finer circles, but they were invariably treated as regional eccentricities. Bubbas can fish, but they can't make art. As Nick Tosches wrote about the elite New York media in his 1977 book *Country: The Twisted Roots of Rock 'n' Roll*: "It is so much more comfortable, so much more acceptable, to dislike rednecks than blacks."

This general disregard (hillbilly until proven innocent) was only intensified with the formation of the defining organ of music journalism, *Rolling Stone*, which dramatically inflated the stakes surrounding pop music. Founded in 1967, *Rolling Stone* christened, then perpetuated the idea that rock 'n' roll could save the world. Country didn't share that ambition and wasn't perceived by outsiders to have it inherently. Among other things, country was for old people and the counterculture establishment believed pop culture belonged to the young. Also, country music was thought to be too traditional, too sentimental, and—let's be blunt— too retrograde, both politically and socially. *Rolling Stone* did cover the occasional country artist, usually proto-rockers like the Outlaws or Tanya Tucker. But by and large, the magazine never welcomed Nashville artists into the fold. None of this would have been so bad—after all, *Country Music* magazine, founded in 1972, never covered Led Zeppelin—if it hadn't been for *Rolling Stone's* far-reaching power within music and journalism circles.

The unprecedented success of Nashville in the nineties presented a new threat to New York editors. The two chief reasons the media had ignored country music—region and class—suddenly seemed even sillier than they had before. First, the South no longer fit its own stereotype. From pariah region—hot, isolated, poor—the South had become the chief engine of growth and change in American life. Spurred by a growing population drawn by service jobs and air-conditioned suburbs, the South began to redefine America—from politics (Clinton, Gore, Gingrich), to cable television (TBS, TNT, CNN). Second, the class makeup of the region had changed. Garth Brooks, as *Rolling Stone* noted with glee, had graduated from college with a degree in advertising. Vince Gill's father was a lawyer; Trisha Yearwood's the vice president of a bank. At least two performers—Mary Chapin Carpenter and Marcus Hummon—grew up overseas when their fathers were transferred abroad . . . one for *Life* magazine, the other for the Agency for International Development. Nashville artists had come a long way from the days of coal miners' daughters.

Predictably, this cleansing of country's dirty-fingernail past, coupled with Nashville's up-to-date sound, made it even more threatening to the high-minded rock set, who suddenly realized that many country artists and their fans had grown up at the mall, just like them. The pop elite needed a new reason to ignore country, which it found soon enough: Country, they said, had sold out its roots. This marked a dramatic shift. Suddenly writers and critics who had never cared for Nashville decided they didn't like current country music because it didn't hold up to the old, which, of course, they hadn't liked in the first place, but now decided to embrace since it was no longer knocking at their door. An odd sort of retro, PC-chic set in. Old country artists were like Indians—oops, Native Americans—who had been ousted from their rightful territory by a bunch of marauders. Though this view is obviously silly—all facets of American culture kick out the old in favor of the new—it was still widely perpetuated in the media. The ultimate message to Nashville was unavoidable: You might as well stop aspiring, you'll never enter our club.

Meanwhile, all the excitement coming out of Nashville had another effect, which is that the less self-righteous, more populist press (*People, Life, Time, TV Guide, Cosmo*) began getting interested in Music Row. The reason: It was news. The fundamental rule of the press, as told to me at my first reporting job in Japan, is to tell us something new, something that surprises us, something counterintuitive. At its most basic, the media take what's small and build it up; take what's big and tear it down. In the early 1990s, country music benefited dramatically from a startling confluence of trends that made it a phenomenal story: You think country music's Southern; well, it's national. You think it's twangy; well, it's rockin'. You think it's old-fashioned; well, it's hip. And, most importantly of all, you think we're kidding; well, we can prove it. The evidence arrived in late 1991, and the person it aided most of all was the fresh-faced cowboy from Yukon, Oklahoma.

From the moment he set foot in Nashville, Garth Brooks understood two fundamental realities about being a recording artist in the nineties: one, the importance of creating an image; two, the importance of communicating that image to the public through the media. Garth set about mastering both sides of this equation with stunning efficiency, and once he had conquered the first part—making himself into a cowboy—he moved on to Phase II—making himself into a media phenomenon. That began in earnest in 1991.

The widely held view on Music Row is that good songs matter more than anything. Have a hit song ("get it in the grooves," as the saying goes) and everything else will follow. The cold reality is: This view is mistaken. Music is important. It is the ticket that can get an artist over the threshold and into the room where the ultimate winners are chosen. But once artists cross into that room, only a few still manage to thrive. Why? The answer, to a large degree, is myth. To compete in the marketplace of personality in America—"Entertainment Tonight," "The Oprah Winfrey Show," "The Tonight Show" monologue, the cover of *People*—one must have not only good music, but something interesting to say. The best artists at this game—the Rolling Stones, Dolly Parton, Madonna, Garth Brooks—develop images that are uniquely suited to their times. By doing so, they don't react to the media, but push the media in new directions.

The first major episode in Garth Brooks's full-court seduction of the press occurred in the spring of 1991 and was, in all aspects, a classic of the genre. "They should teach Garth 101," the head of Epic Records told me. If they did, "The Thunder Rolls" would be a classic case study. In early 1991, Liberty Records decided to release "The Thunder Rolls," the first cut off Garth's second album, as his next single. Written by Garth and veteran songwriter Pat Alger, "The Thunder Rolls" is a haunting, ominous song (it takes place on a moonless summer night, with a storm movin' in) about a man returning from a place he should never have been to a house where a woman waits. Introduced by a gravely growl of thunder and highlighted by occasional claps before the choruses, the song, reminiscent of "The Night the Lights Went Out in Georgia," ends with the man arriving home and the woman smelling strange perfume. An earlier version, recorded but unreleased by Tanya Tucker, has the woman reaching for a pistol, but Garth omitted that verse. His version ends with the sound of a steady stream of rain and a clear sense of impending doom.

The song, though affecting, is hardly controversial. The controversy flowed entirely from the video, and, though Garth would later deny it, the flare-up was carefully, systematically planned. The four-and-a-half-minute video was shot in Los Angeles at a cost of $130,000, an unheard of sum on Music Row at that time. Directed by Bud Schaetzle (who went on to direct many television specials, including Wynonna's with Bette Midler), it featured the husband, played by Garth in a wig and cheesy beard, committing adultery, beating his wife (in front of their child), and then being shot by his wife at the end. "My goal," Garth said at the time, "was to make this man hated so much that every person in America wished it was them pulling the trigger." The first person to object, though, was his wife Sandy, who made Garth promise to do no love scenes. "But then they got to California," Sandy told writer Alanna Nash, "and Bud said, 'But you've got to—that's going to pull it all together.' He called me later that evening, and we had a pretty good-sized fight over that." Sandy also objected to the scenes showing spousal abuse. "We had a tremendous argument on the child watching," she said, "and he took it out. But then he put it back in [because he said] it's missing that one side of darkness." Sandy protested by refusing to watch the final video.

Garth, though, knew he had a media bonanza on his hands. Backstage at the ACM (Academy of Country Music) Awards that April, he anticipated the controversy. "I am gonna get a lot of conflict over this video because it's about real life," he told reporters. "But I swear I just came to make a video." The video aired six times a day, for two days, as CMT's Pick Hit of the Week. Then CMT pulled it. "I'm yanking the thing," said director of operations Bob Baker. "We are a music channel. We are not news. We are not social issues. We are not about domestic violence, adultery, and murder. Our obligation is to protect our viewers." As a TNN spokeswoman echoed when that network banned it as well, "It's a great video, but it doesn't offer any help or hope to anyone in an abused situation." Leaving aside the fact that none of their other videos "offer help or hope" to victims of whatever situation, these comments constituted a startling expression of the paternalism of country's ruling class. Also, coming in the wake of country's once proud tradition of songs about hardship, they were a sad expression of how caution in Nashville had overtaken passion.

Their actions, predictably, had the opposite effect. The episode quickly turned into the best thing yet to happen to Garth Brooks's career. It instantly separated him from the blandness of Nashville and aligned him with the real-life concerns of his fans. Though critics accused him of "pulling a Madonna" (the previous fall, MTV had refused to show her suggestive video for "Justify My Love"), the press still couldn't resist wading in. Garth and Jimmy Bowen were waiting. Bowen even hired additional staff to drum up support from women's shelters in a project termed the "Garth Brooks's 'Thunder Rolls'/Family Violence Campaign." It worked brilliantly. In the last week of May, for example, all four major television stations in Dallas ran stories on the controversy. Both newspapers chimed in as well. These, naturally, prompted more airplay, which, in turn, spurred consumers to rush out to stores. Make no mistake: This is how revolutions in taste are made. Five years later, a woman at Fan Fair in Nashville waiting in line to meet Garth told me she had always hated country music until her husband made her watch "The Thunder Rolls" video. She liked it, went to buy the album, started listening to country radio, went to see Garth in concert, and later decided to come meet him in person. All because of the press, because of the video, because of the song, because, ultimately, of the artist himself, who had the savvy to master them all.

On May 25, in the midst of the "The Thunder Rolls" hullabaloo, *Billboard* quietly announced that it was introducing a new system of calculating album sales. Instead of relying on reports from store clerks, many of whom knew little and cared less about country, *Billboard* would begin taking sales figures from a small New York company called SoundScan, which claimed to count the *actual* albums sold from bar codes scanned in stores (in reality, they took a limited number of scans and estimated the rest). The week the system came out, it became apparent that Nashville had been selling far more records than anyone realized. *No Fences* moved from sixteen to four on the pop album chart, which includes music from all genres.

The big impact, though, came in September when Garth released his third album. *Ropin' the Wind* debuted at number one on both the country *and* the pop charts, the first album ever to achieve this feat. Coming amidst an extraordinary attention surrounding the Judds' Farewell Tour, for which Garth occasionally was the opening act, the album had generated 2.6 million advance orders. *Billboard* ran an article about fans lining up, not only in Houston and Lincoln, but also in Sacramento, California, Mankato, Minnesota, and—in a fit of journalistic flourish—Cambridge, Massachusetts, at the Harvard Coop. Harvard students had become familiar with Garth's music, a clerk reported, hearing it at bars and parties. "The other albums to open at number one this summer have all been by hard rock/metal bands," wrote chart analyst Paul Grein, "Skid Row, Van Halen, and Metallica. Such groups appeal to young, active music buyers who are more apt to find the time and inclination to buy an album in its first week of release than are the older, more settled country and pop fans—or at least that has been the conventional wisdom. Brooks's socko debut suggests that it's time to recognize that country fans can also be active and committed."

Nashville reacted to the event by treating it as the Second Coming. After years of feeling like distant cousins, country suddenly found itself at the center of the entertainment universe. "Sure, Garth Brooks is great for Capitol Records—you bet he is," Jimmy Bowen declared at a hastily arranged celebration at CMA headquarters. "But Garth Brooks is also great for all of Nashville and all of country music." This time he was right. In the next year alone, Garth was on the covers of *Time, The Saturday Evening Post, People, Forbes,* and *Entertainment Weekly,* which rated him first among male singers, above Michael Bolton, Bruce Springsteen, and Axl Rose. Even *Rolling Stone* felt obliged to run a report on Garth's dominance, with reporter Rob Tannenbaum crediting his success (and Nashville's in general) to upgraded recording techniques and to a general decline in pop radio. In a telling exchange, the article expressed surprise that Garth cited Kiss, Boston, and Styx as influences. "Doesn't Brooks worry about honoring such discredited groups?" the magazine asked. To which Garth was quoted as saying: "I don't think anybody's gonna come out and give me flak for that, because it would only be showing their ignorance in what is good music." In twenty-five years, few musicians had so directly challenged the canonical wisdom of *Rolling Stone.* But it didn't matter; Garth himself was on his way to becoming canonic on his own terms. That year he was invited to sing the National Anthem before the Super Bowl. His records started setting records themselves—20 million, 30 million, 40 million and counting. Asked on Larry King to explain his success, Garth's answer was a classic mix of his little-boy and big-boy selves and a knowing nod to timing. "God and SoundScan," he said.

But there was one more thing behind's Garth's success, he gave good answers—deep, dark, fascinating answers—when reporters asked him questions. He became a sort of antihero and superhero mixed up in one. In the fall of 1992, in the midst of his glorious press run, when most artists would have started playing it safe, Garth released a song that not only threatened his standing with many of his fans, but also severely traumatized his family. "We Shall Be Free" was a gospelly, church-picnic Bill of Rights listing Garth's views on when the world would be free: "When the last thing we notice is the color of skin / And the first thing we look for is the beauty within." The song, written with Stephanie Davis after Garth was caught in Los Angeles during the 1992 riots, came with another of his save-the-world videos, featuring Paula Abdul, Burt Bacharach, Harry Belafonte, Michael Bolton, John Elway, Whoopi Goldberg, Jay Leno, Nelson Mandela, Martina Navratilova, Elizabeth Taylor, Lily Tomlin, and General Colin Powell. It was the largest gathering of odd celebrities since *It's a Mad Mad Mad Mad World.*

And, of course, it got reams of publicity, all the more so after Garth announced that one line in the song, "When we're free to love anyone we choose," could be interpreted as an endorsement of gay rights. Many radio programmers were horrified, a response that only worsened after Garth told Barbara Walters that his sister was a lesbian, a shocking and unnerving revelation for much of his family. "He has no edit on his mouth," his comanager Pam Lewis, told me. "I used to say to Garth, 'You should really get a good minister or a shrink because you're telling the world what you should be telling your counselor.' He outed his sister on network television. Caused incredible trauma to his family by doing that. Got a lot of press out of it. But at what cost?" Even more telling: After the interview was taped (but before it aired), Pam asked Garth if she could ask Barbara Walters to edit out his remarks about his sister's sexual orientation; Garth declined.

Garth, as Naomi once said of herself and her daughter, had begun using his interviews as a form of therapy. It was a way to play out the tension between his two selves: GB and Garth Brooks. In the wake of the "We Shall Be Free" episode, for example, *Interview* magazine asked Garth if he'd ever had a man come on to him. "No guy has ever pulled the shitty guy thing by grabbing my ass, if that's what you mean," he said. But he did have a related experience, he said. "There's a guy back home that I sincerely love. He works with us. Rumor has it [that] he's a homosexual. I ran into him one night in a club. We always hug each other, so I'm hugging him, and I'm standing there talking to a bunch of people, and he sits down next to me. We're talking, and all of a sudden I feel this—what he's done is reached down and grabbed my hand. So we're sitting there actually holding hands at the bar. And there are people watching me, making me feel real uneasy about it. Then, all of a sudden, I think: 'Which is going to bother you more? People seeing you holding this guy's hand, or how he's going to feel if you pull your hand away?' Not breaking that guy's heart or insulting him in any way means so much more to me than anybody's opinion about me . . . People that you care about, you try to take care of, and the image takes a backseat."

Not always, though. Indeed, the longer Garth rode the wave of press, the higher he climbed on what seemed like an endless assent of attention, the more consumed he became with his own image. He began to complain about the demands on his time. He began to talk about losing his will ("I have more money than my child's grandkids are going to be able to spend," he said). And, in 1992, when Sandy had a difficult pregnancy, Garth shocked the music world by announcing he would consider retiring if his career interfered with his family (though little known at the time, he was involved in intense renegotiations with his label, which made his remark an effective bargaining tool, if nothing else). In 1993, when *Rolling Stone* finally put him on the cover, Garth admitted, in his most confessional interview yet, that he feared he might have come too far, too fast. "Joe Smith from Capitol was at my house a week ago," Garth told interviewer Anthony DeCurtis. "He pulled me and Sandy over, and he said, 'You got all the money in the world. Make some time for yourselves.' As soon as Sandy walked out of the room, I pulled Joe over. I said: 'Joe, what do you do when you've lived your whole life with stomping the guy's guts out who's in competition with you, just try and knock the *shit* out of him, get number one, do whatever it takes to stay number one—what happens when you feel yourself falling from that because you have a child and a family? Where does the killer instinct go?'"

Smith's answer hit Garth hard. "There's going to come a time," he told Garth, "when people love you more for what you've done than for what you do." "Right then," Garth said, "Dan Fogelberg flashed in my mind. I love all his older stuff, but I'm not buying his latest stuff. Then Bob Seger came to me, *bam*, and all these groups that are gone now, like Boston. I just recently picked up a CD of Kiss's *Destroyer* and loved it just as much the second time—but I haven't bought a Kiss album in twelve, thirteen years. So that hit me. But it did just the reverse of what I thought it would: My shoulders kicked back, my chest stuck out, and I said, 'Well, I'll be damned if that's going to happen to me for a while.' "

But, of course, it did. Once he reached his peak, Garth experienced a backlash. The main reason—other than the basic law of stardom, which is that people get tired of you after a while—was Garth himself: what he said, what he did, and what he recorded. Simply put, a change came over him. Garth became preoccupied with his own mortality. The red-cheeked boy who had seemed so happy to be playing cowboy in public gradually gave way to a road-weary adult who seemed determined to rewrite the rules of stardom. This was Phase III of his career: trying to outsmart death.

Premature death had always been a theme in Garth's life and work. During his senior year in college, Garth was almost killed in an automobile accident, and two friends from that period—Jim Kelley, a guitarist and track buddy; and Heidi Miller, a onetime roommate— later were killed in accidents. Garth's first album was dedicated to their "loving memories." In addition, many of his best songs—"If Tomorrow Never Comes," "Much Too Young (to Feel This Damn Old)," and "The Dance"—are about the consequences of dying young. Even his videos have often contained images of fallen heroes and slain leaders.

Now, though, there was a difference. Garth no longer merely cele-brated martyred heroes; he tried to turn himself into one. Garth's first two albums—produced between 1989 and 1991—contained a wide vari-ety of songs, but none that could be considered preachy. His songs were "from the heart," as he said, and they sounded that way, whether one liked them or not. It was not until the video for "The Thunder Rolls" that Garth began urging people to *do* something rather than just *feel* something. "That video did in two days what I hoped it would do in its lifetime," Garth said, "making people aware of a situation which unfortunately exists in our society and causing them to discuss it, some-times even heatedly. That's what I want my music to do." The truth is, that's not at all what he once said he wanted his music to do. Instead, with success, Garth had embraced the sixties ideal of what a pop star should be: He began believing his music could change the world. Wit-ness "We Shall Be Free," an application for saintdom of the highest order. Witness the video for "Standing Outside the Fire," featuring a battle between parents over whether to mainstream a Down's Syndrome boy.

Eventually he didn't even wait for music as a platform. In 1993, Garth asked Pam Lewis if she would set up a meeting between him and President Clinton so they could discuss world peace. Afterward they would hold a press conference. "You're going to look like a fool," she told him. "It's a very noble idea. Jesus Christ tried it. Mother Teresa tried it. Neither of them was very successful. Why do you think you're different?" The following year a former Disney employee approached Garth with the idea of starting a National Kindness Day. Garth took it one step further and said why not begin a World Flag Day in which one day a year every country would fly the same flag. This idea he actually did take to the Oval Office, where he met President Clinton and Vice President Gore. When nothing came of it, he moved on. "I want to make 1996 the Year of Peace," he told TV Guide. "In that year, during all my foreign concerts, we will try to reach the world's leaders to encourage them to establish peace on Earth." The idea, he said, was to create a Peace Chain. "For every day there is no war on Earth, we will add another link to the Peace Chain." In a little over five years, Garth had gone from wanting to be John Wayne, a grand enough ambition, to wanting to be Mahatma Gandhi.

None of this would have been so bad (just a bright-eyed boy trying to do good) if it all didn't seem a bit wacky. Garth was behaving as he thought a pop star should behave, but because of his own political naïveté—at this point in his life, he had never voted in a presidential election, he told me—his gestures came across as more self-promoting than self-sacrificing. More damaging to his career, though: This new strain of martyrdom began to infect his music. In Fresh Horses, which was recorded during this period, Garth often sounds like a man deeply unhappy with his standing in life. The album begins with a reminiscing song in which Garth longed for the days "when the old stuff was new." Other songs returned to his pet themes of suicide and early death. The album's signature cut was "The Change," written by Tony Arata and Wayne Tester, which Garth said was his most meaningful song since "The Dance." In it the narrator complains that he cannot change the world around him, but that he's going to keep on trying so the world will not change him. "As long as one heart still holds on / Then hope is never really gone."

"Why does Fresh Horses sound so stale?" Michael McCall wrote in the issue of the *Country Music* magazine that came out just days before our trip to New York. "Perhaps Brooks overexamined himself, retooling the songs until he stole the life from them. Perhaps he felt the strain of superstardom and attempted to make each tune an epic venture that chimed with importance. As the bestselling artist of the nineties, Brooks no longer competes with other Music Row performers. Instead, his peers are Whitney Houston, Michael Jackson, Mariah Carey, Michael Bolton. Those performers tend to stretch for the everlasting anthem, the Big Statement, trying to blow every horn with every song. Brooks now follows their lead." Even worse, wrote McCall, Nashville's most influential critic, none of the new songs were memorable. " 'If Tomorrow Comes' and 'Unanswered Prayers' discussed subjects as big as death and fate, but they did so in intimate terms, and Brooks's performances were beautifully understated. 'The Dance' gained strength from how subtly Brooks unfolded its melodramatic message. But Brooks apparently no longer trusts his fans to pick up on such quiet strengths. Now he blares his stories with sweeping musical buildups and histrionic vocal performances, and he milks every topic for high drama rather than gentle reflections."

Perhaps the most surprising consequence of Garth's change from gee-whiz kid to grandstanding grown-up was that he began to succumb to a completely new flaw: an arrogance toward the media. All during his run as cover boy, Garth and the media were perfectly in sync: The media wanted access—and controversy; Garth wanted exposure—and controversy. As long as each side served the other, both were happy. But once Garth realized he couldn't control the press (that it had its own agenda—*surprise!*), he felt used. When a reporter from *People* got his wife to discuss the details of his affairs (Garth said she was tricked; the reporter denies it), Garth was furious. "My respect for *People* magazine is totally through the floor," he told me. *Life* magazine promised him a cover story, he says, but later switched it for a photograph of a fetus. Garth accused the reporter of being cowardly and weak. "My respect for *Life* magazine is through the floor, too," he said. He swore off working with both publications, along with *TV Guide, USA Weekend,* and most other magazines. "All these people are bullshit artists," he told me. "I don't know why they have this ego thing. I mean, would I work with *USA Weekend* if it wasn't for a cover? Give me a break." Finally he developed a rule. "If these people want to work with Garth Brooks again, it's simple. They're going to have to bend over, take it like I did, and then, once the tables are even, we'll sit down and say, 'Okay, this is what we want.'"

For those publications he did agree to work with, like *Esquire*, which hired me around this time to write a story about Garth, he insisted on a written guarantee that he be on the cover and on maintaining approval over the photographs. For country publications, Garth agreed to be interviewed only if he could supply art and only if he could approve the story before it ran. This degree of control was nothing new by Hollywood standards, but in Nashville it was revolutionary. The one difference is that Garth had always been smart enough, subtle enough, and, above all, invincible enough to keep such maneuverings hidden. "Don't let the audience see the puppet string," is rule number one among publicists. But now Garth's domineering was seeping into the open. "Brooks seems insecure about life at the summit," *USA Weekend* wrote in its 1996 preview issue, which featured a drawing of Garth on the cover. "Big-guy handsome at six-foot-one, he worries about his weight and receding hairline and rigidly controls his public image. He wouldn't pose for a cover photo for this issue—and his operatives, when told that an illustration would be used instead, demanded approval (they didn't get it, of course)." When the press regains the upper hand, it's a bad sign for an artist.

Worst of all, Garth kept feeding the story that his career was in a tailspin and that he was spinning out of control with it. It wasn't intentional—Garth was "just saying what's on my mind," as he liked to say—but still it hurt. After country radio rejected "The Fever," the second single from *Fresh Horses,* Garth went on "Larry King Live!" and announced the song was dead, as if his fans really cared about chart position. When sales of the album plummeted (after debuting at number one, it dropped out of the Top 5 within two months), Garth told the *Los Angeles Times* that if people didn't buy his album he would be forced to retire. "If the record and ticket sales don't tell me that I'm stirring things up or changing people's lives, then I think it's time for me to hang it up," he said. "You want to be remembered at your best. You don't want to be a trivia question on some cheesy game show in twenty years and see the contestant get it wrong." He added, "If someone says I'm only saying that because of ego, I'm not sure they're wrong. I'm not sure why I feel this way. But when you stop connecting with people, maybe you're in the wrong field. Maybe that's what God's trying to tell you."

The net effect of all these actions was that Garth—through his music, his actions, and his public statements—had completely remade his own image. The once confident, bright-eyed, bushy-tailed singer had now been recast as an unsure, perhaps unstable, and, at times, self-obsessed individual. And with that transition came a sea change in the public adulation surrounding him. The narrative of the first half of the decade, the rise of Garth Brooks with Nashville riding his coattails, soon gave way to a dramatic new story, the fall of Garth Brooks and the fizzling out of country music. (Since Garth had accounted for a fifth of all country sales, when his sales started to wane, so did the industry's, falling 12 percent in the first half of 1996.) It was, as my first editors in Japan might have noted, news. And it was in this suddenly newsworthy climate that Garth came to New York in the winter of his discontent and talked about his nagging obsession with death.

"Yes, I guess you could say I'm reckless."

Garth was sitting, lounging actually, on a small upholstered bench in a tasteful suite of the Parker-Meridien Hotel in New York. He'd been doing nonstop talk shows since his appearance with Regis and Kathie Lee and now, safely out of public view, he'd quickly removed his baseball cap and sneakers and plopped down on the bench. There was little light in the room. No radio or television. The only two things left turned on were his mind, which grinds continually, and his voice, which he alters with near Shakespearean precision from naïf to sinner. On occasion his mind and voice meet in tandem: this was Garth Brooks, the confesser.

"I'm certainly reckless onstage," he said. "It's part of pushing the envelope."

This was an interview, a chance to enter that arena that is, in fact, a lot like therapy: questioner and questioned; prober and probed. It was an arena in which Garth had always excelled.

"You do some things in entertainment, and you wonder, 'Why hasn't anyone else done that before?' " Garth said. "Then you see it on film and you go, 'The reason nobody else does that is because somebody could have got hurt.' I was that way with throwing cymbals into the crowd. I thought the cymbal would always land flat. Sure enough, the first piece of film I saw it flipped, spun, and went straight down on this guy. I went, 'Damn! I could have killed that guy.' "

This sense of danger was always lurking just beneath the surface with Garth. When I first met him, I was struck by his outward boyishness and good manners. There was something magical about his sheer enthusiasm.

Earlier that day, Mick told a story about a friend of theirs, a former Wild West performer from Oklahoma who claimed to have built the first recreational vehicle. "Even if it's not true, *so what . . . ?*" Garth said when the story was finished. Then he leaned back and said, "I love to believe." For someone who had tried to make himself into a Hollywood-style icon, this comment was extremely revealing. In Garth's mind, fiction was often higher than fact.

But Garth's other side, his dark grown-up side, was also real. At the moment it was even dominant. Perched at the summit of his career, Garth was deeply scared. He was scared of his power. He was scared of the responsibility of being a superstar. Earlier I asked him about a woman who had called a radio show some weeks before and asked if he wasn't ashamed that, as a Christian, some songs on his new record seemed to "promote" sex. Wasn't that a burden? I asked. When he was younger, Garth hoped merely to express himself. Now, with stardom, he automatically became an advocate for positions he took. This was the flip side of becoming an icon: Everybody suddenly paid attention. "Believe it or not, I have thought about that question day and night," Garth said. "And I know what the answer is. The answer is, very simple: That's me. My albums talk about God. My albums talk about sex. My albums talk about death."

But that didn't make it any easier. If anything, that mix of emotions—life and death, sexuality and spirituality—were at the heart of what Garth was scared of. He was scared of his own mind, of the conflicting emotions that swirled around his head. He had once even considered suicide, he told me, to "overcome the demons." At the heart of his problem was the fact that Garth knew he could go no higher.

"There's a basketball game called 'We-They,' " he said. "Greatest game I know. I learned it in the last two years. Two or three guys, and it's you against them, only they don't exist. What happens is, if you miss a shot, *they* get points. If you lose, you've beaten yourself." He raised his eyebrows. "Fair game," he said. "Fair as fair can be. And that's what music is: It's you and the crowd on the same team, and if you lose, if those people walk away unhappy, you've beaten yourself. We're at the point now where we have to disappoint people in order to lose fans. That's a great position to be in, but it's also scary. If we lose them, we have no one to blame but ourselves."

"Do you think you're going to lose them?"

"If I could have a term that I would love to be called, twenty-four hours a day," Garth said, "it would be 'underdog.' "

"But you're not an underdog," I said. "You're a superhero."

"It depends on how you look at things. Now is the time we're starting to hear the comments at the award shows: 'He'll be gone by next year. He can't keep it up. This year's sweep is next year's shutout.' It feels good for people to say it because I can come back and go, 'Okay, boys, let's strap it on. Let's don't even celebrate tonight. Let's hit the workout room. Let's push ourselves to the limit.'"

And there it was, the "limit." The "next level." The "Zone." Garth, sounding more like an athlete than an entertainer, had hinted about this place for years. But I'd never quite understood what he meant. That spot—that *feeling*—had always remained elusive, and yet it seemed to be the very essence of why someone as powerfully successful as Carl Lewis, or Michael Jordan, or Garth Brooks seemed to be so much better than everyone else, and yet never seemed happy unless they were at work. Why was that? Surely it was more than money, more than ego, more than power. It must be something almost divine, something they could not live without, even if the quest for that one thing became the source of their demise.

"I remember the first time I heard a song of mine on the radio," Garth said. "I almost ran over the person next to me. I was on I-40, right before the 65 split going north. I heard it on WSM, popped it over to SIX and they were right in the middle of the pedal steel solo. I'm all over the place, man. Honking, screaming. People are backing away from me. I don't care. Hit me with a flame. Drop a bomb on me. I'da lived through it."

"What was the feeling?" I asked. "Is it that you wanted to die?"

"It was the sense of being untouchable."

"Do you still have it now?"

"Sometimes," he said. "I can play basketball. I'll be on one of those situations where it's, like, give me the ball. I don't care if I throw it up eighty feet out, ten feet out, it's in. I just know it. It's got to be some kind of athletic thing, because in my life I've heard people talk about the Zone, but I've never heard anyone describe it in the same form as I felt it as Michael Jordan did in an interview once, where he talked about the hoop being *this* big . . ." He formed a circle with his arms.

"So how often does it come?"

"It comes mostly in music. But it happens in your kids, too. You know what's going to happen before it happens. And—*boom!*—you're there, and you're handling it well. And you know what's going to happen.

Your wife's going to go, 'I'm so proud you're their father.' You just know it. It's like you can see it coming before it happens."

"What is that?" I asked. "Art, life, love, God?" I had heard other people say that interviewing Garth was like being drawn into the Zone yourself. Now I knew what they meant. This was Garth Brooks, the competitor: at the height of his power, at the edge of his sanity.

"I don't know," he said, "but I know it only happens on things that I really give a shit about. Is it a passion? Is it a total focus? Is it that thing that your dad always told you, 'Son, if you apply yourself, you can do whatever you want.' Is that *true*? You don't know how bad I'd like to know. You know they talk about knowing what ninety-five percent of the human body can do, but only about five percent of what the brain can do. Because there are times when I feel as if I can do anything. There are times I feel I could hang from the rig, let go, and not drop. There are times when I have flown miles, over trees, canyons, and water. And seen it all.

"Rupp Arena, Lexington, Kentucky, I got into something I didn't know what it was. There's this thing they call the purple thumps you get when you're hot. You see these purple veins, you get real winded, and every time your heart beats, those veins seem to glow in your sight. In Lexington, I got the purple thumps. But I kept pushing and pushing and pushing. And I walked over into this realm, where everybody slowed down. I could see the ripples in their shirts, the sweat in their eyes. I could see the fillings in their teeth. I could be sitting there and out of my peripheral vision see someone lifting their hand and before they did it—*boom!*—I'd be pointing at them. It was the coolest thing. I had all the air in the world. I could hold a note for an hour and a half. Then everything jumped back into speed, like film sometimes does, and all I could think about was getting back there somehow, back into that Zone."

"Is that internal," I asked, "or does it come from them?"

"I don't know where that comes from, although something that extraordinary probably has to come from a higher being. But that's what I mean when I say, 'How can I find out?'"

"Are you addicted to that?"

"If the answer is, is that where I want to spend my time, yes. I do. Of course it would get old because it's safe. You know everything before it's coming. But it's also false because you get a false sense of confidence that you can do things."

Garth was distant now. His eyes were out of focus. He was lying on the bench.

"And is that a place you were once or a place that you can be again?"

"I don't know. It was at the end of the last tour. We'll see if I can get there again." And then Garth suddenly jerked back into focus. He looked up from the ground and turned toward me. "But I have a deal with God. 'Please give me the sign when this is over, so I can step down with class.'"

TEN

THE PARTY

In Nashville, you can tell a lot about the status of artists' careers by the types of parties their labels throw for them or that they throw for themselves. On a warm day in early spring, MCA Records and MCA Publishing threw what would turn out to be one of the more pivotal parties of the year for Wynonna Judd, whose first single in two years, "To Be Loved by You," had just limped to number one on the country singles chart.

The party started out unexceptionally enough. The four hundred or so regulars on the Music Row party circuit eased their Range Rovers and BMW 750iLs into the valet lane of Trilogy Restaurant, slid their cellular phones into their pockets, and slipped into their most practiced expression of casual self-importance. "Oh, hi!" Kiss, kiss. "How are you? Love those boots!" Step back and admire. "Hey, congratulations on your artist reaching the Top 20 . . ." Lovingly touch the arm. "Yes, that's right. We just went gold . . ." Demurely smile. "Sure, we sold double platinum last time, but you know how tough it is right now . . ." Raise the eyebrows, grimace a bit. "Oh, no. I'm not worried. For me, it's all about the *music* . . ." Lean forward, put hand over the heart. "Well, then . . ." Peer around anxiously, pretend to spot a colleague, head directly for the bar.

The process of hyping a recording artist—to the media, to radio, and to the all-important industry buzz machine—is a tradition that goes back generations in Nashville. For much of that time, it was a fairly casual operation. "Tammy Wynette would have a gold record," Bill Johnson at Sony told me, "and we'd decide to have a party. We'd call the producers, the managers, a few members of the press. We'd all gather in our conference room with a few beers and bottles of Jack. Tammy would come in, she'd know everybody, and we'd just sit around and talk." Not surprisingly, the only party of any grandeur that anyone seems to remember was one held at the now-defunct Spence Manor Hotel when the Outlaws' album became the first Nashville recording to receive platinum status.

The attention to hype slowly built in the 1980s when more money started flowing into Nashville and when labels started realizing that, when done properly, stroking works. More importantly, when everybody else is doing it, stroking is required. The first target for propaganda, of course, was radio. Labels began arranging so-called showcases—essentially aboveboard payola—in which they fly radio programmers to exotic locations (Lake Tahoe, Key Largo) and pay for them to play golf and get massaged during the day, then get drunk and listen to music at night. The second target for seduction were voters on award shows. In 1983, a minor scandal erupted when a Sony blitz to encourage members of the Academy of Country Music to vote for its artists actually resulted in a sweep of the show. A third target for promotion was the press. "I'm sorry to say this, but I started it," said Susan Levy, formerly a publicist at MCA. She was speaking of the process of sending high-gloss, Hollywood-style mailings to reporters. One campaign she coordinated for the Mavericks, the hip, neotraditional band from Miami, was so successful with its mix of glitzy, four-color fold-outs and intentionally obscure photographs of naked baby bottoms that other labels soon followed. "Now I just wish I could stop it," she said.

By the mid-nineties, these mailings had mushroomed to such a degree that any writer who had ever listened to a country song would receive buckets of unsolicited mail. In one week leading up to the CMA Awards, for example, I received a giant simulated restaurant menu from Vince Gill's Lonesome Cafe; three mailings in brown envelopes from Tim McGraw printed with the words CONFIDENTIAL: FOR YOUR EYES ONLY, the last containing a cassette, à la "Mission Impossible," that included directions to a cocktail party and the warning "This tape will self destruct in ten seconds . . ."; and, finally, a giant pink hatbox from Reba McEntire, which, even in a town world-renowned for its tackiness, was widely considered to be the most outlandish thing ever mailed from a major record label. On top of the box was a photograph of Reba, prancing barefoot on a beach, showing off her recently shorn locks. Inside the box were a few postcards of Reba in various terry-cloth concoctions, the lyrics to her new single on a piece of fake parchment, and an invitation to an open house at her new $22 million office complex (nickname, "Vatican City"). And at the bottom of the box was a plump pillow of white cotton, on the edges of which were a few flakes of scented wax and in the center of which, like priceless artifacts from Cleopatra's tomb, were four perfectly curled locks of Ms. McEntire's hair.

And I don't even vote for the CMA Awards.

The changes going on in country music mirrored the larger changes going on in Nashville. Even after tripling in size from a $700 million annual business in 1990 to $2.1 billion in 1995, country music was still only the third-largest industry in Nashville. The town's two other economic behemoths, religious publishing and health care, grew just as quickly during this period. As a result, Nashville became a boomtown in the 1990s and the poster city for a back-to-roots movement that seemed to be taking hold in America. Nashville, as V. S. Naipaul wrote in *A Turn in the South,* was the home of a "new order leading no one knew where."

Nashville has always been a city of the middle. It's in the middle of a state considered to be in the middle of the Mid-South. Its local products have always been middle-oriented—Maxwell House Coffee, Shoney's, Captain D's, Cracker Barrel, Service Merchandise, Dollar General, Nissan pickup trucks, Ford windshields, and, of course, country music. On the right, Nashville did give rise to the reactionary manifesto *I'll Take My Stand,* and on the left Reverend James Lawson of Fisk was a leader of the civil rights movement, but basically it's a town of political pragmatism: Lamar Alexander on the "right," Al Gore on the "left." Both, of course, are radically middle. Dinah Shore came from Nashville, as did Oprah. John Tesh was a newscaster in town. Pat Sajak was a weathercaster. Today, just on the cusp of 1 million people, it's a town that is large enough to have helicopter traffic reports during rush hour, but small enough so that most traffic lights start blinking at 11 P.M.

The sudden gold rush surrounding country music in the early 1990s would change Nashville forever—and not in ways anyone particularly anticipated. In the past, Hazel said, when hillbilly singers struck it rich, "the first thing they did was buy a Cadillac, the second thing they did was buy a diamond ring, and the third thing they did was leave their wives because as soon as they got money they had young chicks chasing their ass." But in the new age of college-educated country, as soon as stars got money, they started to invest: Reba McEntire bought a fleet of trucks and airplanes, Travis Tritt became a coowner of an indoor football team, and Naomi Judd bought her old hangout of Maude's, poured in $2 million to redecorate it as a "California-themed" eatery, and named it, after herself and her two daughters, Trilogy. As always, she was onto a trend. Though Nashville had never really been a food town (primarily because selling liquor by the drink had been outlawed until the 1960s), suddenly a spate of Hollywood-style restaurants opened up: Sunset Grill, Bound'ry, Cakewalk, Nashville Country Club. In 1994, drawn by the wave of tourists, Hard Rock itself opened its first restaurant in a city of under 1 million people, and, a year and a half later, Planet Hollywood followed.

These developments gave Nashville an air of hipness and excitement completely at odds with its "Hee Haw" image, but completely in sync, I came to believe, with the rapidly evolving South at the time. When I first moved to Nashville, I expected to find perhaps a few nouveau riche, but mostly swarms of WASP-like preppies. I dressed accordingly, in blazer and penny loafers. But in my first week of meetings on Music Row, the only necktie I saw was mine. In fact, in most of the conversations I had, I was the only Southerner in the room. While many music people dressed in country casual—boots, jeans, and polo shirts (no hats)— a surprisingly large number were decked out in three-button, free-hanging Armani suede and blunt-toed, high-heeled Manolo Blahnik leather. This trendiness only heightened in the years that followed. Vera Wang opened a boutique in Nashville. Hummers started popping up on the streets. And, most surprisingly of all, in a town once defined by Johnny Cash and Willie Nelson, *men* started going to hairdressers. By my count, Nashville in the nineties had no fewer than *four* one-named hairdressers, including mine, Yon, from Iowa.

"Since when did men start getting **hairdos**?" asked the Nashville *Scene,* the hip alternative weekly, in its breathless **bolded** style. "Since when did men start having **hair that even needed to get done**?" The paper, which had recently reported on Nashvillians who leaked gossip to the tabloids, went on to report that the Caesar haircut, first made popular by George Clooney, had taken over Music City. **"Brian Williams was the first one around here**. Hard to believe **a banker** would do such a thing, even if he *is* on Music Row. Before long, it grew on everybody. MCA Head **Tony Brown** debuted **his new clip job at the Masked Ball**. Then the other day **at Trilogy**, there was RCA bigwig **Joe Galante**, at the head of the table, showing off his new head of hair. Nobody could pay attention to **anything he was saying**. They were just wondering, 'Can you really ask **a grown man** about his haircut?' " I can think of no better example of what came to be called the New Nashville than the fact that Alan Jackson, one of the last bastions of tradition in country music, the man who still eats bologna sandwiches and insists on having a red velvet cake backstage at every concert, thanked his hair-dresser, Riqué, on his new album. Perhaps this should have come as no surprise: Riqué, as I learned later, has his own publicist.

All this faux Hollywood glamour eventually started attracting actual Hollywood types who were eager to be a part of the excitement, but without the threat of earthquakes. In the span of just under five years in the early 1990s, an amazing and probably historically unrivaled trans-fer of musical power took place between Los Angeles (and, to a lesser degree, New York and London) and Nashville. This new class of immi-grants included Steve Winwood, Peter Frampton, Kim Carnes, Janis Ian, Michael McDonald of the Doobie Brothers, Garry Tallent of the E Street band, Bernie Leadon of the Eagles, John Kay of Steppenwolf, Al Anderson of NRBQ, Mark Farner of Grand Funk Railroad, Bob Welch of Fleetwood Mac, and Al Kooper, the deitific keyboard player on several of Bob Dylan's albums in the 1960s, who was pictured in *Nashville Life* magazine in an article on eyewear makeovers wearing $240 Armani flip-top shades. "Kooper's light-sensitive eyes are his valid excuse for rarely being seen without sunglasses," the magazine said. "For what he's doing, hearing is probably enough." Nashville's lingering stench of hay bales notwithstanding, the new cast of "Hee Haw" would sound distinctly like Woodstock.

Not only graying rockers, but real-live genuine movie stars started hanging out as well. "Notes from Hollywood on the Cumberland," Her-bert Fox, who used to write for "Hee Haw," gushed in a local society rag. "That's Bette Midler grabbing a bite to eat. *Single Guy* Jonathan Silverman is mistaken for Vince Gill on a hotel elevator. Luke Perry lives just up the road. Charlie Sheen drops in at Buddy Killen's Easter Seal benefit . . . The rich and famous just keep popping by—Sharon Stone, Ben Vereen, Martin Sheen, Liza Minnelli. There's one of the Beach Boys. Forgive me if I don't know which one. I'm just getting used to living in Tinseltown." Even Jay McInerney, the literary laureate of 1980s excess, married a Nashvillian who was, as Belle Meade wags pointed out, some years his *senior* and set about nesting in Music City. In a stunning expression of the glam-family values of the times, Jay and his wife Helen Bransford then participated in what will long be remembered as a brilliant metaphor for the bright-lights, gone-country nature of Nashville in the nineties: They procured human eggs from a friend, fertilized them with Jay's sperm, hired a waitress to carry them to term, delivered healthy twins, and then, in their own modern version of a country song, wrote about it in *Vogue.*

Though transplants like Jay often had some difficulty adjusting ("Hey, Suze, the bagels suck down here. And they don't have 'Meet the Press' either . . ."), at least they had themselves, all of which made for some singular evenings in what was still middle Tennessee after all. One night after watching a concert of standards by Raul Malo, the glib, genius Cuban-American lead singer of the Mavericks, I found myself at an impromptu party of recent Nashvillians around his (rectangular) swimming pool. Manuel was there, with his usual *muchacha* on his lap, along with James House, the roots-rock singer. Also Melanie, the zodiacally inspired makeup artist late of Paris, and Frank Callari, the Mavericks' manager and a DJ during the cocaine days at Studio 54. For hours, the assembly writhed to the sounds of Cuban cocktail music, smoked hand-rolled contraband cigars, and finished off nearly a case of Veuve Cliquot, the champagne of choice for James Bond. At 3 A.M. Karen Essex, the former Hollywood film producer who had just written a book about 1950s pinup icon Bettie Page, who also turned out to be from Music City, turned to me and said, "Now, *this* is the New Nashville."

Among country artists, few people encapsulated the glamification of Nashville—the mix of old-fashioned Appalachian roots with trendy L.A. style—better than Wynonna Judd. Wynonna was born in Kentucky, but grew up in Hollywood. She spent her teenage years in frilly dresses pitching lye soap on "The Ralph Emery Show," then spent her twenties in red leather pantsuits blowing the roof off David Letterman. Her mother was pregnant and married before she left high school; her sister, before reaching thirty, had starred opposite Robert De Niro and had an affair with Matthew McConaughey, the last on the set of their film *A Time to Kill.* Wynonna was, above all else, a star. "I can walk into the offices of CAA with any artist on the *Billboard* 200 and nobody will give a damn," John Huie, the head of the CAA office in Nashville, told me, "but if I walk in there with Wynonna, everybody's on the phone in two seconds saying 'Wynonna's in the house, Wynonna's in the house,' and the next thing I know, they're all pouring out of their offices and acciden-tally 'bumping' into us in the hall."

The process of transforming Wynonna into her own free-standing artist began in earnest in 1991, even during the Judds' Farewell Tour. Having moved to MCA from RCA for a better deal, Wynonna was in a delicate situation, trying to retain the tradition she had established with her mother while also trying to establish herself as a fresh new voice. To help, she chose as her producer Tony Brown, who was not only the president of MCA but also the public face of cool on Music Row. The son of a hellfire Baptist missionary who grew up to play backup piano for Elvis Presley in the seventies, then went on to get profiled in *GQ* in the nineties as the discoverer of hipsters Lyle Lovett, Nanci Griffith, and Steve Earle, Tony Brown was an anomaly in Nashville. *Entertain-ment Weekly* regularly placed him on its Power 101 list and he even made an appearance on the *Newsweek* 100. Closer to home, he was also the first person since Minnie Pearl to successfully bridge the chasm between the Grand Ole Opry, where he once played with Emmylou Harris, to Belle Meade, where he moved in 1994. "Tony Brown's always the first person at the Swan Ball," Catherine Darnell, *The Tennessean*'s society columnist, told me. "He's also the one person on Music Row whom all the Belle Meade ladies wish they could get their claws into."

Tony Brown was the perfect producer for Wynonna. If Allen Reyn-olds's gentlemanly philosophical nature, mixed with his penchant for grandstanding, was ideally suited for Garth Brooks, Tony's flamboyance, coupled with his own larger-than-life personality, was uniquely suited for Wynonna, the still-unrealized "Female Elvis." Tony was a pert, dap-per man in his forties who drove his Mercedes 600 SL a little too fast and ordered his wardrobe by the carton from Barneys. Forbidden from seeing movies as a child, he had always been fascinated by the excesses of celebrity, which is why he coveted the job with the real Elvis. "To have watched him in those late years, sometimes looking good, most times looking really bad and being pitiful . . . I wouldn't have taken *any*thing for it," he told *GQ*. "I just wanted to be around Elvis." After the band had been dismissed, Tony would wait around the house for a glimpse. "And, sure enough, at three or four in the morning, he'd walk through in his underwear, his hair sticking every which way, and get a drink of water. He'd stop and say two or three words. After he went back to the bedroom, I'd get up and go back to the hotel." More than just basking in the nectar of fame, Tony went to school on Elvis, learning to bring rock grindings into country. As Don Was, the producer of Bonnie Raitt, said at the time: "The lines between rock 'n' roll and country are really starting to blur. Nobody understands that better than Tony. I don't think of him as a country producer—I think of him as Elvis's piano player. When you view him in that perspective, a lot of things make sense. Because Elvis is the guy who *really* knocked down those walls."

More than any other artist of her generation, Wynonna seemed to decimate those walls further in the seemingly open-ended years of the early nineties. Her debut solo album, *Wynonna,* released in early 1992, was a sweat-dripping, adrenaline-inspiring combination of unbearably painful pleas for strength and lustful declarations of independence—from jilting lovers, crippling memories, and, one can't help thinking, her mother. Years later, it still stands as the most exciting album to come out of Nashville this decade, combining a genre-busting blend of bluesy wails, country caresses, and gospel stand-up-and-shout hallelujahs, all done in the most quixotic, unpredictable voice that seems impossibly suited to all three of its major styles. From its background musicians (Will Weeks on bass, the rare African American among Nashville musicians, to Don Potter, the born-again acoustic guitar player) to the songwriters (among them, Dave Loggins, the sweet-tongued writer of "Please Come to Boston"), the album is a chamber of commerce brochure for the maturing dimensions of country music. Considering Tennessee's cross-state legacy of bluegrass, blues, and country, Wynonna is perhaps the best example of an artist who began in one tradition, Appalachian Mountain, expanded to incorporate the other pole, Mississippi Delta, and in the process redefined what could happen in the middle, a transcendent form of Mother Earth Southern music.

The biggest testament to the music's power is that critics, who weeks earlier had been lining up to lampoon the excess of the Judds' Farewell Tour, now scrambled to anoint Wynonna as the savior. *Rolling Stone,* in a rare lead review for country, called *Wynonna* "the most important release by a country artist this decade . . . powerful, stirring, ennobling." *The New York Times* called it a "faultless nineties country album . . . without an extraneous note or languid moment." Coming in early 1992, as the Garth-inspired media frenzy was just kicking in around Nashville, the album was received as heralding the arrival of the substantive, pop female superstar that the industry needed. "More than any record since Randy Travis and Dwight Yoakam steered country music in a new direction in 1986," James Hunter wrote in his *Times* review, *Wynonna* "demonstrates that a country performer can explore vibrant pop, deep gospel, and straightforward rock and still make sense even to country traditionalists. Taking neither the strict heritage of honky-tonk nor the relative freedom of singer-songwriter rock as her guide, Ms. Judd manages to say that, as the word about forty years of country music spreads, it can be something else as well."

MCA handled the album launch perfectly, forcing Wynonna to speak only about the music with no tearful interviews about her mother. "The strategy," Susan Levy explained, "was to establish her as an artist, not a tabloid figure." All promotional material coming out of MCA referred to her as "the lead singer of the Judds," not the daughter of Naomi. "Wynonna Judd was born in 1964 in Ashland, Kentucky. She moved to California at the age of four . . ." and so on. The celebrity press would come later. Wynonna made her solo debut in January during the heavily watched American Music Awards (which consistently draw more viewers than the Grammy Awards) with the funky contemporary ballad, "She Is His Only Need," about a man who loves his woman so much that he regularly goes "over the line," buying her gifts that he can't afford on his credit card. The following day the song was overnighted to radio and Wynonna was on her way. Advance orders totaled 600,000, twice as much as the previous Judds' album. Released that March, *Wynonna* went platinum in a week, temporarily derailing Garth's *Ropin' the Wind* from the top slot in the country album charts it had held for six months. By year's end, *Wynonna* had become the bestselling studio album by a female in country music history, supplanting *It's Your Call* by Reba McEntire.

As successful as *Wynonna* was as an artistic statement—independent, but still respectful of her mother (Naomi cowrote one of the cuts and sang background vocals on another)—it also raised a new crop of questions. First, strictly speaking, the album wasn't very country—there was no fiddle or pedal steel on any of the cuts. Second, by adopting the one-named approach to celebrity, Wynonna seemed to be likening herself to Cher or Madonna, both larger-than-life pop divas. By her second album, *Tell Me Why,* these trends were even more pronounced. The lyrics were more confident—"I used to be your do-anything, custom made love slave," she insisted, "but that was yesterday"—but the music moved even farther toward rock, blues, and soul. There were dozens of moments of Aretha Franklin in *Tell Me Why,* but only a few reminders of Dolly Parton. The album's most emblematic cut was "Girls with Guitars." Written by Mary Chapin Carpenter, already the spokeswoman for gutsy, suburban women, "Girls with Guitars" is an equal rights anthem for girls, who, despite the pressures to become debutantes and cheerleaders, fall in love with rock 'n' roll, playing Jimi Hendrix in the cellar with the amp turned up. "Get your money for nothin' and your guys for free," Wynonna sings at the end in a wry reply to the Dire Straits line and a playful reminder of Kitty Wells's answer songs. The difference was that all Kitty Wells's songs were written by men. Thirty years later, not only was "Girls with Guitars" written by a woman, an Ivy-educated, foreign-language-speaking woman no less, but it celebrated the fact that once fragile honky-tonk angels could now face the world on their own devilish terms.

Wynonna could certainly understand that. In the span of just under two years, she had completely remade her musical life, converting her vagabond upbringing and tortuous adolescence into the grist for a new type of American female singer, one who knew all parts of I-40—its origins in mountain hollows, its escape through the West, and its pursuit of fulfillment in California. In the process, Wynonna joined Mary Chapin Carpenter and Trisha Yearwood, two other contemporary country singers whose mix-and-match styles captured the spirit of the new mobile America. Together, they began to redefine the type of woman country music idolized, away from the exaggerated divas of the past—Barbara Mandrell, Dolly Parton, and Reba McEntire—and toward a younger, more contemporary woman. These women did not fight the women's liberation movement as their elders—Loretta Lynn, Tammy Wynette, even Naomi—did, but they did have to fight the internal ghosts that those struggles left behind. As they found, in that fight the worst enemy was usually yourself.

Inside Wynonna's party at Trilogy, the star was still absent. The several hundred guests glided from room to room, exchanging bites of gossip and nibbling on polenta cakes, marinated duck egg rolls, and pieces of tuna sushi. From its eclectic, nouvelle Southern menu to its heaping, silk flower arrangements, Trilogy was a vivid reminder of the Judds. The décor had all the trappings of one of Naomi's old dresses: pale pink walls, frilly paintings of Greek goddesses in flowing scarves, and everything—*everything*—perfectly in place. And Wynonna, as usual, was running late. "I was at a party for Bryan White just the other day," Hazel said, "and he stood at the door and greeted everybody as they came in. Can you imagine Wynonna doing something like that? *Puh-lease.* She'll be forty-five minutes late to her own funeral."

When she did arrive, the ceremony was short. Wynonna accepted the accolades of the crowd and the warm wishes from her hosts. The publishing company handed out certificates to the writers. The label handed out plaques to the promotion staff. And everyone posed for pictures. Throughout, Naomi scurried around the back of the room, fidgeting with chairs, picking up crumbs, and generally acting as if nothing was quite right. When she spotted Hazel, she pulled her friend aside. "Hazel," she said, "how old was Dolly when she lost all her weight?" Considering the surroundings, Hazel was shocked. "Naomi, when was the last time you looked in the mirror?" she said. "Don't you realize what's wrong with Wynonna? Having a beautiful mama is the hardest thing in the world. Wynonna is pretty, but she ain't no China doll. And that's hard on her. I grew up at the knee of a beautiful mama. I know. All her life people have told Wynonna that she's not as beautiful as you are. I was thirty-three years old when the first man told me I was beautiful. That was Bill Monroe. And I wept. I suggest you just back off." Naomi, stunned, went about her business. When the ceremony was complete and the television cameras hurried forward for a few words, Naomi was waiting to intercept them. That evening it was her face that appeared on the news.

From the day Wynonna went out on her own, the question of whether she could escape her mother had dominated her life. Even as Wynonna was creating some of the strongest, most independent music ever made by a woman in Nashville, in her mind she was still the unworthy girl her mother had tormented as a child. When Wynonna did get around to talking to the press late in 1992, the stories she told were heart-breaking. About her fragility (*The Washington Post*): "My strength was always as Naomi Judd's child. My identity was not Wynonna, child of God; it was Wynonna, child of Naomi Judd." About her insecurity (*People*): "I go between one minute feeling like I can conquer the world, and the next minute wanting to call my mom and have her come get me." About her guilt (*People*): "When I get to feeling sorry for myself or down from too much pressure, I stop and think how she must feel. When I'm up there singing and look over at her sitting at the monitor board looking up at me, I think how hard it must be for her to let go, not only as a mother but as a professional partner."

As she told James Hunter in *Us*: "Going through this year without her has been really hard. My mother has always been the focus of what the Judds did. While she would be the cheerleader onstage, I would hide behind the drums, acting goofy. Now I don't have Mom anymore to cut up my meat. It's scary when you go, 'Oh, my God, I have to use

my mind.' " As the magazine dryly noted, "This isn't the sort of realization that usually dawns on a successful twenty-eight-year-old woman."
Naturally, the rest of the world sensed her weakness. "I had this lady
walk up to me in a store," Wynonna said. "She said, 'How's your mom?'
She went through this whole thing: 'Tell her I said hello, that I'm so
worried for her.' Then she looked right at me—I kid you not—and said,
'Can you make it without her?' I thought, 'Oh, God, is this the beginning
of how it's going to be?' "

The answer, of course, was yes. Even in retirement, Naomi dominated Wynonna's life. As Ken Stilts, who managed Wynonna as a solo
act, said: "As strong as Wynonna was and as talented, she was equally
weak because she never had to think for herself. Naomi always did that.
Naomi practiced total mind control over Wynonna. When she was not
in her daughter's presence, she would joke about how she could control
Wynonna, how she was so much smarter than Wynonna." During one
of her first nights on the road by herself, Wynonna decided to try sleeping in her mother's old bedroom on the bus, which was larger than hers.
When she walked into the sleeper, she found that Naomi had left a few
of her belongings, namely the tiara and scepter she had worn as part of
her persona as "Queen of Everything" (the nickname had come from
Wynonna). "I didn't know what it meant," Wynonna said, "whether she
was turning over her crown to me or whether she was leaving it there
to remind me who was queen." Her decision: "I went on and slept there
for a while, but woke up and went back to my room."

At the end of her *Us* interview, there is an ominous story. Wynonna
is giving a triumphant concert, vamping with her background singers at
Fan Fair 1992, the first without her mother. In a time of almost over-
bearing insecurity, it is a moment of touching confidence and hopeful
independence. That night, though, Wynonna went home and listened to
her answering machine. There was a message from her mother. "She
said she was sitting at her farm by herself," Wynonna recalled, "and that
Larry called her from the show. She could hear me singing in the background. She burst into tears. She felt removed. She had this sense of
'I'm not on the bus. Oh, God, I'm here by myself.' " Back at home,
Wynonna said, it was quiet, "almost that horrible quiet in a funeral
home. It was strange and unnerving."

Wynonna's reaction to these feelings was typical. First, she ate.
"Food is such an emotional thing," she said. "When I'm feeling like I
really want somebody to hold me like Mom used to do, rubbing my feet
or patting me on the leg, I want to eat all the foods my mom gave me

when I was a child." Her weight, she said, was her "protection from the outside world." Then she sought love with various men—on the road, at home, in the studio. Finally she did something more reassuring, though ultimately more corrosive to her reputation. On May 11, 1994, Wynonna walked into her offices in Franklin and announced that she was expecting her first child. The father was divorced "Nashville businessman" Arch Kelley III, a local yacht salesman with a well-known dating past who looked like Robert Redford with a beak nose. He had met Wynonna six months previously in first class on a flight from L.A. At the press conference, Wynonna made her intentions clear: She would keep the baby and take time off the road. "Of course, I am unwed at the present time," she said with her dog, Loretta Lynn, by her side. "I've made my decision based on what I think is right for me." In fine Judd tradition of piling crisis on top of trauma, Wynonna cited as one of her reasons for leaving the road that she "never had time to heal" after her mother retired. "I want to do the things I've been craving to do as a normal person."

Though it was perfectly "normal" in 1994 for a celebrity to have a child out of wedlock and though it was certainly "normal" in the context of her family, the public reaction to Wynonna's announcement was anything but ordinary. "I felt like I was going to be crucified," she said. "Getting pregnant didn't quite fit in with the Judds' saga." If anything, the experience showed how entertainers had become moral barometers of American life. "I felt so responsible for so many people," Wynonna said. "The fans with sixteen-year-old daughters who called and wrote letters said how important I was to their daughters. For a lot of unwed mothers, I became sort of a hope seller. For the Baptists, though, I became a disappointment." This, of course, was always the unspoken burden of country stars—role models for the flyover country—but now that responsibility was coming back to haunt Wynonna. She had lost control of the public narrative of her life: The little girl who had grown up before the world, who had sung so powerfully even though she was shy, who had persevered even after her mother retired, now revealed herself to be an adult not quite in control of herself. The "Princess of Quite a Lot," as Naomi called her, seemed to be squandering her goodwill. In humiliation, she retreated to her farm.

A year and a half later, when she finally reemerged (predictably, twelve months later than first promised) with a new album, a son, a second pregnancy, and a royal wedding, Wynonna fully expected to lay claim to the role long dangled in front of her as her destiny: the "Queen of Everything." Only now, it turned out, she faced a different threat—

no less daunting than her mother, yet somehow more pernicious. It was the new symbol of Nashville, having just replaced the cowboy hat as the most beguiling—and, to many, reviling—icon of country music in the late 1990s. It was also the thing Wynonna most feared.

It was the belly button.

One month before Wynonna's party at Trilogy, another event took place at the restaurant that inadvertently served to mark Wynonna's overthrow as the poster woman of Nashville. Shania Twain, the twenty-nine-year-old Canadian bombshell whose public image was designed by John and Bo Derek and whose second album, *The Woman in Me,* was produced by her husband Mutt Lange, the South African-born mastermind behind Def Leppard and AC/DC, was being honored for sales of 5 million records. By reaching that plateau, *The Woman in Me,* a mix of bubble-bath self-caressing songs and funky I-am-woman-hear-me-roar numbers, officially topped *Wynonna* as the bestselling album by a female in country music. As a result, Shania Twain, a native of Timmons, Ontario, whose mother and stepfather had died in an automobile accident, leaving her to raise her two younger brothers, became the focal point of an entirely new movement: the advent of the country babe.

Unlike her more traditional elders, Shania exalted in her Victoria's Secret looks and "Baywatch" build, appearing on the back of her album in a cowboy hat and a pale blue halter top, thereby directing attention to her stomach and puckering belly button. In no time, that belly button became the talk of Nashville, spawning criticism, attention, and, of course, imitators. Soon women up and down Music Row were unbuttoning their blouses and flaunting their midriffs in lieu of their Manuels. One young woman, twenty-year-old Mindy McReady, even donned a belly button *ring.* "Women come up and want to talk about my music," she told me, "but men want to see my belly button. Can you believe it?" Yes, and so could her label, RCA, which celebrated her achieving platinum status with a party at Planet Hollywood at which they presented her with a platinum navel ring. A month later, the adornment was shown in a close-up on "The Tonight Show."

More than cosmetics, the ascendance of Shania marked a dramatic change in the culture of Music Row. Sex had arrived in mainstream country music, specifically *female* country music. For generations, sex had been an underground topic, even among men. As Nick Tosches wrote in his book *Country: The Twisted Roots of Rock 'n' Roll,* songs about eroticism, or "smutsongs," were actually more prevalent in preradio country music. Jimmie Rodgers once sang a song called "Pistol Packin' Papa" with the line "If you don't wanna smell my smoke / Don't monkey with my gun." Roy Acuff followed with the voyeuristic lyric "I wish I was a diamond ring upon my Lulu's hand / Every time she'd take her bath, I'd be a lucky man." Even clean-cut Gene Autry once sang of dirty deeds: "Now you can feel my legs / And you can feel my thighs / But if you feel my legs / You gotta ride me high." These sentiments all but disappeared in midcentury. As country became a radio phenomenon, sex was mentioned only in the Freudian, closeted sense. Not until Kris Kristofferson wrote his Grammy-winning 1971 hit "Help Me Make It Through the Night," which was taken to number one by a woman, Sammi Smith, did sexual liberation make it into the open. Still, as Hazel said, "Most people probably didn't even know it was about sex."

Instead, women in Nashville mostly clung to traditional roles, embodied by Tammy Wynette's 1968 classic "Stand by Your Man." As Billy Sherrill, who wrote the song with Tammy, explained: "After being barraged by women's lib and the E.R.A., I wanted it to be a song for all the women out there who didn't agree, a song for the truly liberated woman." Even strong-willed Loretta Lynn, who had such hits as "Don't Come Home a-Drinkin' (with Lovin' on Your Mind)" and "The Pill," about birth control, was devoutly traditional. In her autobiography, she tells a hilarious story about meeting the "Queen of Liberation" on "The David Frost Show." "I was back in the dressing room and this gal started cussing and arguing something terrible," she wrote. "I didn't know this woman from Adam. She was running on about women's rights. I said, 'Isn't it awful what you have to put up with in your own dressing room?' and she smarted off at me, and we were really going at it." Later, after Loretta had completed her interview with Frost, the same woman was summoned onstage. "And I said, 'Oh, my God, it's her,'" Loretta wrote. "It's that Betty Friedan." When Friedan started talking, though, Loretta dozed off. "If I'm not interested in what somebody is saying, I let my mind wander," she said. "I must have closed my eyes for a few seconds because all of a sudden I hear David Frost say to me, 'What do you think of that, Loretta?'" Loretta virtually jumped out of her chair, thereby giving the audience its biggest thrill of the night and inadvertently offering up Nashville's response to women's liberation.

Not until the 1990s did country artists seem to wake up to the open sexuality of the rest of the country. Just as Garth, building on Randy Travis and Dwight Yoakam, had come to embody the new sensitive man, so Shania, building on Wynonna and Mary Chapin Carpenter, had come to define the new power woman. As she sang in her defining anthem: "Any man of mine better be proud of me / Even when I'm ugly he still better love me." For her—and millions of her fans—the message now was "Stand by *Me*." But her success—and those of her bright-eyed followers—also raised a question: Why, after years in which Nashville's central rule was "Women Won't Sell," did women not only start selling, but suddenly start dominating the charts?

The answer, ultimately, has to do with the changing nature of country music. First, Nashville's younger audience now included more women—as much as 60 percent in some surveys. Second, because female artists had fewer role models in country, these artists were forced to draw from a broader range of influences than their male counterparts, which helped them to make music that was more modern and more daring. In a conversation I had with Deana Carter, the first of the post-Shania power cheerleader blondes, she didn't even mention a country artist among her influences, which included Steely Dan, Fleetwood Mac, Bruce Springsteen, and Rickie Lee Jones. "Sometimes I want to be so pop, it gives me a cavity. I know it's sappy, but I *love* Bread," she said, referring to the seventies smooch rock kings. Third, these women were helped by a new crop of female songwriters who delved into topics men had long shunned. Deana Carter's breakthrough hit, "Strawberry Wine," co-written by Matraca Berg, tells the story of a seventeen-year-old girl who loses her virginity to an older man. "He was working through college on my grandpa's farm / I was thirsting for knowledge and he had a car." Hazel called it the first song about teenage free love.

All of these changes in tone and style posed an indirect threat to Wynonna, whose spiritualism and "big woman" body suddenly seemed at odds with the times. But the bigger threat she faced was the personalities of these women: Each was fiercely determined to succeed. "I know what I want," Shania said. "I see no reason why I shouldn't go out and get it. If I use a little sex appeal and I have a little fun, what's wrong with that? The key is maintaining control and knowing when to draw the line." As Deana Carter, who looked uncommonly like Marcia Brady, echoed: "Sure, I may look like a dumb blonde, but people respect me in this town. Maybe it's just the will to succeed. Maybe it's because there's never been a question of failure in my mind. And maybe it's dignity. I've never slept with an executive in this town, and believe me, they have tried for years. But they look at me with respect because I told them to kiss my ass in a heartbeat."

This was an attitude—telling Music Row to shove it—that Wynonna never had before, but that she was just beginning to realize she needed.

Back at her party, Wynonna was fed up. Though the hundreds of people gathered in Trilogy were there for *her* and though the dozen or so television cameras huddled at the bar were waiting to hear what *she* had to say, she didn't feel like talking. "I didn't want to come tonight," she muttered. "I hate these kinds of events." Minutes later, after the entire room surged around her in a cartoon version of a press mob, she'd had enough. "Let's go," she said, ducking under a cameraman and tugging my arm.

A few steps from the bar, she opened a door onto a secret lounge that was tucked in the back of Trilogy. This was Naomi's Room, a dimly lit salon with mod purple and red velvet sofas, a lava lamp, and the general air of a futuristic Las Vegas getaway—Frank Sinatra meets Buckaroo Banzai. "This is where Luke Perry likes to come to relax," Wynonna said. With her manager and publicist scurrying behind her, Wynonna seemed in no mood for relaxing.

"Okay, I have some questions for you people," Wynonna said. "I want to know what's wrong." What followed was the Judd version of a tirade. Wynonna was upset that her newest album, *revelations,* was not selling at the pace of her previous two. The album, an emotional mix of spiritual ballads and heart-wrenching love songs, had started off strongly, but, with mixed reviews and a slow-climbing single, had dropped from the top of the charts. At first Wynonna blamed this slippage on her label. "They are satisfied with just two million in sales," she said. "I want to know why we can't sell four million records?" Then she blamed it on the media. "I know who I am," she said, "but nobody else does. Why am I not in *The New York Times*? Why am I not in *Rolling Stone*? What are the editors of these magazines saying? Tell me. I want to know."

No one, of course, said anything: The scene was too unreal. Paula Batson, Wynonna's publicist and a former MCA executive from Los Angeles, sat quietly in a chair across from Wynonna. John Unger, Wynonna's manager and a preppy, Princeton-educated lawyer, sat next to his client with a resigned expression of "I've seen it all before." "Wynonna is very much an *artiste*," he told me later. "Very intense. Very emotional. Very passionate about her music. But also very confused. She's been on a mythical journey. A journey of finding her own identity.

Of growing up. Of leaving and going out into the wilderness and trying to find out who she is."

"Maybe it's my producer," Wynonna continued. "I told Tony Brown, 'You don't need to be my producer *and* be the main guy at MCA.' Everyone's getting really greedy. When I was making this record, I wanted to bring Eric Clapton in to play guitar. Tony said we just didn't have time. Well, let's make time. Hell, I'm paying for it anyway." Then again, maybe it was all the new women in Nashville. "It's getting gross out there," she said. "It's such a T-and-A thing. They're making dresses now to where the accent is on showing the belly button *and* the nipple. What's next? Designing little hearts around the pubic area? It's depressing. It goes straight for the libido and bypasses the heart."

This scene too was depressing. It reminded me of the moment in Robert Altman's film *Nashville*, when Ronee Blakely's diva-esque character, said to be a composite of Loretta Lynn and Tammy Wynette, faints onstage and retreats to her bed, where she lies brooding in a fever-induced rage, lashing out at the evils of the industry. Wynonna too was capable of such zealous rages—what about my spirit, how about the music, where is the love? But then suddenly, from this tirade, a pause. Wynonna, in a rare moment, turned her gaze on herself. Might *she* be the one to blame? Might *she* have done something wrong? Her voice dipped to a quiver when she confronted this possibility.

"I'm *trying* to organize my life," she said, her voice almost pleading. "I'm studying Stephen Covey. I'm working with Deepak. I'm looking for a personal trainer I can turn my life over to. I'm trying to decide what part of my life is for family, what part for fun, what part for interviews. Is that what the problem is? Is it me? Is it because I'm fat? Do I have to show my belly button and make sexy videos?"

The moment felt like an epiphany, as if Wynonna was waking from a deep sleep.

"Nashville used to be such a refuge," she said. "People felt this longing to go somewhere where they could be appreciated. Now, if we're not careful, we'll turn into another New York and L.A., where everyone's just a commodity." Her voice was growing in strength. She sat up for the first time. "I want women to be independent as much as anybody," she said. "I want women to be appreciated. But there's no reason that just because I don't show my belly button, I can't be strong or in control. Someone called my hotline recently and asked if now that I had a husband and a child was I going to become boring and fade away. I wanted to pick up the phone and call that wench. No *way* am I going to become boring. I'll ride my Harley home from the hospital with my baby over my shoulder if I have to. I'll organize a convention of bikers and ride naked through the desert. I'm not going to become boring, and I'm *not* going to go away."

ELEVEN

THE SINGLE

On the third Monday in April, the most important woman in Wade Hayes's life—or, at least, his career—was staring into a blue-and-white computer screen in a spacious first-floor office overlooking a rain-puddled 16th Avenue South. She was also, by virtue of what came over that screen, one of the most powerful women on Music Row. And at the moment she was pissed.

"That bastard!" Debi Fleischer said. "He's not supposed to do that to me . . ." She paused for a second to consider her options, then blurted, "I must get Jack on the phone."

She pivoted from her screen for the first time in an hour and punched a speed-dial entrant on her telephone. Within seconds, a faraway ringing could be heard from the speaker. When it became clear that Jack was not going to answer, Debi Fleischer, Jack's boss, the vice president for promotions at Columbia Records, and a woman hailed by several dozen gold records on her wall (not to mention a few stuffed animals in the corner) as the "Queen," began preparing a pointed message for his machine.

"So, Jack," Debi began at the tone. She was a round, usually jolly woman in her don't-ask forties with straight black bob hair and an chronically infectious grin. She had a laugh that could lighten the labyrinthine offices of Sony Music (Sony had two divisions in Nashville: Columbia and Epic), but also, when needed, a sarcastic bite that could weaken even the most pompous male bombast in what was still one of the most testosterone-infused corners of the entertainment business, country music. If she was a queen, it was the Queen of Hearts—smiling, smiling, then "Off with your head."

"I'm sitting here looking at BBS in Syracuse," she continued, "and I don't see *Wade* on this list." On her screen, the weekly playlist from WBBS, a radio station in upstate New York, had just appeared. It was one of 183 such lists announcing spins for the upcoming week. Each list would appear over the next several hours on the live, on-line tracking system kept by *Radio & Records.* At the end of the day, the spins from those lists would be tabulated, weighted according to the size of the station, and ranked in order of popularity, thus determining the top seventy-five country singles of the week.

"Their adds are thirteen spins: Shania Twain," Debi said. "Okay, I can see that. Ten spins: Marty Stuart and Travis Tritt. Well, maybe. But five spins: Keith Gattis. Can you explain that last one to me? I thought BBS didn't add records that weren't by stars. And I thought they didn't add those records until they made Top 10. Let me know what *you* know about this situation, and what you don't know . . . find out."

Debi Fleischer was not having a good day. Still two months—or, in radio terms, eight weeks—away from the release of Wade's new album, *On a Good Night,* Debi had chosen this day to release the album's first single, the rockin' Bubba anthem "On a Good Night," to the 2,642 radio stations that program country music full-time. Record companies release singles several months early to radio in an effort to build up momentum before the albums go on sale. To add even more umph to the release of Wade's new single, Debi, a native of Burbank, California, who had worked as a road manager for Crystal Gayle before turning to record promotion in the early 1980s, had decided to transfer the song to the two hundred or so stations that really count (the ones that report their playlists to the chartmakers at *Billboard* and *R&R*) by digital satellite transmission, or DGS, precisely six days before.

"It's more expensive," she explained a week before the experiment, "but right now it's not been done a lot, so it's another way of saying we think so much of this artist that we're doing this special thing. We're giving you his new single in this high-quality instant form so that as soon as it's unscrambled you can rush it on the air." The idea was to create so much buzz around the song that it would smother the other half dozen records also going for adds that week. "We want to be the most-added record," she explained. She also hoped to achieve "breaker" status on the *R&R* chart, meaning 60 percent of the stations had added the song. And what, for her, would be the best-case scenario? At least 120 adds, she said, and a debut in the thirties. And the worst-case scenario? Fewer than seventy-five adds and not charting at all. "That would be a disaster," she said.

Disaster struck early in this case. The satellite transmission system failed miserably. Most stations, it turned out, didn't know they had such a system, and those that did know they had it couldn't figure out how to make it work. Debi, in California at the time, immediately ordered her staff to overnight the single to all two hundred reporting stations via UPS. UPS, though, lost the shipment. By the time a backup shipment arrived the next day, Friday ("Oh, and my consolation is that I'm supposed to be getting the whole thing free," she said), Wade had already missed most of the planning meetings that would have allowed him to be added to most stations on Monday. "On a Good Night" was having a bad week.

Working overtime, Debi and her staff of regional promoters had managed to stabilize the situation by Monday morning, cajoling promises from at least fifty stations that they would add Wade to their lists. By midafternoon, the question had become whether they could regain enough momentum to overtake Alabama's new single, "Say I," when the window for submissions closed at 5 P.M. CST. The mood in her office was tense, with a constant flow of executives, staffers, and even managers (Mike Robertson had come by earlier; Wade was on the road in Michigan). Throughout, the screen remained the center of attention of this minidrama, which was being carried out in real time in offices just like this one up and down Music Row, as well as at radio stations from Los Angeles to Long Island.

As station after station submitted its playlist, Debi would click on it and scan it with animallike ferocity. "Ooooh, Greenville just added Wade at thirteen spins!" she trumpeted. "Plus Ray Hood. Who's Ray Hood? That must be a coerced add." Every now and then, something unusual would pop up and she would get titillated. "Look, KSCS just dropped Garth," she cooed at one point. "I'll give them five minutes to fix *that*." And sure enough, five minutes later, KSCS resubmitted and Garth's "The Change" was back on for another week. "Looks like somebody wanted some tickets or maybe a jacket," she said. "But I tell you, it won't go Top 15. The song's just too pompous." The whole enterprise reminded me of a giant game of on-line chicken. I could think of no segment of American business, with the exception of Wall Street, where so much was riding on a set of actions being played out simultaneously on dozens of computer terminals around the country. And unlike trading stocks or bonds, the central players in this enterprise changed their actions right up until the last minute.

By four o'clock, "On a Good Night" had regained something of a healthy momentum: WDSY, Pittsburgh, twenty spins; KFMS, Las Vegas, seventeen spins; WWZD, Macon, five spins. For someone unaccustomed to the geography of country music, the array was mind-boggling: Wade was added in upstate New York (WFRG, Utica) and Southern California (KIKF, Los Angeles), as well as Anchorage (KASH), Akron (WQMX), and Amarillo (KGNC). He was playing in Peoria, as well as in Elvis's hometown, Tupelo, Mississippi. He was also playing, most crucially of all, on WUSN in Chicago, the largest country station in America with just under 1 million listeners, and on WSIX in Nashville, the most important station in the country, since that's where Music Row gets its information.

At the top of the chart, this day would bring little excitement. Despite an early push, John Michael Montgomery was clearly ahead of Faith Hill, who herself was facing a late-day challenge from Brooks & Dunn, whose "My Maria" was the hottest song all year and would eventually stay perched at number one for three weeks. At the bottom, a battle was developing among Wade, Alabama, and Shania Twain, whose record wasn't officially going for adds until the following week, but because of the energy surrounding her career was being added *early* by dozens of stations, particularly big-city stations that carry more weight in the complex point system that eventually determines chart position.

Debi was sweating for every station she could get, and ones she didn't get, she took personally. When WHOK in Columbus, Georgia, added Alabama and newcomer Keith Gattis (both on RCA) over Wade and another of her acts, the band Ricochet, she was visibly upset. "The fact that they added those records means I consider us screwed. They'd add Keith Gattis over Ricochet and Wade Hayes? Excuse me? If they called me and asked for something right now I'd say, 'Fuck you. When you add my record, I'll give you something.'" Minutes later, she got that opportunity. Jack called back from the Northeast region and reported that RCA had promised WBBS in Syracuse a free show from Keith Gattis the next time he's in the area. Also, Marty Stuart and Travis Tritt were going to be playing in the region in the coming weeks. "And who's going to be their opening act?" Debi asked archly. Jack didn't know. "Ricochet!" she boomed. "Now get back on the phone and tell them if they want any tickets, they better reconsider their attitude." Smiling, smiling, "Off with their heads."

At just before five, there was a rush of final stations. Several members of Debi's staff had gathered in her office, along with the president, Allen Butler, and VP, Paul Worley, who had earlier approved Wade's material in the studio. It took Debi several minutes to slog through the last few lists: Boston, Detroit, San Antonio, Tampa. As she was doing so, the Southwest rep, working out of the Nashville office, came sprinting into the office. "I just got Oklahoma City!" Debi looked at her watch: Just under the deadline. At precisely 5:06, she pulled up WTDI in Charlotte and let out an audible squeal. The mass of anxious executives surged forward as if she was about to go into labor. "We're going to be the most-added record," she began to chant. "We're going to be the most-added record!" The half dozen or so people burst into applause and reached for the bottle of Jack Daniel's stowed safely under the table. A bag of chips magically appeared, along with a quart of Debi's favorite salsa from the San Antonio Taco Co.

At just after six, a small *ding* emanated from the terminal and the chart numbers materialized on the screen. John Michael finished at number one, Brooks & Dunn at number two. The Mavericks cracked the Top 10 for the first time with "All You Ever Do Is Bring Me Down," and Wade Hayes, with a hard-earned seventy new adds—eleven more than Alabama and thirteen more than Shania—debuted in his first week of release at number forty-five, with a bullet notation indicating upward momentum. Debi spread her hands out and bowed her head. "It's good to be queen," she declared.

Since its inception, the popularity of country music and any claim it may have of being "America's Music" has been based almost entirely on radio. Though Nashville in the nineties controlled a sixth of the album sales market, it held, for much of the decade, almost a *third* of the radio market. The 2,642 stations that programmed country music at middecade constituted 1,600 more than the next closest format, news/talk. According to the *Simmons Study of Media and Markets,* a whopping 70 million Americans listened to country radio in 1995. Though these numbers would trickle down in subsequent years, the fact remained: Country dominated radio airplay every year in this decade.

But while this muscle has bolstered Nashville's clout, it has also led to an unfortunate situation on Music Row: Radio now holds most of the power. Because country labels have found few ways to inspire consumers other than radio and because country consumers have been content to buy almost entirely what they hear on the radio, what radio decides to play almost single-handedly decides what will succeed. As a matter of business, this transfer of power is probably unrivaled in American music (rock, for example, and rap are much less dependent on radio). As a matter of art, this abdication of power has been nothing less than disastrous.

Radio has always been central to the success of country music. Before World War II, records were still rare. "Radio was a livelihood," historian Charles Wolfe noted in an essay in *Country: The Music and the Musicians,* "record making was an exotic novelty." Bill Monroe and his brother were so busy playing live at radio stations in the early years of their career that they actually threw away requests from record companies to go into a recording studio. "We finally went up to their studio in Charlotte," he said in an interview, "but we told 'em we didn't have much time, that we had to get back in time to play a school that night." There were several reasons for artists' reluctance to make records: Record royalty rates were almost nonexistent, recording quality wasn't very high, and records themselves weren't played that much on the radio anyway. Indeed, many record companies actually fought to keep their records *off* the air, fearing nobody would buy them if they could hear them free. This began to change in the late 1940s when disc jockeys (the term was coined earlier that decade) began playing more country records. Len Ellis, a disc jockey in Chicago, one of the founding members of the CMA, told me he had to beg his station to give him an hour to play country records, which he then went out and bought himself. Eventually record labels, realizing this potential bonanza, started issuing so-called DJ copies of their records and mailing them to stations at no charge. A new era was dawning.

With the spread of records came the birth of the charts and the beginning of efforts to control those charts. In 1944, *Billboard* magazine began compiling its first weekly chart of the most popular country records, which it determined by polling honky-tonk owners and asking which songs on their jukeboxes were played most frequently. In 1948, it merged the country list into one called Folk. The following year *Billboard* started a separate chart to track radio airplay, which it termed Country & Western. Though the name changed in 1962 to Hot Country Singles, the methodology stayed the same: Chart position was determined by the combination of radio airplay *and* record sales. In 1987, sales were dropped from this equation and a new system introduced that electronically monitors all reporting stations to determine what songs they *actually* play, instead of ones they promise. (The album chart, by contrast, is still based on sales.)

Billboard's decision to monitor actual spins had a curious effect on Nashville. On the one hand, the chart was now undeniably accurate; on the other hand, it could not be so easily manipulated. As a result, labels began putting more importance on *Radio & Records* magazine, which had been founded in 1973 as a rival to *Billboard* and which still had a chart that relied on stations *reporting* what they planned to play. I once asked Debi Fleischer why she bothered with *R&R* and its expensive reporting system when *Billboard* was so definitive. She grinned. "It gives us a chance to flex our muscle."

That flex comes at considerable cost. In the early days of country radio, the relationship between the labels and the stations was much less formal. "We'd call up the disc jockeys and ask 'em to play the record," Loretta Lynn wrote in her autobiography, "and most of 'em did. Those boys have always been on our side." By contrast, before Wade Hayes even released his first album, Columbia Records spent close to $250,000 flying him, a manager, and an executive to dozens of cities around the country to kiss the rings of the programmers (and hug the receptionists) who would determine his fate. Even now, after Wade had four Top 10s, the effort spent courting radio was amazing. For the previous month, Wade woke up at dawn wherever he was on the road and telephoned radio stations, urging them to play his new record. "At first you think that once you leave the studio the record's out of your hands," he told me several days before the single went for adds. "Now you know that that is only the beginning. I spend more of my time talking to radio than I do writing songs."

Not only time, but money. Wade and his team spent alarming sums hiring *independent* record promoters, or "indies," on top of the dozen people on Debi's staff who were already working the record full-time. For "On a Good Night," this included eight indies, each at a cost of close to a thousand dollars, and each paid for by a different entity: Wade's label, his management company, his publishing house, even Don Cook, out of a special budget he had for such expenditures. As of early May, the second week of the first single off Wade Hayes's second album, there were probably two dozen people around the country whose primary purpose was cajoling the 183 stations that reported to *Radio & Records* to add "On a Good Night" to their playlists.

At first it worked.

After the original DGS/UPS debacle, "On a Good Night" built up a substantial momentum and looked, for the time being, like it would be a hit—solidly making it into the Top 10, and possibly even getting a shot at number one. The song got fifty-one adds the second week among *R&R* stations and moved from forty-five to forty bullet ("bullet" is the term the charts use to designate that a record has more spins one week than the week before). Now that he was on a total of 121 stations (or above 60 percent), Wade also achieved "breaker" status the second week. His record was being spun a total of 1,457 times among *R&R*'s 183 stations. "It was a very good week," Debi announced at the end of the day into her automated telephone that would relay her weekly message to her field representatives. "Wade had more adds than LeRoy Parnell. Ha, ha, ha! But there are still tons of stations he doesn't have. A ton of stations. I want forty adds next week."

She got thirty of those adds (plus six more for her Ricochet song, "Daddy's Money") and all in all was feeling bullish. "Not too bad," she told her staff. "Ricochet we will move from thirty-two to twenty-seven bullet. We still have eleven stations not playing our record. Each of you needs to do everything you can to change that by next Monday. Wade Hayes, with thirty adds and forty-four upward moves, will take a nice move from forty to thirty-three bullet, jumping Linda Davis, Trace Adkins, and Kenny Chesney. While we did have a good week there, I know there are way too many stations that should be on this record that aren't. I expect you to be getting tough. I expect you to be getting mean. I expect you to be getting adds. We're looking good. Don't let this momentum in any way slow down. We have two hit records. These are both on their way to being number ones. Only people who are idiots can stop us. Remember, 'Columbia Proud!' "

By week four, though, she began detecting trouble. "There are clearly some stations that are *not* getting Wade Hayes," she told me. "Meaning they're never going to play it?" I asked. With Debi's permission, I had gotten into the habit of dropping by her office late on Monday afternoons to experience the last hour or so of this weekly race. Sometimes it was cheery in her office, sometimes gloomy. Either way, by seven there were usually a dozen or so people crowded around her conference table drinking beer and nibbling on chips with her beloved salsa. "It doesn't mean they're never going to play it," she said. "With a rising star like Wade, they'll have to." "What does that mean?" I asked. "If they don't like it, why play it?" "Because I'll kill them," she said. "If they have a competitor in the marketplace, I'll start giving them promotions. If they want my artists to drop by their stations, I won't let them. They may not think they need me, but they'll learn." And learn they did. Wade had fifteen new adds that week, leaving seventeen stations still not on the record. He moved from thirty-three to twenty-nine bullet. The number of weekly spins now totaled 2,674.

In week five, she began to run out of steam. "It's a brutal day," she said when I arrived. "A wasteland. I can't wait to get out of here. All things considered, we didn't get killed. All our records at least moved up. But there are a lot of labels I'd rather not be right now." And why the sudden blizzard? "It's the competition," she said. "The game. All these labels competing—there are as many labels as there are slots, for God's sake—and all these favors needing to be returned. Reba got dropped. Vince got dropped. Wynonna got dropped. And all were fixed a few minutes later. That's the power of MCA. There are stations that decided they weren't going to play those records—that their audience didn't like them—but still they kept them. No one wants to be the station that kills the record that ruins the star's career." At the end of the day, Ricochet moved from twenty-one to eighteen bullet, and Wade hobbled from twenty-nine to twenty-seven bullet.

Finally, in week six, she hit a wall. It was early summer by now, just weeks away from the album's release. The buildup around the office was increasing, and nobody wanted to launch Wade's album on the back of a floundering single. Debi, with Don's backing, decided to take action. She took out radio advertisements in several cities, rewarding stations that were on the record and penalizing ones that weren't. She told her regionals to let the stations know that they weren't getting ad money because they weren't playing Wade Hayes. One station, KBQQ in Houston, announced they would add the record the following week. But that seemed as if it would be too late. For much of the day, Debi feared she would lose her bullet, an act paramount to taking out a billboard on Music Row announcing, WE LOST THE RECORD! In the end, though, she was saved—barely. "Okay, gang, a miserable week," Debi declared at day's end. "But not as bad as it might have been. Ricochet moves from eighteen to seventeen bullet, and Wade will go from twenty-seven to twenty-six bullet. We picked up two of our holdouts, but we have

eight to go. No small task, but we're no small people. Go out and get
me what we need."

That would prove to be even harder in the weeks to come because
now, six weeks into its life, "On a Good Night" would be subjected to
a new, more potent external force, one that Hank Williams and Jimmie
Rodgers could never have anticipated, but one that was rapidly changing
the course of country music.

Beginning in week five of Wade's single, Debi began getting underground faxes from friends at radio stations around the country indicating
that "On a Good Night" was not popular with their listeners. What they
sent were reports with names like "Power Fax" and "Radio Heat" that
listed every song in circulation at the moment with statistically averaged,
demographically balanced information that claimed to reveal exactly how
many times fans wanted to hear these songs on the radio in any given
week. It was, as I would discover, the most important trend in country
music—market research—and perhaps the clearest indication that Nashville had joined the headlong rush in America to relinquish control of
pop culture to social science.

Basically the system works like this. Several dozen companies around
the country are hired by radio stations to tell them what to play. "What
we bring to the party is the radio audience," according to Joe Heslet,
the president of Marketing/Research Associates in Seattle. "Our clients
are radio stations who want to make sure that they're playing songs their
audience thinks are hits. Before that, it was pretty much a dartboard.
The programming director picked the songs and if he liked the record
it made it on the air." "And how is what you do any different?" I asked.
"We let the audience throw the darts." Working out of a phone center
in Fresno, California, the company polls one hundred country music
fans every week in various markets and asks them to rate anywhere from
thirty to forty songs. In order to get one hundred people (the calls are
made between 5 and 9 P.M.), they must usually call around five hundred
people, whom they screen by playing a tape that includes Alan Jackson,
Reba McEntire, and Garth Brooks. If the person says that's their favorite
kind of music or that they listen to it often ("It's much easier to find
country fans than, say, alternative rock fans," Joe said), they are desig-
nated as "P1's" and move to the next step.

The caller then plays the listener six seconds of each song, usually
the title and a hint of the chorus—"I've got friends in low places . . ."
or "On a good night, I can hop in my truck . . ."—and asks the listener

to rate that song on a scale of one to five: five meaning they like it a lot; one meaning they dislike it a lot. (The six seconds is meant merely to "remind" listeners of a song they already know.) These answers are then averaged, converted into a number from one to one hundred, and ranked according to popularity. The company then faxes this ranking to all its clients. I can think of no better example of the change in country music from a regional to a national form than the fact that a company in Seattle, making calls from Fresno to listeners in, say, Shreveport, decides whether fans in New Haven will hear the new song by an artist from Bethel Acres. Ultimately, going country to such a degree may have cost country music its sense of place.

Why such reliance on pseudo-science? First, competition. With the number of labels in Nashville ballooning from six in 1990 to thirty-one in 1996 and with the number of radio slots holding steady at between thirty and forty, more songs were competing for each slot. Second, with more stations being sucked up by corporate mergers, the importance of getting immediate return on investment has only increased. "Unfortunately, art and business conflict," Joe Heslet explained. "When you've invested $235 million in a radio station and you have investors, they don't necessarily care about the artistry of the music. The goal of keeping a radio listener tuned to your station another five minutes is entirely different than a label's need to get that song played one more spin a week." In this battle for the soul of country music, the research won. Label executives not only release songs they expect will get past the research (instead of the ones they like artistically), they even have started researching songs *before* they put them on the albums. After Reba McEntire released an album of cover songs (recut versions of old hits) that bombed, she sent all her material to Marketing/Research Associates to have them decide what songs she should cut. "We've picked a lot of number one songs this way," Joe explained. And wasn't he at least a little bit concerned that this might undermine any integrity remaining in the genre? Couldn't some songs, like those by a superstar, Garth Brooks, or a rising star, like LeAnn Rimes, be good for a station to play even if they didn't test perfectly? His answer: "Nothing ever helps the record industry, the artist, or the station by playing a bad song."

In reality, though, the research has only *increased* the number of bad songs: not because the research is done, but because the industry has so completely capitulated to the research that it now creates songs it knows will pass its need for blandness. As one poster in a Music Row publisher advised high-minded songwriters: BUY ART. DON'T WRITE IT.

"Pop music manipulates pop radio," Tony Brown complained to me. "Country radio manipulates country music." Sadly, one of the best examples I know of this dumbing down of country music involved "On a Good Night." For weeks leading up to the release of the song, what most on Wade's team were saying privately, but what none had the conviction to say publicly, was that "On a Good Night" was not a very good song. It was catchy, with a "screaming up-tempo" beat to it, as Wade described it to me. But lyrically, it's astoundingly banal: "On a good night," it says on a list of serial male fantasies, "I could put on my hat / Head down to the honky-tonk and dance"; but on a real good night, it adds, "I meet a woman like you."

On an album full of mature, thoughtful songs (including "Hurts, Don't It" and "The Room," which had all the feel of a Song of the Year contender), "On a Good Night" was clearly the most insubstantial cut. Even sadder, everyone knew it. "Well, I know what you mean," Don said when he first played it for me in his office and I failed to jump out of my chair (Don was one of the song's writers, along with Paul Nelson and Larry Boone). "But its whole purpose in life is to set up 'The Room' on radio." "Frankly, it's not a number one record," Debi confided to me at the beginning of its run. Even Wade, when I asked him what he thought of the album, said he didn't much care for "On a Good Night." "I do think it's going to be a hit on radio, though," he said. "It's a real summertime song. Generally there will be a bunch of those that come out, but we're coming out with one first."

So why did all these people, people with good artistic sensibilities and with a profound belief that Wade, more than most of his contemporaries, could establish himself as a serious artist, agree not only to release "On a Good Night," but also to make it the first single, the name of the album, and the linchpin of its entire marketing strategy? Because they were scared. Because they knew, from research, that radio likes up-tempo songs more than ballads. Because they knew, from listener surveys, that Wade's last song had been slow and that they needed to establish him as a "fun" artist to inspire people to purchase the album. Because they knew, from reading the charts, that everyone else got away with releasing fluff, so why shouldn't they? "I read the research," Wade told me. "I know that people are just now beginning to associate my name with my face. So what we do now, at the start of this album, will have a long-term impact on my career." On this, sure enough, he was right.

The early research on "On a Good Night" confirmed what everyone had known all along: It wasn't a very good song. The first week it was tested by Marketing/Research Associates (week five in the life of the song) it scored a 51.4, putting it at the bottom of the thirty songs tested that week and barely making the fifty-point cutoff needed to be included at all. The next week it dropped to 41.9 and fell off the list entirely. This was the week that "On a Good Night" almost lost its bullet and ended its run on the chart in the mid-twenties. But then the odd reality of country radio kicked in: If people hear something peppy and hard-driving long enough, they will eventually grow attached to it. This is why research encourages up-tempo songs instead of ballads. More people like ballads, but more people *dis*like them as well. Invariably, songs that test very high also test very low: This is the curse of good art and the biggest flaw of relying on research. It promotes songs that are homogenous, not songs that are inspiring. As Joe Heslet told me: "Familiarity is the key to success in radio. I've been in this business for twenty years and the one thing we know for sure is that if people don't like a song, they're going to punch the button. And if your competitor is not playing songs they don't like, the listener will stay there until they do. So the theory behind radio is if one in five people is likely to bail on a song, I have to ask myself what the benefits are to playing it." In other words, the goal of a radio programmer is not to play songs that people *love*, but to play songs that people don't hate enough to switch the station.

"On a Good Night" certainly passed that test. It wasn't a bad song. It was fun, and its hard-driving fiddle and thumping guitar did put a little bounce on the gas pedal. Eventually the research did detect that: Though few people liked it enough to give it a five, few people *dis*liked it enough to give it a one on the research scale. In fact, the number one song in the country at the moment, "My Maria," like many megahits, had a high number of both fives and ones. Accordingly it would "burn out" faster on the radio, meaning it would scorch to the top of the charts, stay for a few weeks, then get dropped precipitously, presumably making way for more mediocre songs.

"On a Good Night" built more slowly. After losing its bullet in week six, it scored a 53.4 on the research scale in week seven and regained its momentum, moving from twenty-four to seventeen bullet ("We picked up two of our holdouts," Debi said. "We've got to get the rest . . ."). The following week it moved from seventeen to twelve bullet ("KIKK in Houston said they'll be with us next week. We're Columbia, damn it. We're hot. We're proud . . ."). By week nine, just at the time when the album would go on sale, the single was just where Debi wanted it wanted to be: poised to go into the Top 10. Debi had worked her sleight of hand. Ricochet, following Shania, would have its chance to go number one in early July. Wade Hayes, it now seemed likely, might also have a chance, several weeks after that. "We've done our job," Debi said, assessing her sudden turn of good fortune. "We turned the song into a hit. The only thing is," she added, "we can't make people buy records."

TWELVE

THE POLITICS

The stage of Limerick's Theatre in downtown Dublin, Ireland, was black. A single white light reached down through the dark, where a man from Miami sang a song, cut in Nashville, about a woman from Havana who escapes to America. "I'd like to dedicate this song to my aunt," the man said, "and to everyone who's died trying to make that trip."

And then from this somber scene came a voice, the most haunting voice to emerge from Music Row since Roy Orbison thirty years ago. There was a bit of flamenco hot pink in it, a hint of South Beach blue, but at this moment it was a single color: an angry, bleeding red. "This ninety-mile trip / Has taken thirty years to make / They tried to keep forever / What was never theirs to take / I cursed and scratched the devil's hand / As he stood in front of me / One last drag from his big cigar / And he finally set me free."

On a chilly night at the height of the "Gone Country" era, the most unusual—and arguably the most daring—artist to come out of Nashville in the 1990s was standing on a stage in downtown Dublin, wearing a bright purple Manuel jacket, and pouring sweat from his rock-idol bangs. Raul Malo was an unlikely front man for a country band. The son of Cuban-American émigrés, Raul and his band, the Mavericks—four guys from Miami with funky facial hair, no cowboy hats, and an affinity for Vegas underworld clothing (picture Quentin Tarantino doing a makeover on the "Rat Pack")—jolted Nashville out of its protective cocoon by making Fidel Castro, his counterpart John F. Kennedy, and the entire suburban lounge culture of that era the centerpiece of a new brand of retro-revival. They also cussed, smoked dope, dissed their colleagues, threw wild parties, and, in the process, completely redefined what it meant to be country stars (actually, they called themselves "rock stars") at century's end. By doing so, they also made a point: Cowboys aren't the only heroes in Nashville; rebels have a cause, too.

" 'From Hell to Paradise' is about the Cuban struggle in particular," Raul, the stout, often brooding leader of the group, told me later that night. It was the first day of their maiden swing through Europe, and we were talking about whether country music had become "America's Music," as it liked to call itself. "But when we started touring," Raul continued, "I met a lot of folks who came over as Polish immigrants, or Jewish immigrants, or Italian immigrants, who came over at times of oppression in their countries, and that song touched a nerve for them. Here was this little song about my family that turned out to be about America."

Being about a new kind of America is what made the Mavericks notable. Their music—a smarter, hipper version of the Nashville Sound (country without the twang, but with an edge)—seemed to test the boundaries of what constituted country in the nineties. On the one hand, the Mavericks were a perfect embodiment of all that was different about the contemporary South: They were young, outlandish, and spoke a foreign language. (The final verse to "From Hell to Paradise," in which the woman promises to return home, is sung in Spanish.) Politics pervaded their aura. "I remember having a pretty good understanding of how the democratic system works," Raul said, "because I remember my parents discussing how different it was in Cuba." On the other hand, their highly contentious style (one cut on *From Hell to Paradise* took on religious fundamentalists like Jim and Tammy Faye Bakker) was a direct challenge to Nashville's conservatism. That they chose to make their home in Nashville, instead of New York or L.A., seemed to foreshadow a broadening of country's base. "We wanted to play music that we loved and only loved and nothing else," Raul said of the band's original manifesto. "We wanted to play *country* music because it was wide enough and growing enough that it could take us in. Nashville still had Lyle Lovett, Steve Earle, and Nanci Griffith. We figured if they could do it, we could, too."

At first, they couldn't. For all their initial efforts at bringing the voice

of Little Havana into Middle America, the Mavericks' sloganeering and
dark message songs turned off radio programmers and failed to interest
fans. "There were times I just thought, 'I can't believe this isn't going
to happen,'" said Robert Reynolds, the bass player, who is married to
Trisha Yearwood. So they adapted. With their second album, *What a
Crying Shame,* produced by Don Cook, the Mavericks took their fascination with the past and broadened it to an entire look and feel that mixed
the grit of Hank Williams and Roy Orbison with the panache of Frank
Sinatra and Dean Martin. By doing so, they finally hit the essence of
country music: its longing, its sentiment, its romance. In the process,
they also uncovered a previously hidden craving in the new suburban
fields of country music: While some consumers might escape their five
hundred channels and TV dinners with cowboy hats and horseback riding, others clearly prefer an edgier version of the past—swanky pool
halls, red velvet blouses, his and hers cigars. The Mavericks exploited
this craving with a style that was part white trash (So what if Elvis had
dirt under his fingernails?) and part bourgeois (It's cocktail hour!): You
can ride Harleys and play blackjack; you can eat pesto and still embrace
Nashville. In the genre-bending spirit of the 1990s, they had created a
new form of music: martini country.

"I remember when we were trying to decide the album art for *Music
for All Occasions,*" Paul Deakin, the drummer, told me of the band's
latest album as we walked through a drizzle to a private nightclub where
the Mavericks, in the selfless name of sowing Nashville's seed, would
field the adoring pawings of their fans. "We were thinking of the time
of our parents." The record, a sort of Valentine to the music-to-make-
love-to era, had campy art and dancing typography on the cover and
inside a photograph of the band pushing lawnmowers across a "Leave
It to Beaver" front lawn. "When they were our age, they were listening
to Ray Conniff, Peggy Lee, and stuff like that. They were all alcoholics
back then. Real liquored up."

"Of course, we're all products of those fucks," added Robert.

And they were. Instead of poor children growing up in the fields of
the South, the Mavericks were four children of the middle class (even
Raul's father had become a banker in Miami) who were singing about
their own sense of nostalgia, even though their longing for the past was
vastly different from that of, say, Garth Brooks. "It's a highly unique
thing that's going on with the Mavericks," explained guitarist Nick Kane,
the son of an opera singer. "As long as I've been in the business, there's
no rule that says that just because you're talented, and just because you

work hard, and just because you give a shit, you're going to be successful. There's a part of society that was ready for this kind of country music."

Which is what's most surprising of all.

Of all the misapprehensions at loose in the world about country music, perhaps the most persistent is that it's the music of racist, redneck Republicans. Certainly all these groups have a home in Nashville. "Redneck" is still a term of pride in many country songs, like Joe Diffie's parody "Leroy, the Redneck Reindeer." Racism has been a theme in country music for seventy-five years, including Hank Williams, Jr.'s 1988 anthem "If the South Woulda Won." And Republicans have thrived in Nashville for decades. Richard Nixon, feeling the heat of the Watergate scandal, opened the new Grand Ole Opry in 1974 by singing "God Bless America" and "Happy Birthday" to his wife Pat. George Bush, facing reelection, attended the CMA Awards in 1991 with his wife Barbara, then asked Wynonna to sing at his renominating convention. Four years later, Travis Tritt did the same at the behest of Newt Gingrich. "I'm a fan of his," Travis told me of the congressman from his former hometown. "The thing I've always respected about him is that he's not a politically correct politician, and that gets him in hot water. I'm proud to say he's a friend of mine."

But while country may give voice to Southern conservatives, its origins and fan base are far more diverse. Liberalism, for one, thrived in the fields of the South. "Most of the Southeastern audience for country music comes out of the Appalachian tradition," Bob Titley, the manager for Kathy Mattea and Brooks & Dunn, pointed out to me. "And big-dollar, big-government, New Deal programs have done a lot for those people." From Woody Guthrie's *Dust Bowl Ballads* to Johnny Paycheck's "Take This Job and Shove It," the ideal of the little guy fighting big business has long been important to country music. Beginning in the 1960s, though, economic liberalism was overshadowed by social conservatism in the South as Republicans exploited widespread Southern support for the Vietnam War (and opposition to civil rights) to build a base in the region. Pro-war anthems like "Ballad of the Green Berets" and "Okie from Muskogee" drowned out protest pleas by Gram Parsons and Emmylou Harris. Liberals still had roots in Nashville, but conservatives carried the day, especially since many of the economic assistance programs put into place by the New Deal had worked, thereby eliminating the chief source of liberal protest.

The South, in other words, was no longer filled with rural workers fed up with the system; it was filling, instead, with suburban middle-class workers with their own set of problems. By the 1990s, Nashville finally turned its attention to addressing *their* concerns. And though some claimed this amounted to an abandoning of country's "roots," in fact it was a belated recognition that the country *itself* had abandoned its roots. For country music to explore a place other than where the nation was at the time would have been the biggest abandonment of all.

Instead, Nashville took up the plight of the middle class, as unheroic as that might seem. The result was a completely new set of themes. If, as Public Enemy's Chuck D once famously asserted, rap music is the "CNN of the ghetto," country music, in the nineties, became the CNN of the suburbs. The news from that front was surprising. On the surface, politics themselves seemed to be of little interest to middle-class country music listeners. Gone were the days when Johnny Cash could rail against mistreatment of Native Americans in "The Ballad of Ira Hayes" or Merle Haggard could discuss the darkness of his own prison life. Gone, no doubt, because those children of the Depression no longer seem relevant to today's descendants of Cheez Whiz and Pop Tarts. Instead, hit songs in country were now almost exclusively about interpersonal relationships: people fighting people, instead of the elements. In many ways, country became the pop music parallel to the confessional, self-referential talk-show boom on television (though without much of the deviance). Even old beer-in-the glass staples, like breakup songs, developed a pampered middle-class feel to them, a freshly buffed coat of self-esteem. Terri Clark, for example, had a smash hit with a song that listed all of the things she'd rather do than see her ex-boyfriend, including washing her car in the rain, checking the air in her tires, and "straightening her stereo wires." "I'd love to talk to you / But then I'd miss Donahue."

When issues were raised in the music, they invariably related to domestic turmoil—alcoholism, child abuse, spouse abuse—rather than such external issues as class, race, or poverty. Anxiety became internal in country music. The threat was not across the ocean or even on the other side of the harvest, but on the street where you live, in what Mary Chapin Carpenter called this "House of Cards": "On the surface, it looked so safe / But it was perilous underneath." The most striking example of this trend was Martina McBride's defiant hit "Independence Day." In the song, named 1995 CMA Song of the Year, a battered woman decides to burn down her house with her husband still in it. "Throw the stone away / Let the guilty pay / It's Independence Day."

The song, like many in the new social landscape, was written by a woman, Gretchen Peters.

Still, despite a few powerful songs like "Independence Day," country music in the nineties by and large recoiled from engaging the world. It was the genre's biggest weakness. Though the music certainly reflected the new middle-class reality of its audience—on a three-hour trip back from Knoxville not long after our journey to Ireland, I heard no less than *three* songs on country radio that were set in airports—it seemed to have little understanding of where those planes were flying. I can think of no song I ever heard on the radio, for example, that mentions the name of a foreign country—and I don't mean Mexican food. ("From Hell to Paradise," a sterling exception, was never released as a single.) Instead, just at the moment its audience was at its broadest, country music became more narrow-minded than it had ever been before. What passed for depth in most country songs was best captured in a Tracy Lawrence hook, which said that if the world had a front porch "like we did back then," the world would still have problems, "but we'd all be friends." Even some of the stalwarts of thoughtful music eventually succumbed to navel gazing. Mary Chapin Carpenter, who produced such searing songs as "House of Cards" and "He Thinks He'll Keep Her," lightened in her advancing years to "Shut Up and Kiss Me" and "I Want to Be Your Girlfriend." Eat enough TV dinners and drink enough martinis and even the hardest edge becomes fat.

Which is exactly what happened to the Mavericks. When Raul, Robert, and Paul came to town (Nick met them there) in the early 1990s, they represented everything country music abhorred but needed: They were arrogant, brash, and iconoclastic. After winning their first CMA Award, the Mavericks were asked if they listened to the music of their competitors, like Blackhawk. "Nope," Raul answered bluntly. "And you know what? We won, so fuck 'em." It was certainly honest. But by the following day, disc jockeys around the country had listeners call into Blackhawk/Mavericks face-offs and Raul was forced to apologize. Music Row's conservatism had begun to rear its head.

"When I was young, I wanted to change the world," Raul told me the day after the Dublin concert. We were sitting in a hotel brasserie in Belfast eating fish-and-chips and pâté de campagne. "I was angry. I was young. I wanted to be Bruce Springsteen."

That morning the band had made the four-hour trek to the North, where an uneasy calm had settled over the civil war.

"I think there was a certain naïveté about me back then," Raul continued. "I was naïve enough to think that I could make a difference. I wrote that song 'Children' about child abuse, and we played it for a while. We played 'End of the Line' about Jim Bakker. And you know what? It didn't change a fucking thing." His hair plopped over his face again. Around his neck an elaborate cross jiggled against his black corduroy shirt. He looked like a biker. "So I kind of changed myself. Instead of worrying about changing the world, I decided I'm going to take care of myself and my family. Instead of worrying about whether the other guy is smoking or not or eating granola or not. Do whatever you want. Let's not kill each other, and I'll raise my family."

In that way, Raul and his band became a perfect expression of the New Nashville, where angry young men seem to come to grow old. Whereas once he seemed revolutionary, Nashville's answer to the Rainbow Coalition (as a first-generation Cuban-American, he was the first Hispanic performer to have notable success in country since Freddie Fender rode "Before the Next Teardrop Falls" to a brief run in the seventies), now he became nostalgic. Music might not be able to save the world, but it could capture the idealized America he had envisioned as a child. "I'm a huge fan of the early sixties," Raul told me, referring to the music of his idols—Bobby Vinton, Herb Albert, Keely Smith, Perez Prado—as "the music of dorks." "I would like to bring back some of the values we had then, some of our goals. People seemed to have a hope back then. But it has to be subtle, not slapping you across the face, like my earlier material. That's the difference. We are 'entertainers,' after all."

Accepting that equation is what kept them in Nashville. Others who pushed the envelope in country—Lovett, Griffith, k. d. lang—were ultimately pushed out of town, left to peddle their music from the chancier realms of Los Angeles or New York. For better or worse, those artists refused to subscribe to the mainstream demands that country radio imposes and that country labels demand. That was the fatal flaw of Nashville in the mid-nineties: Having achieved the breadth of market it had sought for so long, it no longer seemed capable of accommodating the breadth of its own audience. That the Mavericks, even the defanged Mavericks, were able to persevere through this system, mixing their rock 'n' roll heroics with a retro-country message ("I want to tell people to lighten up," Raul told me. "Have a drink. Have a smoke. It's not going to kill you. Relax . . ."), offered a faint ray of hope: Perhaps Nashville might be able to produce a music broad enough in its origins to reflect the country as a whole.

Later that night, in a windowless club in Belfast, I watched with heart-pounding delight as six hundred people stood shoulder to shoulder and shrieked their way through the Mavericks' crotch-grabbing version of Johnny Cash's seminal hit "Ring of Fire." For a second, I thought my eyes would burn, the energy in the room was so intense. And for an instant, I believed that Nashville might, someday, realize its dream of producing "America's Music."

That feeling didn't last long. A month after the Mavericks made their triumphant tour through Europe, Cleve Francis prepared to take the stage of the legendary Birchmere in Alexandria, Virginia, for his valedictory concert. "My lifelong dream wasn't to become a musician," Cleve told me just hours before the show. We were sitting in a diner across the street from the club, eating salads and vegetable soup. Five years earlier, the forty-five-year-old Cleve had left his cardiology practice in Virginia and moved to Music City for a shot at becoming the first African American country artist in over a generation. He was handsome—a shorter version of Denzel Washington—with a gentle voice and a cropped goatee.

"My lifelong dream wasn't to become a doctor either," he said. "I just wanted to be successful. My family was poor. I had no intention of being poor. My family was uneducated. I had no intention of being uneducated. My family had never left the state of Louisiana. I had every intention of traveling. I guess you could say I've done everything I ever wanted . . . Oh, yeah, except be a successful country music artist."

For a time, it seemed as if Cleve might achieve that dream. After bankrolling an album on an independent label in the late 1980s, Cleve forked out $25,000 to shoot a video for the album's first single, "Love Light." Released in April 1990, the same month as "The Dance," the video for "Love Light" surprised everyone by landing at number nine on CMT's Top 10. Jimmy Bowen, by then at Capitol Records, saw the video and offered Cleve a deal. "Bowen deserves a lot of credit," Cleve said. "He sat me down and said, 'I don't know if this is going to happen. It's been a long time since a black man pulled this off. But you sound genuine to me.'" On the strength of the video alone, the album, *Tourist in Paradise*, earned advance orders of 100,000, an astounding figure for an unproven artist. Even more importantly, the press lapped up the story of the doc who chucked it all for his dream. There were pieces in

The New York Times ("A Physician who Heals with Both Science and Art"), the *Chicago Tribune* ("Two Practices Make Perfect"), and *USA Today* ("Cleve Francis Is Feeling His Musical Pulse"). *People* magazine, topping even the best pun headline writers in America, ran a two-page story called "Country Doctor" ("Singer-cardiologist Dr. Cleve Francis can break your heart *and* mend it!"), which included a picture of Cleve in a lab coat over a hospital gurney holding a stethoscope to the heart of a six-string guitar.

With all this momentum, he seemed like a sure bet. The album's first single, though, failed to make it past the forties on the chart. His second single stalled in the thirties. Through two more albums and seven more singles, none would go higher. "I don't want to sound like sour grapes," Cleve said. "I know I had several strikes against me. I was older. I was a doctor. Many people thought this was a lark. But I also know that most people in control of the industry are between forty and fifty, and that's an age where race comes into play. The fans didn't mind. I've got rooms full of letters that prove it. But I talked to club owners who wouldn't book me, fair organizers who wouldn't have me, and radio programmers who said they wouldn't play my music because they thought it was a joke: a black guy singing country music." He paused to catch his breath. "That has to do with a lack of education," he said. "We simply have a whole generation of people out there who don't know the history of country music."

In one sense, Cleve was right. The history of Southern music is full of stories about the influence of blacks on traditional white music. In a TBS series on country music that aired not long after that night, legend after legend in Nashville's pantheon stared into the camera and proclaimed their debt to blacks or black music. For some, the link was to the black church; for others, the blues. For still more, like Johnny Cash, Bill Monroe, and Waylon Jennings, the connection was to the entire culture of rural blacks that permeated the lives of rural whites. "I had this friend in the flats," Waylon told me, referring to the black part of town where he delivered ice. "He taught me to move all my strings up on my guitar and put a banjo string on the bottom. That way you could push the strings and make a more bluesy sound. One day I was talking to my dad. I said, 'Daddy, what would you think if you took the mix of country music with blues?' He said, 'It might be good. I know a lot of people who like blues.' It wasn't that long before I heard Elvis."

Today, thirty years after many of those legends were at the height of their popularity, the country music community is more removed from the African American community than at any time in its history. Despite three decades of improved racial integration in the South, as well as a decade of booming sales and audience growth, at the time of Cleve's farewell concert there were not only no black artists with major record deals in Nashville, but also no black label executives, only one or two black songwriters, and only a handful of black backup musicians. With Nashville's grip on the radio market and its prominence in the sales market, country was, by default, the largest segregated corner of American music. The question—*why?*—had haunted me from the first time I met Cleve and, by the time of his concert in Alexandria, had led me to perhaps the saddest recognition of what country music at century's end had to say about America.

The easiest explanation for Music Row's white face would be to dismiss the industry—and its audience—as racist. Even a cursory examination of country music, though, suggests this isn't so. First, country's audience is far broader, far younger, and far better educated than it was in the past. If the American public will embrace Dolly Parton's version of "I Will Always Love You," Vince Gill's duet version with Dolly, *and* Whitney Houston's version, it's hard to believe they also wouldn't embrace it coming out of the mouth of a black country singer. Second, the executives who run Music Row these days are no longer country bumpkins. "I'm not saying that bigotry and racism don't exist in our business," said Tim DuBois, the head of Arista and a former official with the Federal Reserve Bank of Dallas, "because they do. But I guarantee you that if there were a marketplace for a black artist, and if there were a talented person out there, few people in this town wouldn't sign that talent." In Nashville, the only thing more powerful than black or white is green.

Instead, the real reason for the absence of blacks in country music is more subtle, though no less troubling. Although no one intended it, country music has become yet another example of a widening racial gulf in American culture, where blacks and whites choose to watch different television programs, read different books, and listen to different kinds of music. In other words, it's voluntary segregation. In a world of choices, blacks have few reasons to pick country music. This is the conclusion Tony Brown came to after looking for a black artist for years. "It'd either be some black kid trying to sing like Charley Pride," he told me, "only a really bad version of that. Or it'd be somebody who really sings like James Ingram, who decided he couldn't make it in pop music, so he could make it in country. If you're a young person with real talent, you're going to be a pop star because that's the biggest star. To be in country music, you have to love country music or you have to think it's such a stupid industry and you're so much better than it that they won't know the difference."

Naturally, since blacks and whites had fewer options forty years ago, more people shared cultural traditions. Particularly in the rural South, the cradle of country music, blacks and whites lived and performed in close proximity to one another, promoting cross-pollination. "Thirty years ago, you could rarely meet a white Southerner who not only knew a black Southerner, but who loved one as well," Alice Randall, the only African American songwriter to have a number one country song in the 1990s, told me. "For all the racial hatred, white Southerners had black people working in their homes, as baby nurses, as longtime servants, as illegitimate relations." These two traditions, which had developed along separate but related tracks throughout much of American history, collided in the 1950s. The integration of black rhythm and blues into mainstream white culture—first with rock 'n' roll, later with Motown—mirrored many of the changes taking place in the civil rights movement. Suddenly blacks were not only *influencing* American culture, they were *redefining* American culture, particularly music.

Nashville, meanwhile, which has a lower percentage of blacks than almost any other major Southern city (around 20 percent, as compared with over 50 percent in Memphis, Atlanta, and Birmingham; the reason: There were fewer plantations in the area), resisted many of these changes and clung to its more traditional (read: segregated) past. The leaders of the Grand Ole Opry in particular were reluctant to accept black influences into their music. "It was what they called 'n-word music,'" Waylon told me, referring to the word he preferred not to use: nigger-bop. The Opry even banned drums from its stage. By doing so, Nashville turned its back on many of the central features of black music at exactly the moment they were penetrating American culture at large. At a time of youthful rebellion, sexual revolution, and black empowerment, country set itself up as America's lily white, above-the-waist popular music. "One can find musical sources for that split," Jim Ed Norman, the head of Warner/Reprise Nashville, told me, "but I can't help but think there was some sort of sociopolitical context for that split as well. It had to do with what was symbolized, or insinuated, in the rock and rockabilly styles. I can remember riding in a car one time with an uncle, and there was a battle of the radio stations. He was clearly more comfortable listening to country music, and I noticed that any time anything stepped outside the boundaries, an insecurity rose up in him; it was safer and easier not to listen to that music."

No sooner had that split occurred than a complex economic system—record labels, charts, and radio stations—arose to service and, ultimately, to entrench this splintering. Ironically, the most prominent example of that splintering may be Charley Pride. Long before Pride emerged as a prominent country singer in the mid-1960s, blacks had been listening to and performing country music. Negro swing bands toured the Texas circuit. Harmonica player Deford Bailey was an early Opry star. Even Georgia-born Ray Charles put out several groundbreaking country records in the early 1960s in which he recorded country songs to R&B instrumentation. But Pride, a former player in the Negro baseball leagues, was the first to emerge as a superstar, helped by the fact that RCA deliberately withheld publicity photos from his early records. Still, it was not until he joked about his "permanent tan" that country fans accepted him, which they did warmly for two decades. What is striking about Pride's career, though, is that he didn't open a floodgate of African American performers. He was "no Jackie Robinson," as Cleve had put it to me. He was an exception, not a role model. "Jackie Robinson was specifically picked to go into the major leagues because of the type of person he was," Pride told me. "Nobody came to me." If anything, blacks shunned him. "Not many of them bought my records or came to see me," he said.

Twenty years later, that creeping resegregation is almost total. Though many labels scrambled to find the "next Charley Pride," none came close. Instead, the absence of even a negligible black presence on Music Row reflected the new middle-class reality of the industry. Whereas country was once concerned with the plight of rural whites and played directly into the homes of rural blacks, it became, over time, the near-exclusive purview of suburban whites. By doing so, it alienated blacks and became, in effect, the soundtrack of white flight—not angry at the world, just oblivious to it. Of course, there are blacks in the suburbs—about one third of all African Americans. And research suggests that as many as 17 percent of blacks do at least listen to country music. But those audience members are not being felt in the marketplace. "I've seen those data," Tim DuBois said, "but I've also been to the concerts. Those people are invisible in the process."

Which is what's so sad. At exactly the moment America was becoming *more* inclusive of African American culture, country music was becoming *less* so. Eventually even the music itself lost contact with its black roots. "Country basically *is* white music," Tony Brown ultimately concluded. "Why would black people want to sing those straight notes? Why would a black person want to be in a format that gives any white singer who tries to do a little curlicue or deep groove so much grief? I work with artists, like Wynonna, who really draw from black music. But most people can't. Reba would love to feel it, but she just can't, and she doesn't even know it. In the studio, when even a great hillbilly song doesn't feel good, my terminology is: 'It sounds too white.' That means it has no feeling. To me, black music is about feeling, and white music is about no feeling."

That absence of feeling is one of country's biggest drawbacks. As Alice Randall noted: "The black suburban experience is not the same as the white one. I have a child. My child goes to private school. She just spent the summer in France. But my child has been told by people that they would prefer she not be black. People have told her they would not have her over to the house to play. Some soft music that does not have a sense of the edge of life is not going to appeal to her. Her suburban experience is not one of ennui." Perhaps that is the reason for the growing segregation. As the music has become more middle-class and the entertainment business more balkanized, country has assumed a position as the music of American bliss—a touch of regret every now and then, a hint of suburban angst, but, ultimately, an overarching sense of contentment.

As Ed Morris, the former country editor of *Billboard,* summed it up for me: "Country is fundamentally based on the white experience. It's about where whites live, what they read, what they see. And I don't know if you should expect more from art. It's putting a really heavy burden on artists to have them do what the culture at large hasn't done. Sure, country music has failed, but I don't think the nation as a whole has done any better than we have."

Two hours later, Cleve Francis stepped to the door of the Birchmere's spacious dressing room. He was wearing black boots, black jeans, and a black silk shirt. His vest was shimmering purple. Out in the audience, three hundred people, many friends, family, and former patients, crowded around neatly arranged tables with red-and-white-checked tablecloths. The atmosphere was jovial: local hero done good. Most did not know that at the end of the show Cleve would board a rented bus back to Nashville, pack his bags, and return the following week to his stethoscope.

"To be honest, I'm a little relieved," he said. "I know what I have to do. For me, it's not to pursue this dream anymore. There's just no point."

"And tonight?"

"Tonight my thing is going to be just to entertain people. It's not a Lenny Bruce thing where I'm going to go out and be political. I'm going to be true to what I do. History will judge Cleve Francis the singer, the philosopher, the scholar, the doctor."

"And what will history say?"

"That he was good at what he did: singing lyrics, telling stories, and making people feel good."

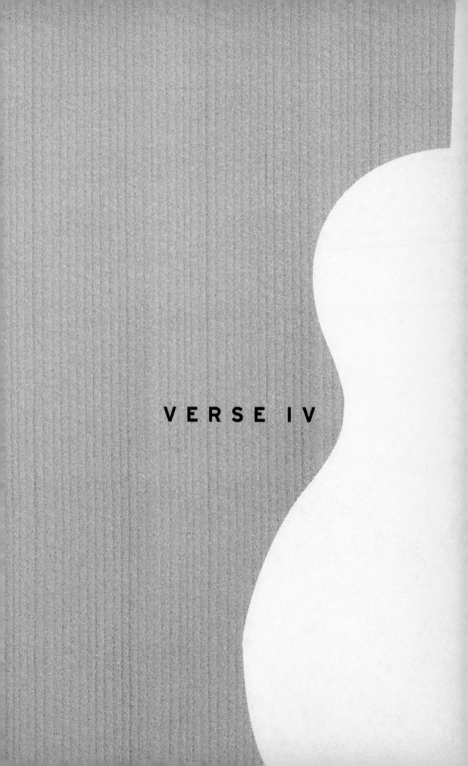

VERSE IV

THIRTEEN

THE STAGE

It was just before three o'clock in the morning when the giant corrugated metal door of Atlanta's Omni Coliseum opened onto the foggy night. Outside, the first of ten brand-new hospital white tractor trailers was idling with its back panels facing the auditorium. Minutes later, the Peterbilt thirty-eight-foot truck inched its way into the building and a team of several dozen men swarmed around its doors. A few of the men did warm-ups. A handful smoked. One drank coffee. They were dressed in blue jeans, torn sweats, and shorts, with Lynyrd Skynyrd T-shirts and Oakland Raiders parkas. They had on black Nike hightops, workboots, sandals. Each was sporting a well-used pair of gloves—woolen, stretched leather, canvas, mesh. And all were wearing a drop-dead I'm-a-professional expression as they freed the locks that turned the handles that opened the doors that were adorned with a frank admonition: NOT FOR HIRE. The only hint of celebrity on the trailers was a lavender circle on the back of each door inside of which was painted an understated purple g.

By the time his platoon of "g-trucks" arrived in downtown Atlanta, Garth Brooks's once meteoric career seemed to be in total free fall. The new album on which he had pinned so many hopes, *Fresh Horses,* had failed to keep pace with his previous releases. Though it had sold 3 million copies in its first three months, it had failed to hold the top position on even the country album charts—a far cry from several years earlier when he held that position for well over a year. Instead, by early March, he was down to a mere 15,000 units a week on the SoundScan chart that once had signaled his rise to prominence and that just a year earlier had calculated his weekly sales at ten times that amount. To add to his troubles, Garth had been roundly pilloried—then ridiculed—for his actions at the American Music Awards. At the Grammy Awards several weeks later, Vince Gill walked to the podium after winning an award for Song of the Year and announced that, unlike Garth, he planned to accept his trophy. At the Nashville Music Awards, a good barometer of industry attitudes, host Gerry House, a prominent DJ, declared that if anyone offered him an award, he wouldn't be fool enough leave it behind.

Then, to add to the growing perception that he was losing his flawless touch, the video Garth released for "The Change" consisted entirely of footage of rescue workers at the site of the Oklahoma City bombing. As a native son of the region, Garth had intended the video to pay tribute to the valor of the rescue workers. Instead, it was widely viewed, particularly in Oklahoma, as exploiting the tragedy for his own financial gain. Irate viewers deluged CMT with requests to pull the video. Garth worsened the situation by trying to strong-arm his label into sending a CD of the song to *every* radio station in America—not just country— and asking them to play it at 9:02 CST on the anniversary of the bombing. When the label refused, saying he'd be mocked for his Jesus complex, Garth took out a full-page ad in *Billboard* on his own letterhead inviting radio stations to telephone his office and coordinate a tribute. Faced with such stubbornness by the still reigning if weakened king of the genre, CMT and hundreds of radio stations did honor his request, though the following week many stations quietly dropped the song from their playlists, making "The Change" the second single from *Fresh Horses* to die a premature death. "Spend any time in Nashville and it's clear that Brooks just baffles the hell out of the local music industry," Melinda Newman, his friend and confidante, wrote in *Billboard*. "People speculate on everything: 'How does he handle his fame so well?' 'Why doesn't he just go ahead and quit while he's at the top?' 'Why does he insist on controlling every facet of his career?' 'Why on earth does he keep talking about being forgotten?' "

The start of Garth's new world tour, if nothing else, was giving him a reason to stop worrying—or, at least, getting him out of the house for a while. But even when the show started setting records (80,000 tickets in Atlanta in two and a half hours, more than Elvis; 88,000 tickets in Washington, D.C., a venue record; 60,000 tickets in Miami, the most since the Grateful Dead), Garth began fearing that deep inside he might even be losing enthusiasm for the one thing he'd always enjoyed most in his career: his performances. After two years of waiting, tonight he would find out.

Once the back doors of the Omni were opened, the crew went to work. The forklift driver, nicknamed "Grumpy Old Man," started removing giant cases of equipment. A team of riggers—"Go-Go," "Charlie," and "The Abomination"—sidled up the twin aluminum ramps and began rolling out silver fiberglass boxes. As soon as the boxes hit the ground, a new crew of workers took over from the first and shepherded the equipment—banded lights, grid motors, spansets, ratchets—onto the floor of the auditorium. The floor was covered in a patchwork of thick cardboard panels that sat directly on top of the ice hockey rink.

Within minutes, the entire crew was working. A few men arranged the portable silver cases along one side of the floor. A few more carried the first of the light trusses—giant rectangular scaffolding structures—and put them on the cardboard floor. Occasionally one would shout or burst into song. A few tripped over a stray cord or patch of exposed ice ("Hey, Mack, get some tape over here before somebody kills *hiss*self . . ."). A steady chorus of loud metallic clanks accompanied the scene. But mostly everyone did his job with quiet determination and a few commands from the small team of roadies who traveled full-time with the show.

"Opening a tour like this is a major deal," explained Debbie Diana, a thirty-something lighting specialist who, along with other members of the Atlanta crew, was making ten dollars an hour for her work, with no guarantee that she would even get to see the show. "We look forward to it. We don't know what the stage is going to look like. In six hours, though, we'll see."

"Is there something special about this particular show?" I asked.

"Well, since Garth started touring a couple of years ago, most country acts are as large, if not larger, than the rock tours going out. Five years ago, country was not like this, with lots of vari-lights and sophisticated equipment. Garth came along and was so influenced by rock 'n' roll that he did country with rock production. I've heard that in this new show there's movable parts up in the truss, with motors and stuff. They've got forty-two motors—that's a lot. They're moving something up in the sky.

We'll have to wait and . . . ooh, look," she cooed, "a giant mirrored ball. I think we have our first clue."

By the end of the first hour, the shape of the rigging was beginning to unfold. The eight-foot trusses were laid out in a giant octagon, with a smaller octagon truss inside it, connected with four-foot-long crossbars as spokes. The whole rigging looked like a giant oven burner, roughly twenty-five feet in diameter, that in time would be raised to the ceiling of the building and used to suspend the lights. This rigging would be the first thing constructed, followed by the speakers, and ultimately by the stage itself, which would rise from the floor and spread its wings from one side of the arena to the other.

At 4:45 A.M. a team of men who had found their way to the ceiling began lowering strands of one-inch garden chain to the floor. One by one, the chains were hooked to the motors that had been affixed to the various joints of the double octagon. As the chains were being attached, the rest of the crew worked furiously. Two men set about attaching strobes to the giant truss. Several more strung bundles of cable around the perimeter. And one lone man with a pair of black jeans that sagged, plumberlike, down his backside had the all-important job of affixing two F-100 Performance Smoke Generators, a machine about the size of a portable vacuum cleaner that contained a heater, a pump, and a tiny fan, into which water and mineral oil were poured and out of which spurted artificial fog. "The ones on the floor go pretty much all the time," he explained. "The ones up in the air we just give a blast every now and then and they produce just fine."

With the rig now pieced together, the pace seemed to quicken. At a quarter after five, one of the riggers started the computer that would run the lights. Minutes later, one of the electricians arrived with a silk-screen frame and a bucket of purple paint, which he used to paint Garth's signature g on both ends of each light rack. And just before six, the first rack of several dozen lights was bolted to the rig. "There are five hundred pars, or fixed lights," the head electrician explained. "Another hundred and fifty varis, or moving lights. All together, about a thousand."

"Any particular color scheme?" I asked.

"The truth is, there are so many moving lights. Everything changes color, changes focus, changes position. It's pretty much the case where your imagination is the only limit."

By half past six, the rig was ready to be raised, but production manager John McBride, who had met his wife Martina when she opened for Garth, decided to hold the ascension until seven, when Mark McEwen would be doing the national weather report for "CBS This Morning" live from the Omni floor. Jolly, friendly, almost giddy in his early morning element, Mark positioned himself at the edge of the floor, and, at precisely 7:01, received his cue from New York: "I wonder if I might quote a Garth Brooks song," he said, " 'I'm much too young to look this damn old . . .' " He chuckled at his happy TV pun. "So here we are, getting it all together. Garth is going to open his first tour in two years right here at the Omni tonight. We'll be talking to Garth later on this morning, but shall we take a look at the weather . . ." Mark was not two words into his speech when McBride gave the word and the giant rig, with cables dangling and lights shivering, slowly began climbing to the sky. A burst of applause went up from the crew. Mark himself glanced briefly at the creation. And fifty feet away, to no one's particular notice, a bleary-faced man with sleep in his eyes walked underneath the giant metal door, past the last truck waiting to be unloaded, and gave a simple, satisfied salute to the stage.

"Okay, folks, here we go."

It was a little past 7:30 A.M. when Garth ambled into the backstage holding room. He had changed clothes in his dressing room, then lumbered across the hall to where the rest of us were waiting: his two publicists, his brother, and the makeup lady who had been hired for the day by CBS. He was not in a good mood.

"I knew I never should have touched it," he said to no one in particular. He greeted everyone with hugs and shakes, then plopped down in the chair. "It'll probably be a year now before I'm normal again. I wanted to surprise my wife. And look what happened . . ."

He removed his black GARTH BROOKS WORLD TOUR baseball cap. His hair, normally graying and thin, was a vivid peanut butter color—a little bit darker on the top, a little bit orange around his ears.

"Oh . . . ? No . . ." The makeup lady was fumbling. "It doesn't look . . . *bad*," she finally said. She was lying.

"I cut it as close to my head," Garth said, not buying it for a second. "You get out in the sunlight and this thing takes off like it's on fire."

"Who did it for you?" the makeup lady asked.

"Somebody who does Sandy's hair. When she finished, she said, 'Oh, my God. It's red . . . !' And to think, Sandy loved the gray."

"And what was wrong with that?" she said.

"I don't know. Ten-year anniversary. I look at those wedding pictures. Sandy looks the same. I thought I would try to do something. That week she was gone, I thought I would surprise her. And now look . . ."

With his makeup applied, Garth stepped outside to examine the rigging. Immediately his executive instincts took over. "You got your ratchet up?" he asked one of the workers. "How 'bout those plugs?" It was that side of Garth, his Barnum-esque showmanship combined with his Bailey-like business savvy, that had forever changed the dynamics of country stage shows. In the early years of Nashville, country artists rarely traveled outside the South. Performers would play the Opry on the weekend, then drive to small-scale venues within a day's commute of Music Row. In time promoters began packaging Opry legends and taking them on the road, like the tour with Minnie Pearl that played Carnegie Hall in 1947. By the late 1960s, country music had grown to a level where some name performers—like Johnny Cash—could begin to venture into larger arenas, but for the most part country artists were unable to compete with pop artists. All that began to change in the late 1970s with the emergence of country as a mass-appeal music. Though most country artists still made the bulk of their money playing state fairs and annual rattlesnake jamborees, a handful of stars—Kenny Rogers, John Denver, the Oak Ridge Boys—were beginning to show real clout in drawing fans to arenas. Unlike outdoor festivals, which are known as "soft tickets" since fans get the added lure of the fair or the carnival, arena sales are considered "hard tickets" since all you get is the concert.

One reason fans had long been reluctant to see country artists in concert is that they didn't do much onstage. For generations, country singers were simply that: singers. They stood in front of a microphone and sang. While rock shows had been remade in the 1970s by everyone from the Rolling Stones to Kiss, country artists were content to plant their feet, lower their hats, and drip their tears into their proverbial beer. More than anyone else, Garth Brooks recognized this disparity and set about designing a show that would mix the best of the arena acts he remembered seeing as a child with the country acts he so desperately wanted to emulate. Simply put, he yanked country into the age of the arena.

"Originally I was scared to move around too much onstage," Garth told me, "I remember one night in 1989 we were opening for Chris LeDoux, this real cowboy, and I told my guys that we would have to curtail what we normally did onstage because we didn't want to overwhelm the people. We went out there and put on this nice, quiet, gentlemanly show. Then all of sudden Chris LeDoux comes out. They turn out the lights and—*boom!*—he comes catapulting over this hay bale and lands in the middle of the stage. I said, 'Geez! Look at that!' From then on, I started understanding that whatever the audience does out there, I can do back to them, and they'll send it back to me ten times bigger."

Within a year, Garth had moved from being an opening act to head-lining his own gigs, and he quickly began adding more elaborate stunts to his act: catapulting from trapdoors, swinging from ropes, even dan-gling from ladders he had suspended from the rigging. As with so many other areas of his career, he managed to combine pop pyrotechnics with country sincerity. Following the lead of Madonna and Michael Jackson, for example, he employed a headset microphone, but in his case he suspended it from the brim of his hat ("What people don't realize," Garth told me, "is that the hat actually blocks out the light and makes it easier to see the fans.") More importantly, Garth realized that these theatrics would work even better on television. In 1992, he began what would become a series of televised specials on NBC—in effect, hour-long commercials—that brought his stage show to millions of fans who never had to leave their living rooms. Having successfully moved from "soft ticket" to "hard ticket," he then rewrote the rules even further by moving to *no ticket*: All you had to do was sit at home and watch. Instead of diminishing sales, though, the exposure made Garth's show a must-see. Not just in the South, but in cities from Miami to Seattle to San Diego, and, beginning in 1994, abroad: Australia, Germany, Spain, the Netherlands, Ireland, England, and Scotland. It was that tour, his most recent, which was foremost in Garth's mind that morning in Atlanta.

"To me, this isn't the first day at all," Garth said when I asked him if he had any opening day rituals. He was wearing a black jumpsuit and, to cover his errant hair dye, a black baseball cap from the current tour. The makeup had removed most of the weariness from his face. A few drops of Visine had eliminated most of the puffiness from his eyes. "As soon as I got in the dressing room," he continued, "the same thing was happening as in Aberdeen, Scotland. There's no hot water, no food. It's freezing. I got in there and felt that cold water and thought, 'This ain't the first day, it's just the next day.'"

The only thing this was the first day for, he said, was the stage and the show. And even the show was similar to the past, though without many of the pyrotechnics he had used in his television specials.

"You've got to remember," Garth said, "if you've seen us on TV, you haven't seen us live. TV is what Schwarzenegger does. You do things you don't do elsewhere. You burn things. You toss cymbals. You fly. We did that once. It was in Dallas, at Texas Stadium, and it cost us 1.7 million dollars to film three nights. We couldn't afford to do that every night. I wouldn't *want* to do it every night."

In fact, they almost didn't get a chance to do it at all. That famous show in Dallas, taped and broadcast as the second of Garth's NBC specials, proved almost deadly. In his typical shoot-for-the-rafters style, Garth decided he wanted to fly above the audience. In order to pull it off, though, his tour managers were forced to come up with an elaborate rigging that stretched from the floor of Texas Stadium to the roof. On paper, the scheme worked fine, but no one had tested it. Several days before the taping, while his team was installing the equipment, the rigging buckled. "Everybody panicked," Garth said. "When the thing started snapping, people started jumping. I had friends who were up on it, and they had never been so scared in their lives. There were seventeen we took to the hospital." But still he had a show to do. "So I made this speech to my guys," Garth continued. "I said, 'Look, I've got to go down there and clean it off. Nobody here has to go.' There was this big box of hard hats sitting there. I grabbed one and started going down to the stage. And I turned around, man, and here comes just this trail of those white hard hats, following right down. I see my wife with a hard hat. My crew. It was a good day. Of course it was a bad day—but it was a good day, too."

Garth had that dreamy look in his eyes again—the one he develops when he starts talking about certain moments in his career. It was that part of him that often made him seem on the verge of falling into a trance.

"So as I was saying . . ." He returned to the present. "I'm not sure different is the thing. People come to a concert like this to have fun. They come to forget, they come to scream as loud as they want, they come just to sit and watch. Whatever they want to do. That's what we do. We'll play the old stuff. We'll do some of the old gags. We keep a ladder hanging down in case we want to run around on it. We keep the pit cables down in case we want to swing on those. And we'll see what the people like."

By a little after 4 P.M., the stage was nearing completion as Garth returned to the coliseum for a preshow run-through. Eschewing the stairs, he scampered up the stage as if it was a jungle gym and started prancing from one side to the other. "Look at this baby!" After all the attention paid to the lights, the stage itself was clearly the dominant part of this creation. The heaping piles of steel and metal that earlier had looked like pots and pans had been transformed in the previous twelve hours into a giant elevated stage the size of half a basketball court. Unlike most stages, with their bundles of sound monitors and towering ramps, this construction, which cost $300,000, was surprisingly flat. All the bells and whistles had been suspended underneath the rostrum, along with a series of scrims, trapdoors, and assorted platforms that would come rising out of the floor with a simple glance from Garth (and the press of a button underneath).

To add to this futuristic feel, the stage itself was made of corrugated mesh, like a high-tech freezer shelf. "It's an ample mesh," Garth explained. "Stretched. We've taken the regular mesh and stretched it even further. Then we went and put the monitors underneath it. We had fifty-something monitors last time on a stage that was seventy percent of this size. This year we have thirty monitors, but we cover more ground."

At the center of this massive plateau—and clearly its focal point—was a giant Plexiglas capsule about the size of the Apollo command module that contained the fifteen-piece drum set. Made with half-inch Plexiglas designed for use in 747s and christened the "U.S. Hope" by Garth, the capsule was mounted on a giant hydraulic piston that gave it the ability to rise periscopelike into the air and spin itself around like a giant robotic ventriloquist's dummy gone horribly out of control. "Houston, we have a problem: The drum set is leading a coup." To accent this galactic theme, the capsule had been decorated with a single decal of the American flag that Garth had received from NASA, along with some navy blue spacesuits for the crew to wear. "There are all sorts of names for the capsule," said drummer Tommy Johnson, "the 'egg,' the 'drum dome.' I prefer 'boy in the bubble.' "

Garth liked the "boy in the bubble" for different reasons. From a practical point of view, the bubble, by preventing the drums from bleeding into other instruments, allowed him to record each show on a forty-eight-track recorder for use in a "live" album later. From a showman's point of view, the bubble gave him a toy to play with in his own private playground.

"So what kind of soles do you have on those shoes?" Garth asked.

"Rubber," I said cautiously. After months of hanging around him, I had learned to be wary of such outbursts.

"Then go ahead, run up there," he said. For a second, I thought he was talking about one of the risers that had been raised from the floor. But before I had a chance to decide, he wiped the bottom of his Nike hightops and darted up the bubble in three rapid-fire steps. I followed (though, unlike him, I had to balance myself so I wouldn't fall off). "Now you can see what I've been talking about," Garth said. Indeed, from atop the pod, the view of the stage was stunning. The cold layer of cardboard on the coliseum floor had grown into a glacier of technology. The stage itself—cool, flat, accented with bursts of artificial fog—seemed like something out of a sci-fi opera. The fiddle player was warming up in one corner. A soundman was testing the board in another. With musicians popping in and out of trapdoors, the whole scene began to feel like an audience participatory game of "Chutes and Ladders."

"There's two things I would say about this stage," Garth said. "It's smart, and it's flexible. First of all, the money we'll save because we don't have seat kills"—those seats in the front several rows that usually go unsold because they have obstructed views—"will pay for this stage in its first month alone. Then the whole front section can be removed when we go to Europe, since their venues are smaller. We're just really proud of it."

Proud is an understatement. Earlier, Garth had introduced me to the head of his film crew, whom he had flown in from Ireland to record the show. I asked him the difference between a Garth show and a U2 show, since he had worked with both. He didn't hesitate. "The amount of time Garth spends thinking about it," the man said. "He's obsessed with it."

Indeed. "I hope this show tonight will be the worst one we give all year," Garth said, "but that it will still be a pretty good show. What I hope most is that people walk out of here saying, 'I came here four years ago for seventeen bucks. I had the time of my life. I screamed. I laughed. And I came here again tonight for seventeen bucks. I had the time of my life. I screamed. I laughed.'"

What Garth was most worried about, though, was perception. "One of the big questions that will be in everyone's mind when they sit down here is: 'Has he forgotten country music?'" Garth said. "That's why the second song out of the holster is 'Two of a Kind.' We want to tell them we're still country. We want to make sure these people sit there and go, 'All right, it's the same guy I knew.' And then they're open to new stuff. But if you just bomb them with stuff, they say, 'Ah, well, he's forgotten.'"

Though his show was only a few hours away, Garth could not escape the central worry that had consumed him for so long—what did the audience think of him, did they still like him, would they still enjoy his show? Even given the notorious insecurity of artists, it would be hard to imagine Mick Jagger, say, or Elvis, or the Beatles giving in to such anxiety about whether the public thought they were too pop, too rock, too crossover, too successful. Yet despite Nashville's leap into the American mainstream (or maybe because of it), country artists, more than those in any other genre, are obsessed with this definitional question. For Garth, it was a near-religion. "Our message is this," he had said at a preshow press conference. "We are not country music. We are a *part* of country music. We represent the best we can what we feel our brand of country music is. And we take that all over the world with us."

What he didn't say at the press conference was that having spread that message as far as it could go, he now doubted his own fitness to be its chief evangelist.

"You want to know the thing that scares me most?" he asked me. "It's pretty heavy." He squatted on top of the drum capsule and slid down to the stage floor. "What scares me most has to do with Katharine Hepburn. She made a wonderful speech about 'it' in one of her movies. She said, 'There's just a thing that some women have, and if you've got it, it doesn't matter what you don't have. And if you don't have it—no matter what else you've got—it's not enough.' That's what I'm worried about. I'm worried about getting on top of that piano and going, 'Where is he? Is GB going to show up or not?' And I'll know as soon as I jump off the piano and hit the grating. That's my biggest fear. Sitting on the farm for two years going, 'Is it still there?' "

By just before showtime, that fear had become almost palpable. A funereal hush had come over everyone backstage. The only person the least bit upbeat was Garth's mother, Colleen, who had just driven in with her husband Troyal on the bus Garth bought them for Christmas.

"The only thing I'm worried about tonight is some of his antics," she said. A sprightly woman in a purple jumpsuit and freshly upswept Martha Washington white hair (actually, she could have passed for Hazel), she looked like the biggest bettor—and the one with the saltiest tongue—at the local ladies' auxiliary bridge club. "He always runs up that drum thing in tennis shoes. He's going up there with ropers tonight. He could slip. Or fall. If he does, his finger will go through that stage. That stage is like razors."

Colleen had her own opinions about why her son, the one she

claimed to be her spitting image ("He's bullheaded like his father. Other than that, I see nothing but his mother . . ."), continued to be so insecure.

"All males are weak," she said. "It doesn't make any difference who they are. It takes Jesus Christ, or a woman, or a parent to make them realize that they're not. I think all males truly want to be a hero. What Garth wants to be is what he thinks every man should be: a person of complete strength, a man of his own mind. I think he'll be perfectly happy with himself when he knows he is a good man."

"And is he there yet?" I asked her.

"I think he is so close. But I think he doesn't feel it yet. When I talk to him I say, 'I'm so proud of you. You're such a good man.' And this means all the world to him. What Garth needs to do is stop and say, 'It's not what I've done, but who I am,' and realize he can only do so much. I don't care who you are—John Wayne or anyone—life only offers you a certain level. Garth's reached it. He just doesn't realize it yet. If I could give him any advice, it would be: 'Son, you're there. Be happy and enjoy it.'"

At 8:45, Garth emerged from his dressing room. To ease his surreptitious entrance underneath the stage, he was wearing a navy blue NASA jumpsuit over his traditional cowboy garb. On his shoulder, he balanced his black plastic hatbox. Ten minutes later, the hatbox was empty and the mammoth light rig, which had been lowered from the ceiling to cover the stage, started rising in a cloud of artificial smoke. After several minutes, a white grand piano rose up from the stage. At the keys was a man in a white tuxedo and a white cowboy hat. It was a perfect tableau of Garth's last video, "The Red Strokes," and it elicited an explosion of flashbulbs. But the scene was artificial. Minutes later, the real Garth Brooks, looking like a postage stamp in red, white, and blue, emerged hydraulically from inside the fake piano, sang the line he had written especially for this moment, "Oh, I said a little prayer tonight / Before I came onstage . . ." and opened his arms to the crowd.

FOURTEEN

THE MONEY

By early summer, Wynonna finally acted on the plan she had been formulating since spring. For the first time in her career, she outlined a course of action, marched through it on schedule, and slowly began to wrench control of her often uncontrollable life. First, in the final week of June, Wynonna gave birth to her second child, Pauline Grace, at Baptist Hospital in Nashville. (That evening, in a sadly emblematic gesture, her mother, sister, husband, and son all attended the red-carpet opening of Planet Hollywood downtown, where Naomi said of her daughter's delivery: "It was like pulling a watermelon through a nostril.") The following week, Wynonna dropped her publicist, Paula Batson. Then she split with her manager, John Unger. And finally she prepared herself for the biggest decision of the year: What label would release her next album? Though each of these choices was—or would be—made as amicably as possible (a far cry from a similar round of bloodletting several years earlier, most of which ended up in court), collectively they brought Wynonna into step with the cutting edge of country music—the bottom line. And, for the first time in her life, she finally seemed ready to confront what had long been the first rule of Music Row: Music is more than a gift from God; it's also a source of money.

Or at least it can be.

Since the arrival of Ralph Peer to record Jimmie Rodgers and the

Carter Family in Bristol, Tennessee, in August, 1927, country music has always been a self-consciously commercial enterprise. Particularly in Nashville, the lyrics or instrumentation may have changed to reflect the times, but the essence of what constitutes country music has changed surprisingly little: Country is music that chronicles the country *and* music that the country will buy. If anything, the second half of that equation is more important than the first. More than artists, executives, or even critics, it's consumers who have always decided what constitutes country music. This is the Faustian bargain that Music Row has agreed to: It doesn't matter what music anyone in Nashville might like; if the country doesn't like it, Nashville will change. "Think about this as a definition of country music," Bill Ivey, the director of the Country Music Foundation, wrote in an essay. "A country record is any record a radio station that calls itself 'country' will play and any record that a consumer who considers himself a 'country fan' will buy." That's it, he said of this grimly mercantile vision—no fiddles, no steel guitars, no high lonesome harmonies, no rhinestone suits required. Also, he might have added, no artistic vision. Chet Atkins, Ivey mentioned, when asked to define the Nashville Sound, put his right hand in his pants pocket and gently jingled the change. "That," Chet said, "is the Nashville Sound."

This preoccupation with making money only heightened in the 1990s as Music Row faced a new reality that rapidly transformed Nashville, as well as the South in general: the arrival of international corporations into a once regional subculture. Perhaps the chief mark of the "Gone Country" era was not that country music became popular around the United States, but that it became a vital profit center for corporations all over the globe. MCA Nashville, which is based on Music Row, reports to MCA Entertainment Group, which is based in Los Angeles, which in turn reports to Seagram, which is based in Toronto. Capitol Nashville reports to CEMA in New York, which in turn reports to Thorn International in London. Sony is based in Japan; BMG in Germany; PolyGram in Holland.

Far from merely being a technicality, this change has completely altered the nature of country music. "What has changed is that people don't care about the music anymore," one former senior executive for both MCA and Capitol told me. "All they care about is money. EMI is a financial company, controlling a record company, telling each and every level of that record company: 'Okay, how many dollars are you going to give us this month?' Well, how can anybody project the future of the record industry?" What Nashville offers up is blind guesses. "We have to project by month, by quarter, by year what artists are going to sell. Of course nobody has the answers to that. But then we're asked, 'How come this artist didn't hit that number?' 'Because it's music,' we say, 'not shoes.' Then they respond, 'Then you have to stop spending on that artist.' " Artist development, once the watchword of the industry, is the first to go. "If we think an artist is moving in the right direction, but a single's just not there, it doesn't matter. The patience is gone."

This patience was tested to an even greater degree as more and more corporations witnessed Nashville's boom and swept into town to grab a piece of the action. Though the income the industry generated increased by threefold, it was nowhere near large enough to accommodate the *five*fold increase in product being released. As a result, a disastrous scenario was set into motion: The more the industry expanded, the *less* money most people made. As manager Bob Titley put it to me: "The increase in income drew more companies into the community, which in turn increased competition. The increase in competition drove up marketing costs, which in turn lowered profit margins. The lower profit margins led to increased pressure from headquarters, which in turn made the entire industry more cold-blooded." Put more bluntly, as Debi Fleischer complained to me one afternoon in her office, "It's just *no fun* anymore."

What was "no fun" for executives, though, was downright perilous for artists, most of whom get only one shot at stardom. For them, the new financial equation led to an almost impossible challenge: Concern yourself too little with the business aspects of your career and risk financial disaster; concern yourself too much with business and risk artistic distraction, even collapse.

What is perhaps most surprising for new artists in Nashville, even ones like Wade Hayes who experience out-of-the-gate success, is how *little* money they actually make. In country music, there are basically five ways for artists to make money: records, performances, merchandise, songwriting, and, if they get successful enough, endorsements. Though records are the most visible part of the equation, they usually account for the least amount of money. When Wade signed a deal with Sony Records in 1993, he agreed, as part of the standard new artist contract, to give them the option to record seven albums, plus a greatest hits package. In return, they agreed to release at least one of these albums and to advance him money for the album, his wardrobe, and various other expenses. This advance money adds up fast—an average of $50,000 for legal costs; $200,000 for recording costs; $10,000 for a Manuel-style wardrobe; $5,000 for a cosmetic makeover; $30,000 for the cost of half a video; and the black hole, as much $3,000 a day for tour support to keep the artist on the road. As a result, the day his debut album was released in January 1995, Wade was effectively in debt to Sony Records for around $350,000. Though he would be under no obligation to pay back this money if the album stiffed, he still could not start making money until this advance was "recouped" by his label.

That recouping was done through royalties. The standard royalty rate for a new artist is 12 percent of the wholesale price of a record, with a third of that going to the producer. Most labels in Nashville sell their records to wholesalers for a base price of $10.30 for a compact disc (roughly three quarters of country sales are on CDs; one quarter on cassettes; vinyl records are manufactured rarely). With various discounts, the actual base price of a CD ends up around $8.00, which means the standard artist royalty per CD totals around 99 cents—again with 33 cents of that going to the producer. For cassettes, the total figure is closer to 66 cents. (For the roughly one quarter of country sales that are made by record clubs, that figure is closer to half the retail royalty.) The net result of all these calculations is nothing short of devastating. In its first year of release, *Old Enough to Know Better* was certified as a gold album, meaning it had shipped—though not necessarily sold—500,000 units for a gross of approximately $4 million for Sony Records. Of that, Wade Hayes, the bestselling debut artist in Nashville that year, earned less than $200,000, leaving him still $150,000 in the red to Sony.

For Wade, this situation was made slightly less bleak by the fact that he, unlike many of his contemporaries, was also a songwriter. Songwriters make money through four principal avenues: record sales, radio airplay, television and film licensing, and, if they're successful, long-term residuals. Every song on every album sold, regardless of discounts, earns what's known as a "mechanical" royalty rate of 6.95 cents, which is shared between the writers and their publishers. On *Old Enough to Know Better,* Wade was a cowriter on three of the ten songs, meaning he earned an additional 5.2125 cents on each album sold as a writer (actually, as a writer-artist, he was entitled to only 75 percent of that figure), and an additional 2.5 cents on each album sold as part of his copublishing arrangement, bringing his total mechanical royalty off the album to roughly $30,000. Since mechanical royalties, like songwriting royalties, are kept separate from album royalties, Wade got to keep this money.

In addition, since two of the songs he had a hand in writing, "Old Enough to Know Better" and "I'm Still Dancin' with You," went to number one, he earned a percentage of the roughly $150,000 that usually accrues through airplay to a chart-topping song. In Wade's case, after his publisher and his performance rights group took their cut and after he paid back the advance his publisher had been giving him over the previous several years, his songwriting royalties from his number one records were close to $70,000. So far, *Billboard*'s Debut Artist of the Year had netted $100,000, and that's before his manager took 15 percent off the top, bringing Wade's total closer to $85,000.

The final major source of income for Wade was touring. As a new artist, Wade Hayes spent almost 250 days a year on the road, playing close to 150 dates. Traditionally, it was the road that has provided Nashville artists with most of their income, up to 80 percent in many cases. For baby acts, with minimal airplay, a night at a small club earns them as little as $1,500. An artist with a hit record, like Wade the night I saw him in Richmond, Virginia, earns closer to $5,000. Two hit records could earn an artist a coveted slot on one of the half dozen or so major shows coming out of Nashville every year that are headlined by a superstar: Vince Gill, Reba McEntire, George Strait. Though playing on one of these tours gets a young artist exposure, it pays little more than $7,500 a night. As a practical matter, this rarely covers expenses: $700 a day to rent a bus; $2,000 a day for his band. Add to that the costs of equipment managers, sound engineers, merchandisers, and daily living expenses, and in a year on the road, Wade Hayes grossed almost $1 million, but netted almost nothing. If anything, he ate into his songwriting kitty. At year's end, with a gold album, two number one songs, and a slot on one of the biggest-grossing tours of the season, Wade Hayes earned a grand total of $75,000—not bad for a former construction worker, but still far below the norm for most pop stars (in their prime, Kiss could spend that in a weekend), not to mention NBA players or Hollywood actors. "Sure, it's not that much," Wade's manager said when we discussed this grim equation. "But Wade's happy. It's more money than he's ever made."

As stark as this financial picture was for a young artist like Wade Hayes, if he could make it through the start-up years and gain a degree of stardom, the financial rewards could be staggering. No artist in Nashville history proved that more than Garth Brooks.

From the beginning, Garth Brooks did one predominant thing for country music: He brought the rhythms of pop culture squarely into Nashville. While the short-term impact of this influx of energy and excitement was beneficial for everyone: greater sales, more media exposure, and higher numbers all around (Bob Titley told me he believed Brooks & Dunn were boosted immeasurably by their being placed after Garth Brooks in the alphabetical bins at retail); the long-term impact proved devastating: higher expectations, more fickle audiences, and greater financial pressures for all. Just as Garth embodied the upside of this change, to Music Row, he also came to represent the opposite. As one veteran publicist put it to me: "After Garth Brooks, there was no such thing as a pleasant meeting with a manager, a pleasant meeting with a label, or a pleasant meeting with an artist. Everybody wanted to have the world—and to have it now."

For Music Row, this split attitude toward Garth (Was he the galloping hero saving the town or the evil mogul bringing big-city ways?) was made even juicier by Garth's ongoing shootout with his label head and principal rival in Nashville: Jimmy Bowen. For most of the 1990s, the battle between Garth and Bowen (even friends referred to him by his last name) was the preeminent soap opera of country music and, because of Garth's outspokenness and Bowen's long list of enemies, the source of probably half the gossip on Music Row at any time. Bowen was a burly bear of a man with a grizzly voice, a foreboding beard, and a Greek fisherman's cap perpetually on his head. He was also a relentless self-promoter who was renowned for starting rumors about himself on the golf course in the morning just to see what form they'd take by the time they reached him in the office that afternoon. By December 1989, when he swept into Capitol Records (he eventually changed the name to Liberty to emphasize its independence), Bowen had been a dominant force in Nashville for two decades, having run five other labels—including MCA twice—and having revived the careers of everyone from Waylon Jennings to the Bellamy Brothers to Reba McEntire. He'd also terminated the careers of many other artists by imposing a strict bottom line. "I think Jimmy Bowen is the single most insensitive person in Nashville," producer George Massenburg told one profiler. "He came along at a time when the town was ripe to pillage, and he pillaged what was left. He did increase the recording activity; anyone will warrant that. But he's wrecked the music. He's turned it into a commodity."

Bowen certainly understood the financial potential of music. Born in Santa Rita, New Mexico, in 1937, James Albert Bowen first·entered the music business as a bass player and singer in a college country band, even visiting Nashville with the idea of becoming a country artist. When his band switched to pop, Bowen briefly became a teen idol. Abandoning his singing career, he eventually gravitated to L.A., landing a job as a record executive and producer. He revived Dean Martin's career with a string of hits, including "Everybody Loves Somebody," and went on to produce some Sinatra sensations, including "Strangers in the Night." Disillusioned with Los Angeles, he drifted back toward Nashville in the mid-1970s and reacquainted himself with Music Row. In 1978, Bowen became general manager of MCA, then moved quickly to Elektra/Asylum, where in four years he moved the label from a $2 million deficit to an $8 million profit. In 1985, he returned to MCA, where sales tripled under his leadership. When he arrived at Capitol in 1989, Garth Brooks had just had his first number one song.

At first Bowen and Garth seemed perfect for each other. For starters, they both understood marketing. "When I got to Capitol Nashville, Garth was just having his first big hit," Bowen told me in his breezy, though sometimes self-serving way. "But when I went out into the country to our distribution branches, they said to me, 'You know, we could sell a lot more of that Brooks kid if he had a lot more marketing and advertising dollars.'" Left with little budget, Bowen hit on a solution: He raised the cost of the record by a dollar and gave the retailers the margin for advertising. "That wasn't brain surgery," Bowen said, "that was good business." Next, both men understood the importance of generating controversy. "I remember calling up Bowen during the whole 'Thunder Rolls' flap," Garth told me, referring to the controversy over his video, "and he said, 'Do you realize how good this is going to be for you?' Of course, he was right."

In time, though, the relationship soured. In December 1990, after sales of his first two records had topped 3 million units, Garth Brooks did what every successful artist does: He set out to renegotiate his contract. Because the basic record contract is so one-sided—the label assumes the risk, but also, if the record succeeds, reaps most of the benefit—artists frequently rewrite their deals as soon as they achieve star status. In this case, Bowen agreed that Garth deserved a higher royalty rate and offered him 15 percent, plus a few other long-term incentives. Garth, the budding businessman, was so pleased that he purchased a full-page ad in *Billboard* that showed him reading the new agreement: WHOA . . . !!! WHAT A CONTRACT! CAPITOL NASHVILLE, YOU ARE AN ANSWERED PRAYER. I LOVE YOU. At the time "Unanswered Prayers" was at number one, and forever after—though he would later regret it—Garth had guaranteed that his financial affairs would be a matter of public record and at the forefront of his public persona.

The following year, when his sales continued to climb, Garth wanted to renegotiate again. In most renegotiations, the basic outlines of the deal don't change. The label retains control of the masters (the original recordings), as well as the production schedule of the albums; only the royalty rate grows. Also, the label usually agrees to shell out more money for marketing and pay a sizable advance. Garth, though, wanted more. "I want the Michael Jackson deal," he told Bowen, referring to Jackson's deal with Sony in which the artist gets close to 30 percent of the royalties. Also, he wanted to own his masters, and he wanted to dictate the production schedule of his albums. And he wanted a guaranteed payment every year in the millions of dollars. In effect, he wanted to own his records and basically lend them to Capitol Nashville to distribute for him. The negotiations, which Garth headed personally, lasted for well over a year. Ultimately, faced with the potential ruination of his company and no money left to develop other artists, Bowen turned Garth down. "You don't deserve the Michael Jackson deal," Bowen told him, pointing out that Garth hadn't been a worldwide superstar for thirty years. "This turned Garth stone-cold," Bowen told me. "He was seething. Fortunately he was at the office, so he couldn't explode."

What he did do was go over Bowen's head, first to Joe Smith in New York, then to James Fifield in London. In January 1993, a new deal was reached. The initial news was shocking. Bowen, quoted in *USA Today*, said the agreement was probably bigger than Madonna's $60 million contract and Prince's $100 million contract. By the time the agreement arrived in Nashville, where it was stored in a top secret black loose-leaf notebook that only a handful of people have ever seen, it became clear that its value was even higher. According to several people who have seen it, the deal works basically like this: Garth's records are owned by a company he controls called Pearl Records, Inc. (named for Minnie Pearl). These records are then licensed to Capitol Nashville, which must distribute them on a schedule he determines. Capitol pays for the artwork, which Garth then owns. Capitol, in turn, pays for all marketing and promotion costs. After these costs are deducted, Garth earns upward of 58 percent of the net profit, bringing his take from every album to above $3.00. In addition, in an even rarer move, these terms were retroactive to day one of his career, covering each of the 30 million albums already sold. Finally, in an unprecedented step for Nashville, Garth earns his share of this agreement in annual payments of close to $5 million.

As Bowen would say, it doesn't take "brain surgery" to total up these numbers. Based on album sales alone, Garth Brooks earned close to $150 million in the first eight years of his career. Add to that mechanicals for all the songs he's written, performance royalties, merchandising, and concert tickets, and Garth's gross revenue for the "Gone Country" era was close to $300 million. "Are you ready?" Garth claimed with utter seriousness when I asked him about this figure. "Even *I* don't know how much money I make. I don't have a clue because money has never been that important to me." "Then why drive such hard bargains?" I asked him. "Why spend so much time negotiating business deals?" Garth seemed pained by this question, then replied, "What hard bargain? I just do what I think is fair." (As for the deal itself, Garth, citing a confidentiality agreement, refused to discuss its details. He did say he thought it was justified, considering he was responsible for 90 percent of Capitol's profits and 60 percent of EMI's profits in North America. "I think everybody on this Row deserves that deal. It's a fair artist deal: If you go up and hit so many, you get so much. It's that simple.")

To be sure, any artist's desire to maximize his or her profits is understandable and, given the record industry's track record of bilking artists, even commendable. But Garth's aggressiveness did come at a cost. First, it created a chill between Garth and Bowen, the most effective marketing tandem in Nashville history. As Bowen put it in his 1997 autobiography: "Garth was turning into a control freak, wrapped up in details his people should have been handling. His explosive success and new fame—he externalized it as 'the GB thing'—had distracted him. I sensed a dark, almost self-destructive aura around him." Further, Garth's deal undercut the label around him. "Most artists think they're not getting enough," Bowen told me. "I can understand that. When I was an artist, I thought I was getting screwed bad. But there is a point where it can get lopsided the other way. An artist can make a deal that's so good that you weaken the company behind you, and then you've hurt yourself." The larger problem, Bowen felt, was that the deal was dominating Garth's life. "You know what the biggest damage of that deal was?" Bowen said to me. "The year and a half he took to negotiate it personally. His music suffered. In forty years in this business, I never saw an artist as involved with the contracts, the deals, as Garth. I think he just loves it. Either that or he trusts absolutely no one."

Certainly trust was a factor. As Garth said to Bowen at one testy moment in their relationship: "Trust is on a bus down the road a few hundred miles, but it ain't got here yet." Also, as Garth said to me, he watched in horror as superstar after superstar, including his friend Billy Joel, get ripped off by advisers. But the true nature of the conflict had to do with sales. By 1993, Garth's sales started to dip. From 9 million units in 1991 and 17 million in 1992, he slumped to 5 million in 1993 and the same in 1994. Garth could have blamed his fans for this decline. He could have blamed himself—for making controversial statements or inferior music. Or he could have blamed Bowen. Garth chose the latter, specifically focusing on Bowen's inability to retain Joe Mansfield, the marketing head who had spearheaded Garth's rise. Though Bowen re-hired Mansfield as an adviser, Garth remained upset. "Where Bowen and I got the furthest away from each other was over the letting go of Joe Mansfield," Garth told me. "That was probably the beginning of the end. As an employer, the reason Bowen gave me for letting Mansfield go was very acceptable, but my whole fight was: 'Bowen, you've got four hundred other acts to go to, but this is my one shot, so I've got to do it my way.'"

Initially Bowen was prepared to accept the blame, if only to relieve pressure from his star artist. "I never met an artist in my life who's going to take the blame for failure," he told me. "It's the nature of them. And if you're working with a bigmouth like me, I'm the obvious one to blame." But gradually the goodwill faded. "I can give you fifteen reasons why his sales dropped," Bowen said, "but he can only give you one: me." The first of Bowen's reasons was Garth's mouth. "Garth was the most perfect artist that ever came along," he said. "Then he started doing talk shows, and when you have an opinion, you make enemies as well as believers." Next, his freshness began to fade. In 1994, Bowen did a study that determined that Garth had a core audience of 3.5 million people who would buy whatever he released. New audiences, however, were tuning him out. Finally, as critics noted and as Bowen believed, the new music was simply not as good. "His later albums weren't as strong because he spent so much time negotiating his record deal, plus the book deal, plus the movie deal, plus, plus. He's a brilliant kid, but he was just so busy." Unlike Elvis or the Beatles, Bowen said, who both turned to drugs, "Garth's drug was power, which can be worse."

By 1994, Garth and Bowen were openly feuding. "By then, I didn't feel it was my label," Garth told me. "I didn't set foot in the building for two years." That fall Garth went to visit Bowen at his 9,100-square-foot home on Franklin Road, which Don Cook later bought for $1.875 million. The two got into a sharp disagreement, at which point Garth threatened to get him fired. "We went through this whole thing," Bowen said. "And Garth said he and his people had decided that since his records didn't sell fifteen million anymore, there had to be a scapegoat, and I was it." As Bowen knew, such posturing is not unusual. "What was unusual was that you had a thousand-pound bear doing it." Bowen calmed Garth down, and eventually Garth went back to his office. Ninety minutes later, he returned. "I went to my people and they don't buy any of it," Garth said. "They want me to go to war with the label. And when I say 'my people,' " Garth added, "that means me too." Bowen was startled. "If a song publisher and an accountant know more about the record business than I do," he said, "then listen to them. These are the choices you have to make in life. But I'd be very careful about taking advice from people that want you to go to war when they have nothing to lose and *you* do." "Yessir," Garth said, "that's a good thought. But I gotta tell ya, I agree with them."

When I asked Garth about this story, he said simply, "So what? Going to war with my label? I do that ten times a year. I mean, that's what artists and labels do because they're in it for the money and we're in it for the music."

In the months that followed, though, Garth, by all accounts, took his war even farther: He instituted an underground "dump Bowen" campaign—first to Charles Koppelman, the new head of CEMA in New York, then to James Fifield in London, and eventually to Fifield's boss, Sir Colin Southgate. Eventually Koppelman summoned Bowen to New York and chastised him about his expense account, particularly his habit of playing golf every morning. "Your problem isn't golf," Bowen told him. "It's Garth." The three sides had reached an impasse. Would EMI fire Bowen, as Garth wanted the company to do? Would Garth go to another label, as EMI feared? The answer will never be known because in the summer of 1994, Jimmy Bowen paid a visit to the Mayo Clinic, where he was diagnosed with lymphoma. Bowen decided to bail out. "At that point," Bowen told me, "I said, 'Okay, that will do, thank you very much.' I didn't need to be between the corporate thing in New York, which I always hated, and an artist that big and powerful. That alone can kill you." In December 1994, Jimmy Bowen resigned from Liberty Records.

This rapid turn of events, though tragic, gave everybody a graceful out: Bowen could take his fortune and retire to Maui and play golf.

Garth could plausibly deny having Bowen fired ("I thought it was his illness," he told me) and still come across like a dignified winner. "You got to know, I truly did and still do like Bowen a lot," Garth told me. "I probably dug him so much because we were so much alike."

Still, the underlying reality did not change. Garth Brooks had completely rewritten the rules of business in Nashville and was increasingly acting like a lone vigilante, all the more so since he now had no manager, because Pam Lewis and Bob Doyle, his original managers, were locked in a bitter legal dispute over how much money each one deserved from their partnership. For Garth, the dispute was crippling. By 1995, at exactly the time he was supposed to be concentrating on his new album, *Fresh Horses,* Garth was now managing himself, his recording obligations, his merchandise arrangements, and his international touring responsibilities. In the ageless battle to balance business and artistic concerns, he had completely succumbed to the burdens of tending his own exploding empire. This set the stage for the saddest irony of all. Wade Hayes, after a brutal first year, had earned a mere $75,000 and was, at least for the time being, happy. Garth Brooks, meanwhile, after an explosive half a decade, had earned $300 million and was deeply *un*happy. This was not an expression of music; it was an expression of character.

Wynonna, meanwhile, had the opposite problem. If Garth cared too much about business, she cared too little. Especially in the years with her mother, Wynonna was legendary in her lack of interest regarding financial matters. "I think that you can probably not be any more ignorant of the details of the business side of the career than she and her mother were," John Unger, her lawyer and later her manager, told me. "She just had no interest in that." Wynonna agreed. "Here's what I thought," she told me. "I do the singing, you manage. You call me and tell me what to do, I do it, then I can go back and have my personal life." Unfortunately, that lack of attention amounted to an invitation to rip her off.

After Woody Bowles, the Judds' first manager, stepped down in 1986, Naomi and Wynonna were managed exclusively by Ken Stilts, the businessman whom Bowles had brought in. (Naomi remembers telling him after their first meeting, when she thought he looked like a "mafia don": "Ken, if you decide to work with Wynonna and me, I promise you'll find us infinitely more interesting and rewarding than all your previous business ventures.") Twelve years later, those words would prove prophetic. In early 1994, John Unger, then Wynonna's lawyer, walked into her farmhouse and announced: "So, do you want to know how screwed you are?" With the help of his clients' willful disregard, Stilts had amassed an enormous financial empire on the backs of the Judds. According to a story first reported in the *Globe* (which a judge later determined was "essentially true" in response to a libel suit from Stilts), Stilts and his partner, Steve Pritchard, made $20 million from their association, while Naomi and Wynonna made $5 million combined. As Wynonna described the situation to me: She and her mother believed they owned Pro Tours, Inc., the company that handled most of their affairs. It turned out that Stilts and Pritchard each owned a third, while Naomi and Wynonna together owned the other third. The Judds thought they owned their $350,000 tour bus; Pro Tours owned it and charged it to them. The Judds received an $80,000 bonus in their last year together; Stilts, $250,000. "There was no way to describe how I felt when I learned this," Wynonna said. "Just the audacity of some of the things he did: His sister was my accountant. His lawyer was my lawyer. His best friend did my insurance."

Though Naomi wanted to sue Stilts, Wynonna initially refused. "I just wanted gone," she told me. "I loved Ken. I didn't want to take him through this fight. He was my father figure." Also, as she knew, she had given him control. "When you willingly give away the title of executive producer and you make two hundred thousand dollars from the farewell concert, you're screwed," she told me. Eventually Wynonna did join her mother in a lawsuit, and though Stilts and Pritchard denied they were liable for wrongdoing, they did acknowledge the essential facts of the relationship. The suits were settled out of court.

Either way, the damage had been done: The "Female Elvis" proved to be like her namesake in another sad and crucial way. "This is the part that's really hard to admit," she said. "I had a manager that dominated my every move. I didn't even buy a car without his approval. I was isolated from making my own decisions because I didn't have enough confidence in myself, I didn't think that what I had to say mattered. I'll just let him do it all. His way of dealing with it was he's Colonel Parker, I'm Elvis, he made all the decisions, I did what I was told." And why did she do this? "You place a worth on your life and I didn't feel I was worth very much." Sadly, her wish came true. Though she was two years into her solo career at the time and had the bestselling solo album by any female in country music history, Wynonna, in early 1994, was broke.

Even worse, just as her relationship with Stilts collapsed, her relationship with her record label disintegrated as well. Though it's seldom
noted, the Judds were originally signed not by Joe Galante at RCA, but
by Mike Curb, the maverick businessman from California who later
moved his company, Curb Records, to Nashville. Curb signed the Judds
to a production deal, then agreed that RCA would promote and distribute the duo. Later, when Wynonna moved to MCA, Curb remained
attached. ("Mike's either got the greatest lawyer in the world or he
sleeps with the devil," one MCA executive told me.) As part of that
deal, MCA believed it controlled the Judds' recordings. When Curb
disagreed with this assessment, MCA sued. Curb countersued, alleging
that MCA had excluded Curb from decisions surrounding Wynonna's
solo career. Though both suits were essentially frivolous (and were later
settled out of court), the hostility between the two camps was calamitous.
"It got to the point that everybody was paranoid," a former MCA executive told me. "I remember having nightmares about this. Literally having
nightmares, looking at an ad that didn't have Curb on it and fearing
being dragged into court. It was just pure silliness."

These dealings, of course, devastated Wynonna. "It was like being a
child between two divorced parents who hate each other," she told me.
"You love both of them, but they're so intent on destroying each other
that they don't realize they're killing you too." At the heart of the battle
was the seemingly never-ending question of whether to keep Wynonna
a country artist or cross her over into pop. The latter would mean taking
her videos to VH-1, working her singles to different radio stations, and
generally shedding her Nashville imagery. On one side was Mike Curb,
who wanted to take her pop. "I think Mike Curb had a more expansive
view of Wynonna's potential than MCA did," John Unger told me. On
the other side was MCA, who urged her to remain country. As Tony
Brown, the president of MCA and Wynonna's producer, explained: "Wynonna has a pop voice, but she's part of a country music legacy. Had
she and Naomi broken out as a pop duo, it'd be simple. But they didn't."

John Unger clearly sided with MCA. "I thought this whole country/
pop thing was semantic," he said. "Don't talk about it. Just do it and
nobody will know the difference. You can be a country artist and still
have Elton John come sing with you. And that's cool, as long as it's
perceived that Elton John wanted to come hang with *you*, rather than
you wanting to be something you're not." Indeed, as Garth had shown,
you could be country and still rule the world. As Unger put it, "The
mountain did come to Mohammed."

Wynonna initially resisted this argument. When the head of radio promotion for MCA came to see her and told her she was too rock 'n' roll and needed to get back to her roots, she exploded. "Show me the size of your penis, and then we can talk!" she said. "I thought I was beyond that. I thought I was in that space of not having to prove myself anymore." Her conclusion: "To hell with politics, I'm a heart person. I'm going to do what I want." This conclusion, though liberating, had two immediate conclusions. First, it forced John Unger to quit. "Wynonna just wasn't in the mood to listen," he said. Also, it terminated her relationship with MCA. "They just don't have the vision of what to do with me," she said. Privately Wynonna sent a message to Doug Morris, the head of MCA/Universal, who had come to see her in a secret meeting some months earlier and tried to lure her to his boutique label in L.A. under the MCA umbrella. She was ready to make the move, she told him. A week later, Wynonna became a Curb/Universal artist: a country singer with an out-of-town label; a pop star, basically, with country roots. Elvis.

Wynonna's decision to leave MCA marked the end of an era. Her debut album five years before had heralded a new moment, one in which country, pop, gospel, and blues could combine into a new American music based in Nashville. Now that music was homeless again. Nashville had grown so big, so fast, it no longer seemed capable of accommodating its own genre-bending artists. Not surprisingly, this coincided with a downturn in sales. At exactly the moment Wynonna was contemplating her future, the rest of the country woke up to the fact that Nashville's boom had come to an end. In a little over a month, *Entertainment Weekly*, *USA Today*, even *The New York Times Magazine*, which hadn't run a story about Nashville in years, all wrote articles about the slump of country music. This was the Revenge of SoundScan: The chart that several years earlier had so documented country's rise now equally vividly chronicled its sag—down 12 percent from the previous year, 21 percent in the last quarter alone. Though typical media overreaction (this was the "take something big and tear it down" part of the perennial press equation), the stories did capture Nashville's mood. "We're really greedy right now," Wynonna said, "and when you get greedy, you don't put out your best product. You get into the cookie cutter mentality."

Naturally, this frightened her. As Wynonna prepared to go back into the studio, she realized, if nothing else, that she could no longer wallow in her own laziness. "There is a fight in me now that I don't really know that I've had in my whole life," she told me. "I've fired a few people.

I've started battling back. I'm just saying, 'I'm not going to take it any-more.'" The little girl, it seemed, might finally be growing up. "I got into this business to sing," she said. "And I thought the way it worked was that they sold the records. Now I realize it doesn't work that way at all. If I don't get in there and make decisions, nothing gets done. So now I'm making decisions. I have a plan. And I'm going to make it come true."

FIFTEEN

THE FANS

By late afternoon, they'd already missed lunch, an ice cream break, even a chance to see Shania Twain dance in the rain in her black leather pants on the stage in the infield of the Nashville Speedway. But still they stood, necks craned, on the balding sawdust ground. Still they waited, three abreast, in the long and winding line. And still they thought, even now, they were blessed with such good fortune they could hardly believe their dumb luck.

"I've been in this line since noon," said Chris Lowe, forty-three, of Sioux City, Iowa. "I got up at four-thirty in the morning, waited by the gate until they opened, then went to his booth. But they said he wasn't coming. I was horrified. Then I was standing in line for Wade Hayes when I heard he *was* here, and I ran to this line. I've been here ever since. I told my husband, 'If I have to stay here all day and half the night, I'm going to get his signature.' "

Chris, a guest services operator at Target, was wearing blue jeans and a gray T-shirt that was crowded with buttons from all the stars she had met at Fan Fair: Alan Jackson, Vince Gill, Aaron Tippin, Alabama. This was her third Fan Fair, the annual backstage festival-cum-family reunion in which 24,000 die-hard country music lovers flock to Nashville in early summer for the chance to rub cheeks with their favorite stars. This was also Chris's vacation and was, in a moment, about to become the highlight of her life.

"There's a story behind this," she said. "I had a panic attack problem for over thirty years of my life. I never went anywhere. Even as a child, I never went camping, or to Girl Scouts, or to anything that made me stay overnight. Then a song that Garth Brooks put out, called 'The River,' changed my life. It got me to see that I could get over my panic attack problem if I could just give it to God. There's one line in particular: 'Don't sit along the shoreline and say you're satisfied. Choose to chance the rapids, dare to dance the tide.' That inspired me to leave the house, to come to Fan Fair, and now to meet him." She paused to look up at the tan cowboy hat, now just several paces away. "Because that's the man that put out the song that changed my life."

Twenty minutes later, Chris Lowe reached the front of the line that stretched out through the stockyards of the Tennessee State Fairgrounds, around the corner toward the Nashville Speedway, and out toward the vinyl cowboy boot vendor, so far from the front that fans who were waiting there would not reach their target until sometime the following morning—twenty-three hours and twenty-one minutes after the biggest star in country music history had started signing autographs.

"Oh, my God, I can't believe I'm here," she said when her turn finally came.

Garth Brooks looked at her with the smile of a storybook Good Humor man. "What's your name?" he said. Like her, he was wearing blue jeans and a gray T-shirt. His boots were dusty from half a day in the dirt. Since he arrived, he hadn't sat down, hadn't eaten, hadn't gone to the bathroom. It was perfectly in keeping with the spirit of Fan Fair: two parts dedication, one part obsession.

"My name is Chris."

"Nice to meet you," he said.

"I'm from Iowa," Chris said. She launched into her story. She told him about the panic attack problem, about never leaving the house, about going to see him in Des Moines in 1992. "You mentioned from the stage that night that we should all come to Fan Fair, and I told my husband that night we were going." For two years, she had come, she said, but Garth hadn't. Chris and her husband went home empty-handed.

"Thanks for not giving up on me," Garth said. He started to sign her T-shirt.

"I would never give up on you—*ever*," she said. She started to cry.

"Did you have fun at the show?" he asked.

"It was really a great concert," she said. After taking a few snapshots and posing for a few pictures, she reached up to give him a farewell hug. He would do it hundreds of times on this day alone. She would do it with a half dozen others in the next several days. But at that moment, it was the most important embrace of her life.

"Oh, my God!" Chris squealed as Garth let his arms linger a tad longer around her back. "I'll never forget this as long as I live. I could die right here in your arms."

Certainly the most impressive feature of country music today is the intimate, almost familial relationship that exists between artists and fans. That relationship is so familial, however, that at times it seems codependent, even gothic. Perhaps the South's biggest legacy to country music is a tradition of morbid family closeness. Think of Faulkner's Snopeses hanging out at the food court, eating fried chicken, and simultaneously cozying up to and smothering one another, and you get a sense of what it's like being around most country artists when they come in contact with even their most casual fans.

Increasingly that relationship has been turning violent. Though Robert Altman's *Nashville* ends with an obsessive fan shooting the Ronee Blakely character in front of the Parthenon, not until the 1990s did some younger stars, like redneck balladeer Tracy Lawrence, begin wearing bulletproof vests at their outdoor concerts. Lawrence, himself no stranger to violence (he once pulled a gun on a carload of people who chased him on a highway), was both stalked and mobbed early in his career. "There was one lady who grabbed me when I was getting on the bus after a show," he told *The Wall Street Journal*. "They couldn't get her to let go. She got away with a hank of my hair." Reba McEntire was later the focal point of the bizarre incident in which two young women from Ohio hatched an aborted scheme to meet their idol by taking visitors to the Grand Ole Opry hostage. Garth, meanwhile, who told me he "has more guns than he could count," has people camping out in front of his gate most months of the year, thereby forcing local law enforcement to provide protection. He also regularly received bomb threats at his concerts and even devised a special way for his staff to notify him of such incidents—by courier underneath the stage.

The flip side of fan violence, of course, is fan sex. Musicians have always had groupies, and Nashville, despite its Baptist leanings, has been no exception. Hank Williams was deluged with women, even during his two marriages, and one of his many girlfriends gave birth to his only daughter two days after his funeral. With the Outlaws, Nashville had its first gang of artists prepared to flaunt their indulgences. "The music, the pills, and the women," Waylon wrote in his autobiography, "that was our life on the road. Sometimes I'd screw two or three a night. My road manager remembers me reserving extra rooms in hotels, running up and down the emergency staircase to get from one to the other. Once, in Louisville, I had girls stashed on three floors. I had been awake for several days and was determined to please them all." Twenty years later, those activities had hardly changed. "I literally chuckled to myself when I was reading Waylon's book," Travis Tritt told me. "There's one section where he says, 'I had a reputation for being able to take more drugs and screw more women than anybody in the business.' I just thought, 'That's me, man.' Because I could drink more liquor and screw more women than anybody I knew. And I had 'em lined up." He grinned. "In case you wondered, it is possible to have sex with three people in one night. I know because I've done it—on numerous occasions."

So how did he do it? "It's kind of like herding cattle," Travis said. "I'd walk offstage every night and go, 'You, you, you, you, you, and *you*. I would cull maybe six or seven, rent a conference room in the hotel, and have my guys set it up. Then it's one big party." This part of the evening was for screening the candidates. "That's your opportunity to sit down and talk to them," he said, "find out which ones suit your taste, which ones don't. Then once you get it narrowed down to the two or three, you move those on up to your hotel room. Or, there's a possibility, depending on what the temperature of the situation is, you may put three of them in three different rooms." And what were his criteria for selection? "Obviously, it was appealing to the eye," Travis said. "This is a guy who never had any kind of attention from females in school. And once I did get it, it was nowhere near these gorgeous babes. I mean, these girls looked like they stepped out of *Cosmo*, *Penthouse*, or *Playboy*. And they're there. For the picking. Every night. And they don't want commitment!" All of which leads to the obvious question: Why did they do it? "I don't know," he said. "I don't know if it's because it makes them look more important. Or if they feel it's their only connection to someone famous. It's kind of like the old syndrome of touching the hem of the garment. If they feel like they can be associated with an artist, they know people are going to talk about them. She's going to walk down the street and people are going to say, 'She slept with so and so.' I guess some people find that glamorous. And as far as I was concerned, I wasn't going to look a gift horse in the mouth."

The potentially explosive mix of violence and sex has led to a new growth industry: backstage passes. Invented in the 1970s as a way to standardize backstage admittance to rock shows, passes arrived in abundance in Nashville in the late 1980s when country acts started hitting it big. On Alan Jackson's tour, for example, the singer had seven different kinds of stick-on passes: rectangular for BACKSTAGE, square for PRESS, diamond for MEET & GREET, and triangular, circular, and octagonal for different categories of GUEST. Each stick-on was printed in both red and green versions, which the road manager alternated daily to confuse posers. "We figure we'll be far enough away by the third day," he told me, "so someone won't drive and try to reuse the same color." The ultimate pass, laminated and given only to permanent staffers and, presumably, the artist's mother, was ALL ACCESS. On the Garth tour, this pass was considered so sacred it actually contained a passport photograph of the bearer and a hologram of the artist's trademark. When I failed to have even my temporary pass on display just hours before his opening concert, two burly cops promptly lifted me from the arena floor, carried me up the stairs to the lobby, and tossed me, like a hammerthrow, outside the Omni door. Having been at the building nearly twenty-four hours by that point, however, I knew precisely how to sneak around the back and enter Garth's dressing room from behind. It didn't hurt that Garth's mother knew my name.

The main reason, of course, for all this heavy breathing among security personnel is the presence of so many heavy-breathing fans out there. Backstage passes—stick-ons, that is—have several pejorative nicknames on the road ("knee pads" are one example) growing out of their widespread use as commodities to be exchanged for sex. It's not that the stars need backstage passes to land dates—usually a wink, a nod, and a road manager can get the lucky fan or two backstage. Instead, it's the crew. Many road crews have what they call a "bimbo pass," which offers wearers after-show access only. A crew member ushers a young damsel backstage with the promise of unexplored wonders and a potential spot on "Lifestyles of the Rich and Famous" ("The fan who met her hubby behind the curtain and went on to fulfill his every dream . . ."), only to find that (Oops! Who knew?) the star has unexpectedly departed and the only person left to fulfill her fantasies is the second light rig assistant.

Even journalists have been known to benefit from such refracted glory. I was once flirting with two women at a Brooks & Dunn concert in Nashville. "Oh, you should be talking to my friend," the prettier of the women said to me dismissively. "*She's* single." As soon as she noticed the laminate around my neck, however, she changed her mind. "Wait, you have a backstage *pass*," she cooed. "In that case, I'm a single too."

All of these components of fan management are never more evident than in the largest contact sport in Nashville, the grand ole orgy of country music, Fan Fair. Though it takes place at a location normally reserved for cows and pigs, Fan Fair, like much of country music, is actually something of a sitting duck. Just the idea of thousands of fans converging on middle Tennessee in the middle of summer to eat corn dogs, listen to fiddle playing, and wait in interminable lines for the chance to receive a hug and peck from some Stetson-wearing good ol' boy (or girl) is often too much for cynical observers to pass up. Even if those observers are part of the country music industry, the target is still irresistible. "The real question about Fan Fair," Wade's publicist said to me just days before I attended my first, "is whether polyester conducts heat?" Minutes later, after I told her I was writing about the event, she quickly retorted, "Now, don't go write something snide."

To be fair, there are many parts of Fan Fair that even the most sympathetic person would find extraordinary. One is the nature of the gathering itself. Try to imagine, say, Barbra Streisand or the members of Megadeth shelling out thousands of dollars of their own money to build what looks like an overgrown science fair booth, complete with fake paneling and faux haystacks, behind which they then sit for hours at a time, sign autographs, and receive silk roses with tiny vials of perfume attached. Country artists not only do this, but most participate in what amounts to an internal rivalry to outcountrify one another. Alan Jackson decorates his booth every year to look like a storybook farm, complete with a white picket fence and hanging plants. John Berry builds a simulated front porch. Pam Tillis topped everyone when she sliced a motorboat in half and attached it to the back wall of her booth to draw attention to her upcoming song "Betty's Got a Bass Boat." Her effort still didn't compete with the Oak Ridge Boys, who continue to earn "aaahs" of admiration for their 1982 stunt of arriving at the fairgrounds by helicopter.

Behind this veneer of country excess is a remarkably enduring, even warmhearted tradition. Fan Fair was started in April 1972 as a way to bring country music fans into closer contact with the genre's artists: superstars, unheralded newcomers, and a sprinkling of flat-out eccentrics. Originally held in a downtown arena, the event drew five thousand fans in its first year alone for a mix of concerts and autograph sessions.

The following year it doubled in size. In 1982, to accommodate even more fans, the event moved to the hilly yet barren Tennessee State Fairgrounds not far from Music Row. Since then, it has become both the highlight and the dreaded swamp of the country music year. Artists sometimes complain about the cost (most not only pay for their booths but also spend thousands more to feed their fan clubs at the obligatory picnic), but fans love it. For $90 dollars, they get admission to five days of events at the fairgrounds, as well as entrance to Opryland U.S.A., and the Hall of Fame. At the fairgrounds alone, fans have a choice between an ongoing series of concerts and display halls where they can meet the stars. Attendees end up declaring their allegiance early on. "I can go to a concert anytime," said a lawn-care worker from Kentucky who carried a special scrapbook with over 100 pages of signed portraits of country stars, "but you don't always get a chance to get autographs. You come up here, you get to meet your favorites, and if that means you have to wait eighteen, nineteen hours in line to get there, well, that's how it is."

While this level of dedication is the hallmark of Fan Fair, it also raises a question: Why would someone wait in line eighteen hours for an autograph? What explains this degree of devotion? After leaving the line of people waiting to meet Garth on day two, I drove fifteen miles on I-40 out to Hermitage Landing, the lakeside marina several miles from the antebellum plantation where Andrew Jackson once lived. It was the location Wade had selected to host his first fan club picnic. About four hundred people were gathered on the beach when I arrived, lying on blankets and dancing in the sand as Wade and his band gave a free show on a small stage. The sun was setting on the piney vista behind the drummer. Every now and then, a plane would dip, silhouetted across the orange sky, on its way to land at Nashville International Airport.

"As y'all may know, we have a new album coming out next week," Wade told the sympathetic crowd that included his parents, standing peacock-proud at the rear. Wade was wearing blue jeans and an orange button-down shirt. He had a yellow pin on his shirt that said: ON A GOOD NIGHT: COMING JUNE 25. "We'd like to play you some new material off that album," Wade said. "If y'all don't mind, that is."

He launched into an introduction of "On a Good Night." "As you know, it's been on the charts several months now," Wade said, treating his fans like they were part of his business team. "I pretty much get nauseous when those charts come 'round every Monday. I watch 'em close. And I tell you, the Top 20 is just jammed up right now. We've got so many good folks with new records out. Vince and Wynonna, for example. But ours is still going up. We've jumped another two records this week, even though there's no place to go." The audience broke into applause. "Vince's single went back two places," he continued, "as did Wynonna's. And they've still got bullets, which means they're not through yet." Would he tell them about the number of station adds? I wondered. What about the research? "But they're lettin' us sneak through, which is pretty nice." More applause, this time with screams. "We might just have another hit on our hands!"

After the show, Wade retired to a nearby shed where the volunteer president of his fan club had set up a small barn scene that mimicked his latest video. Wade settled in front of a red barn door covered with horseshoes, posed for pictures, and signed autographs atop several hay bales. He would stay there for the next four hours. Outside, 100 or so people milled around the picnic tables finishing their complimentary meals and trading gossip. The chief stories this night: Alison Krauss had lost fifty pounds on a mysterious crash diet; Faith Hill and Tim McGraw, long rumored to be a couple, had begun French-kissing onstage ("Locking tongues like snakes," one woman told me); and Alan Jackson's wife Denise had stood up at his event at the Ryman and said she wanted to end the rumors that her husband was also having an affair with Faith. "The only women in Alan's life are Allie and Mattie," she announced, referring to his daughters. As one shocked fan from Alabama recalled, "I almost fell out. If you're secure in your marriage, why do you say that?"

Across the grass, I ran into Trish Fuller, the best example I had met of a devoted fan. Trish, forty-five, was the postmaster of Mason City, Iowa. She had shoulder-length black hair, a slightly beaked nose, and a friendly Midwestern accent. On this day, she was wearing her signature Wade Hayes jean jacket. I had first met Trish several months earlier after a Wade concert in Owensboro, Kentucky, a fourteen-hour drive from Mason City. Since that time, she had also seen him in Green Bay, Wisconsin, a six-hour drive from her home, and Severville, Tennessee, a twenty-hour drive. "I got this tour jacket at last year's Fan Fair," she explained. Then Wade had written: "Trish, Thank you for spending my first Fan Fair with me." This year he had added, "To Trish, The First Fan Club Party." She also owned two Wade pillowcases, she explained, a poster, and a set of Wade dog tags, all of which she had purchased from his fan club. Her Iowa license plate said WADE FAN, and when Wade didn't believe her, she brought the license to him, took his picture holding it, and made that into her keychain. Then she put the picture of him and the license on the front of a T-shirt, and on the back she listed all the dates and places she had seen him in concert. It totaled nine states; twenty-three locations; 15,683 miles. "It started when he told me that Trish is also his mother's name," Trish explained. "Obviously he could remember my name. We developed a really special relationship."

"And do you consider yourself a groupie?" I asked.

Suddenly her tone became serious. "There's nothing sexual about this," she said. "This is what I do. This is what's important to me. I sincerely like country music. And country artists, there's nothing like them. You don't see the Spin Doctors meeting with their fans. It's just totally different. And I'm glad it is. It's just really special to me to be able to meet that person whose music I enjoy listening to and whose songs really move me. Then to get to meet them and know what they're like, that's what special about Nashville."

Back at the fairgrounds, it was almost midnight when I arrived. The evening concert had long since finished, and the nightly fireworks had long since been fired. The only people left inside the facility were the several hundred fans who were still in line to get their audience with Garth Brooks. Garth had been signing for fourteen hours and, according to the people in line, still hadn't broken for food or relief. It was the perfect expression of his mind-set at that time: the stubborn hero. Would his fans have minded if he interrupted his signing for a few bites of a hamburger?

More striking, though, was the reaction of everyone around him. Close to four hundred people stretched in calmly snaking fashion from one end of the football-field-sized shed to the other. A half dozen cops stood idly by, drinking coffee and nodding off. Soon it began to rain. The temperature dipped. By any measure, this should have been the hour when tempers started to flare. It turned out to be the opposite: a country version of *Lord of the Flies,* where people formed alliances instead of rabid packs. One group of linemates played charades. Another strapped umbrellas onto their heads and simulated a bullfight by waving ponchos. A group of women used a cellular telephone to call Domino's and have them deliver a pizza. When it arrived, though, they decided it wouldn't be right to eat it alone. "We're doing a Garth Brooks deed," one of the women said. "We're going down the line just feeding the children." Several hours later, a man near the end of the line ran out of cigarettes. He offered to buy one from a neighbor for a dollar. The neighbor, having only three, declined, pleading addiction. The man went to the next person in line, who also declined. "I've got ten, but I'm kind of a heavy smoker," he confessed. Finally a nearby woman had eleven. She sold the first man a cigarette, who lit it and passed it to the second, who passed it to the third, who passed it back to the woman—each of them taking several drags. "Never seen each other before," one observer remarked with utter admiration, "and here they're smoking the same cigarette. It's like a bunch of inmates in jail."

By half past six in the morning, the sky started to lighten and the mood turned to celebration. A reporter from "Good Morning America" set up a camera to do a live remote broadcast and had to urge the audience to act more subdued, as if they were tired from staying up all night. As soon as the light on the camera went on, though, the fans cheered. They were part of a history-making stunt, and they were proud of themselves. Back at the front, meanwhile, Garth was starting to strain. His eyes were narrow and bloodshot. He was leaning against the fence. A cup of undrunk Dr Pepper had sweat through the bottom of its container. At nine, the camera of a woman in front of the line broke, and Garth, exhausted, said, "We're doing the hard part here." Whereas six months earlier, on the verge of his return to the public eye, Garth had signed autographs at the Grand Ole Opry until the middle of the night and seemed, at the time, to be drawing energy from his fans; now, feeling battered and bruised by six months of decline, his effort to push himself beyond what was humanly possible seemed more desperate. Plus, it wasn't working. Instead of Garth drawing strength from his fans, this time his fans seemed to be drawing strength from him.

At the start of the twenty-fourth hour, there were eight people in line and, now that Fan Fair had reopened, a huge throng of gawkers pressed around the hastily formed police line to watch the end of the sideshow. When the last person, Shirley Johnson of Fontana, California, stepped up to have Garth sign two shirts, a poster, and a cowboy hat, a giant roar rose up from the crowd. "I just can't believe I'm here," she said. "I'll never forget this the rest of my life." Garth, by this time, could hardly speak. He gave her one last hug and smiled blankly at the empty space where once the line had stood. He gathered up the notes and wilted flowers that were scattered around his feet and hobbled toward the exit route that Mick had carefully cleared. The transfer of power was now complete. Garth Brooks had nothing left to give; his fans had taken it all.

SIXTEEN

THE LAUNCH

On the last Tuesday in June, Wade Hayes's mother was standing in her cramped, naturally lit kitchen off a clay road in Bethel Acres, Oklahoma, baking a loaf of banana bread. The persimmons and cottonwoods just outside her window provided a layer of protective shade in an unusually arid summer. Over her kitchen table, a single-room air conditioner was straining to cool a three-room area that already, at nine o'clock in the morning, was filling with the hurryings of the day's homecoming. Today was the day that *On a Good Night,* her son's second album, would go on sale from Key West to Puget Sound. Today, his young career would confront its most critical sophomore exam. And today, his first since striking it rich in the manner his father had always dreamed, Wade would make his long-awaited return to the no-stoplight town an hour's drive southwest of Oklahoma City that didn't even have a post office, but that did have what was even more important for this ceremonial-cum-commercial occasion: a Wal-Mart.

"It's a wonderful day," Trish Hayes was saying. Several years shy of her fiftieth birthday, Trish, a trim hairdresser and amateur watercolorist, was having the kind of day about which most mothers dream, but often come to dread. She had gotten up early and done her hair. "It's short, layered, bleached-blonde hair," she explained with professional precision. "I change it every month." Then she went out to feed winter wheat to the small herd of cattle that a friend had given her husband as barter for a carpentry job. "This is the first time we've ever had cows," she said. "Our original purpose was that we didn't have any kids at home and we needed a tax break." Then she set out cleaning the house and baking Wade's favorite dessert.

"It's a wonderful day because he's coming home," she continued. Her voice had an upbeat, knowing tone like Jessica Lange in one of her beleaguered Midwestern heroine roles. She had little of Colleen Brooks's buoyant optimism to her or Naomi Judd's sculpted sincerity. Trish Hayes had lived through several heatless winters as the wife of a ruined musician and the mother of three hungry kids. "This is the first time the town's recognized him," she said. "The mayor's going to be there. All his friends from school. I don't think Wal-Mart's prepared for what's going to happen. People are already standing in line—have been since seven o'clock this morning. Heck, we have an unlisted number, and in the last few days it's just rang and rang and rang and rang." And just as she said that, it rang again. Across the room, her elder daughter, Stacey, put down her four-year-old son and went to answer it. "You see, people are *so* excited!" Trish said. "Wade's put Bethel Acres on the map."

And in the process, she believed, culminated a journey that had been under way for generations. "All my husband's family were musicians," she explained. (Her husband Don was off at work.) His father played fiddle. His two brothers played guitar. Don played guitar and sang. "He's *almost* as good a picker as Wade," Trish said. "Singing, though, they sound exactly the same. Wade was on the radio last night promoting the album and a friend called and said, 'Trish, he sounds just like his daddy talkin'.'"

"And what did that feel like?" I asked.

Trish pulled her hands from the sticky cinnamon batter. "It's like watching your destiny unfold," she said. "I'm kind of awed by it. It's like we've been on a path, and I'm finally getting to see where we're going and why. When you travel through life, you get confused a lot of time, especially if you have a faith in God, which I do, strongly. But when things go so badly, you whine, 'Well, what did I do wrong?' But then years pass and you see why you had to go through that. I've told people, 'You pay for your kids to go through college. We paid for Wade's education in music.'"

Don Hayes and Trish Snow met in Shawnee, Oklahoma, in 1966. Though both had roots in the state, their path to Shawnee, the "Crossroads of Oklahoma," had been circuitous. Don was born in 1948 in Early Mart, California, on the grounds of a labor camp. His mother, a native of Oklahoma, had been raised on Cherokee Territory, though she denied being part Native American. Don's father was born in Arkansas and met his wife when he moved to Oklahoma to become a farmer. The two married, had two sons, then lost everything in the Dust Bowl. In a story taken right out of *The Grapes of Wrath,* the family then fled to California, where they found more hardship than opportunity. While living in a camp, they had a third son, Don. "Again, this is where music comes in," Trish said. "Don and I always loved Merle Haggard's music because it touched so many things in our lives. Merle had a song—'Working Our Way Back Home One Row at a Time'—and that's exactly what his family did. They worked their way back home to Oklahoma by being migrant workers." (Later, when Wade opened for Merle at the Ryman, Merle said to him, "Son, where'd you learn to pick a guitar like that?" "Sir, from listenin' to your music.")

Back in Oklahoma, Don met Trish at the local Sonic hamburger drive-in when both were in high school. Trish had been born in Hugo, a small town south of Oklahoma City, but moved to Shawnee when her father got a job at nearby Tinker Airfield. Shawnee was a boom-and-bust oil town of thirty-five thousand that was home to both Jim Thorpe, the famed Native American athlete, and Dr. Brewster Higley, the writer of "Home on the Range." With little diversion, teenagers spent their afternoons cruising the drive-ins. During one such session, Don, a shy, lanky musician, told his friend Specks, "There's the girl I told you about. I'm gonna marry her someday." "Her?" Specks said. "That's my sister." The two were married the following year, in February 1967, when both were eighteen. They went to work as his-and-hers hairdressers and moved in with his mother in nearby Bethel Acres. Within months, Trish was pregnant. "We were so poor and so stupid," she said. "I stayed pregnant for two years." Stacey was born in 1968, Wade in 1969, Charity in 1973. By then, Don had given up cutting hair ("He just couldn't take the gripey women," Trish said) and went to work in construction.

Don had always played music and, even after getting married, would pick at locals bars once a month. After Merle Haggard won CMA Entertainer of the Year in 1970, Don sat Trish down and said, "What do you think of us trying to do this more seriously?" She was agreeable. "I think we can do that," she said. Trish extended her hours at the salon, and Don formed a band: Don Hayes and the Country Heritage. It included two guitar players, a bass player, and a drummer. In no time, the band was playing four nights a week, opened for Nashville acts who played the area, and eventually cut an independent album, which they sold at their shows. When a small label in Nashville released a single from the album in the late 1970s it briefly grazed the playlists of a few local radio stations. Soon the head of the company promised Don even greater fortune if he left his band and came to Nashville. In 1982, Trish and Don Hayes sold their house in Oklahoma, packed up their three kids, his mother, and their cat, and moved to Tennessee. As soon as they got there, they realized they'd been had: The company didn't exist; the man had been a fraud.

"It was scary," Trish said. "Nashville hadn't boomed yet. It was the worst winter in a hundred years. We couldn't even afford to heat the house. The only thing we had was a rack of wood that the previous owner had left us. Stacey had to get her first job. Wade had to go without clothes. I went to work in a beauty shop." Though the pressure was great on the whole family, Wade, thirteen, still caught the music business bug. "The whole episode was bizarre," Wade said, "but still, it was probably easier on me than it was on my two sisters. First, I'm a boy, and those kinds of things are always easier on boys. And second, I was eager to get in and learn about the music business because country music had always been such an important part of my life." Trish, though, was fed up with music. "I felt it was over," she said. "We shot for it. It didn't work. Let's go home. My kids were falling apart. They were having race riots in the school. Wade was freezing. Stacey had become ill with mono. It was just awful." Finally she told Don, "I'm going home." "Not without me," he responded.

Still, they didn't have enough money to move, so while Trish was in Washington, D.C., visiting her sister-in-law, Don sold the family van for $2,000. He also gave the buyer, a gospel musician, the title. By the time Trish returned, the check had bounced. The musician turned out to be a con artist. And the Hayeses had suffered their second disaster in ten months. This time, though, they were desperate. With the help of a policeman, Don tracked down the con artist, burst into his house, and took the van from his garage while he was in the process of stripping it. The next morning they loaded the kids, the cat, and the grandmother into the van and headed back to Oklahoma. When they got there, Trish sank into a depression, and Don abandoned his dream. "Moving down there in some ways was a bad decision," he said later, "but in other ways, it was something we had to do. We had to find out if I could make it." Also, he added, "it was a real learning experience for Wade."

Once back in Oklahoma, Wade, shy and lanky like his father, couldn't shake the idea of a career in the music business. He'd been fascinated with music since childhood. "I can't remember a time when I didn't want to be a musician," he told me. "I'd drive my parents absolutely crazy, walking around all day singing those country songs." Trish agreed. "Sometimes his veins would almost burst, he sang so hard. But he was always very talented. He learned. He would absorb. Don would play music and Wade would just listen." At ten, Don bought him his first mandolin. "I was furious," Trish said. "It was winter. There wasn't much work, we didn't know how we were going to keep the electricity on, and Don brought this mandolin in. It was the first time I went to my mother's. Went there for two or three days." Wade, though, was transformed. He'd set his alarm an hour before he had to get up, just so he could practice. Returning from school, he'd head straight to his room. The following year he switched to guitar. "We'd never fuss at him," she said, "but he'd drive you crazy. He'd be sitting on the couch next to you watching TV and doing runs on the guitar. You just wanted to bop him on the head. But I thought this would all be worth it someday."

Eventually Wade followed his father's lead and put together a band. "Anything I did, Wade wanted to do," Don said. "If I was building houses, he wanted to do that too. If I was playing in clubs, he wanted to do that." "He was always in tune with his daddy," Trish agreed. "He knew the ins and outs of what we had been through. I didn't have to tell him how hard it was. He knew. But *I* knew that he was a real gifted person who could survive." Wade, though, was scared. "That whole episode in Nashville put such a fear of people in me," he told me. "I'm not a trusting person anyway, but because of that I didn't trust anybody. It's kind of unhealthy." After graduating, he went to college for a few years, worked a few odd jobs, but seemed haunted by the idea of a career in music. "I don't feel like this was something I chose to do," he said. "I think it kind of chose me. I tried for a lot of years to get it out of my system and get those thoughts out of my head. I tried to get through college. I tried to get a regular job and be happy and comfortable. But as I got older, I kept realizing that that wasn't me."

"I think he wanted something secure in life," Trish said. "When you're older, you know that nothing is secure, but when you're young—" She shrugged. "Wade grew up knowing that everybody's parents worked at Tinker Airfield or at GM and had those good paychecks coming in every week, and he thought he wanted that kind of security. But in college, he was becoming more and more unhappy and hard to live with.

He went and did a show with a friend in Nashville and they talked him into moving down there." The week he was supposed to leave, though, his grandmother passed away. "So of course he couldn't leave that week," Trish said. "Then he didn't want to leave the week after that. The longer he waited, the scarier it got. Finally I sat him down and said, 'Wade, if you keep putting it off, these people are going to forget they ever saw you. You're young. Your dad has given you the tools to survive. It's *in* you to try it. I've lived with your dad. And I know you. Someday you're going to try. It's better now while you don't have a wife and kids.' "

It was 1992. That month Ricky Skaggs stood at the podium of the CMA Awards and said, "All of you young musicians that are struggling with your art, you need to go ahead and pursue it because that's what you're called to do." This time Wade got the message. "His dad gave him his best guitar," Trish said. "Wade had saved up four hundred dollars. We had a bedroom suite his grandmother had left him. He loaded that up in his pickup and we cried watching him go down the road. But there was just no doubt in my mind. He'd call me and say, 'Mama, I don't know.' I'd say, 'Wade, you're a Christmas package that's come to this town. You can do it all. You can write. You can sing. You can play. You've been to college. You've got the looks. It's all there. You just hang in there. It's going to happen.' "

And, of course, it did. Within weeks, he was playing guitar at Gilleys'. Within months, he was writing songs with Chick Rains, a veteran song-writer from Oklahoma. And within a year, Chick introduced him to Don Cook. "I remember the first time we went to see him in Nashville," Trish said. "We wanted to see what kind of people he was meeting. Wade introduced us to Chick and Don. They took us to breakfast at the Shoney's on Music Row, and I knew immediately that they were good people. Wade got up to go to the bathroom, and I said, 'I just want to thank you guys for being so good to my boy.' They looked at each other, smiled. 'Oh, we think your boy's gonna be real good to us.' " Heading back to Oklahoma that evening, Trish and Don were giddy. "I just kept saying to my husband, 'It's really gonna happen. It's really gonna *happen* for him.' And Don said, 'It really is.' It was like getting something for Christmas that you really wanted, but you never thought you'd get."

But what exactly were they getting? Today, three years after Trish and Don Hayes first visited Wade and a year and a half after the song he played for them during that trip reached number one, Wade's career seemed, by all measures, to be skyrocketing. He had two number ones, a gold album, was *Billboard's* reigning Debut Artist of the Year, and was returning home as a conquering hero. But from Music Row's perspective, Wade's career was in a surprisingly perilous position. His new album was turned in two months late. His first single from that album was mortally wounded at radio. And while that was happening, Bryan White, his altar boy competition from just up the road in Oklahoma City, had scored his *fourth* number one, had seen his first album go *platinum,* and had leapfrogged over Wade into the position of "It Boy" of the moment. As a result, Wade was actually under *greater* pressures now than he had been at the start of his career. The reason: Having shown he could make money, Wade was no longer the little boy who could, but had become a corporation's chief economic asset. This was the backside of the Nashville dream in the 1990s. We can take you to unexpected heights, Music Row whispers, but in doing so we will put you under unspeakable stress. For Wade in particular, the financial expectations of his company, coupled with his craving to regain his momentum, had conspired to undermine the one thing he most desired from his career: security.

How Wade coped with this new state of affairs was fast becoming the unspoken drama surrounding the release of *On a Good Night.* On the one hand, the signs coming out of Sony Records were universally positive. Sony believed in Wade Hayes and, because of the label's lackluster performance in recent years, deeply needed him to succeed. As a result, even though the album's first single was not performing well at radio, Sony still pushed ahead with its strategy to launch the album in stores with a major marketing onslaught. "Without putting any of my team's butts on the line," Allen Butler, the head of the label, said at a luncheon several weeks earlier touting Wade's "gold-plus" debut, "I think they would all stand up and commit to this on the spot: His new album will be a platinum-plus album." Mike Kraski, the vice president of sales and the man charged with selling as many albums as possible in the all-important first week, was even more blunt in a conversation we had just days before the release. "We have to go out and make a major statement that Wade Hayes is *still* the king of the crop of new and upcoming artists," he declared. "Basically we have to kick Bryan White's ass." Bryan had sold 14,000 units in the first week of his second album, Mike noted. "We need to exceed that."

The plan he concocted to achieve that goal reflected not only the new marketing clout of Nashville, but also the increasingly high stakes of competing in the music marketplace. Country, for all its inroads into the mainstream, is still an extremely price-sensitive commodity. Fully two thirds of country music is sold at a discount merchandiser, with *half* of that coming from either a Kmart or a Wal-Mart. For Wade, who sold particularly well in the Southeast and Southwest, the figure was closer to 80 percent. Anderson Merchandisers, the company responsible for stocking (the industry term is "rack jobbing") most of the two thousand plus Wal-Marts, controls 60 percent of country music sales in many parts of the country. This concentration of power has made these wholesalers some of the most powerful players in country music. As a result, even before an artist goes into the studio, he or she is whisked off on a corporate jet to meet, schmooze, and often perform for distributors. Such performances can be life or death. Garth Brooks is famous for advancing his nice-guy reputation by tirelessly seeking out factory workers and distributors. "He's the only artist I've ever worked with who uses the term 'rack jobber' in conversation," one Capitol executive told me. Vince Gill, by contrast, though considered one of the nicest men on Music Row, nearly torpedoed his reputation at one wholesaler party several years back by cursing employees who weren't listening to him perform, then storming off the stage. Even Wade, certainly a "yes sir, no sir" kind of artist, nearly stifled his nascent career by having too much to drink at his first appearance at such an event.

He recovered, though, and the relationship he developed with major retailers around the country proved critical to his success. Such relationships matter because retailers often rent prime shelf space to labels for a monthly fee. Endcaps (the prime area at the end of display cases), counter space, even floor space—painting an artist's face on the tile— all come at a price, as high as $50,000 a month, which labels happily pay since surveys show that 60 percent of purchases in these stores are *un*planned.

Mike wanted to do something that would make Wade stand out in Wal-Marts in particular. "We went down to Anderson's laboratory store," he explained, "and told them, 'All your space is already allocated. If we gave you an opportunity to make an aggressive statement on Wade Hayes without using up any existing space, would you say yes.'" They laughed. "What are you going to do?" they said. "Hang it by the ceiling?" With that, Mike unveiled a special cardboard display case he had designed. Called a "wing bin," the case, about the size of an ironing board, had a picture of Wade at the top and two shelves for CDs. It was designed to fit in one of the vertical slots normally reserved for the section sign saying COUNTRY. "We slid it in the slot and they fell in love with it," Mike recalled. Within weeks, three thousand such bins, which Sony manufactured at a cost of nine dollars apiece, were slid into place in every Wal-Mart in North America.

While the long-term impact of these promotions, coupled with the warehouse visits and in-store singalongs Wade had been doing for months, might be to make a "major statement" as Mike hoped (advance orders were so high, he upped his initial shipments to 235,000 units), they were also having a more immediate consequence: They were starting to overwhelm the one person they were most intended to help. Originally Wade had welcomed Sony's aggressive stance. "He's an intensely competitive guy," Mike Kraski said, "which is surprising because when you meet him he appears to be very aw-shucks and humble. But he wants to win in the biggest way and he told us, 'Put more on me, put more on me, I'll tell you when it's too much.'"

In the weeks leading up to the release, though, Wade did begin to bend under the pressure. The steady increase in confidence that had been under way for most of the previous year suddenly started to regress. First, at the Academy of Country Music (ACM) Awards in late spring in Los Angeles, Wade was visibly horrified when Bryan White defeated him (and David Lee Murphy) for the New Male Artist of the Year. "I assumed it would be me," Wade told me, "because I had been on more tours, sold more records, and had more hits than anybody in the category. So how could I not win it? It hurt my feelings pretty good." Even worse, instead of gamely applauding (which is the award show custom) he openly grimaced and slumped his head on national television. "I was embarrassed," groused one executive at Sony. "If he doesn't know how to behave, he should stay at home."

Next, in his prealbum interviews, he came across as a man on the verge of a nervous breakdown. "Hayes frets so much about his career that it has actually made him ill," wrote Bob Oermann in *The Tennessean.* "He admits that he's in constant fear that it will all end tomorrow." The article went on to quote Wade as saying he was consumed with the business aspects of his career. "I'm a chart watcher, an everything watcher," he said. "It's to a fault, to the point where I make myself sick physically. I get headaches and start losing weight and stuff like that. When my career started unfolding is when it got bad. I just recently figured out that I'm going to kill myself if I keep going the way I am."

It got so bad, Wade told me, that the hair in his beard stopped growing in places.

Ultimately, the pressure erupted at Fan Fair. Two nights after his fan club picnic, Wade took his turn before the fifteen thousand people gathered in the stands of the Nashville Speedway. It was the night of the Sony show and the entire management of the label gathered on the stage, along with Don Cook, Chick Rains, and Wade's parents. The sight of so many people swaying and singing along to "Old Enough to Know Better" and "I'm Still Dancin' with You" was enough to bring tears of joy to Debi Fleischer's eyes. "I think we have a superstar on our hands," she said. When Wade's set was over, the team of supporters rushed to the side of the stage to celebrate. Wade, though, never showed. Instead of stopping by for a congratulatory hug and high five, Wade ducked into the scrum of security officials and hurried away to his bus. Once inside, he retreated to the back of his cabin, locked his door, and sobbed.

The following day, three days before the album release, Carol Harper, one of his managers, called me at home. "Wade's in a bad position right now," she said. "There are so many people who are counting on him—his parents, his band, the label. He's gone in a little over a year from being a construction worker who had a dream to being the head of a multimillion-dollar corporation. He needs to concentrate on what he's doing right now. Its okay if you go to Oklahoma, but we prefer that you not try to talk to him. He just needs to be left alone."

"I feared this would happen," Wade's mother said. Trish had finished mixing the banana bread and had placed it in the oven, filling her compact home with a storybook aroma. She cleaned her hands and sat down at the table, pausing to retrieve a scrapbook from a crowded shelf of photographs of her son with Merle Haggard, Marty Stuart, Waylon Jennings. She flipped through the yellowing pages—Wade as a boy with his first guitar, in the living room playing with his band, and on the way to his high school prom. In that picture, Wade was wearing a white tuxedo, accented with a pink cummerbund and pink tie. His hair was dyed blond and permed—the high school perk of having a hairdresser for a mother.

"Wade would kill me if he knew I showed you that picture," Trish said. "But it's the only picture we've got from that night. Wade wanted to take Don's Adonis guitar to the prom that night. Don said, 'Okay, you can take it, but I don't want you laying it anywhere. You play it, you bring it home.'" Wade took the guitar in his pickup, and after playing it, drove it back out to his parents'. "And we lived *way* out," Trish said. "So while he was home, he changed into his shorts and then went back. But they hadn't had their pictures taken. Boy, his date was really steaming." She smiled when she realized what she had said. "I guess he's always put his music before his girlfriends."

She closed the book.

"I guess what concerns me now is his spreading himself too thin," she said. "I'm a mom. At Fan Fair last week, I could see he wasn't feeling too good. He seemed distracted, not his usual self. It doesn't surprise me, but as a mother it hurts me. I've seen him sing when he was too sick. But, of course, I've seen his dad do that too."

It was, she believed, part of his legacy. "It's also the pressure of the position," she said. "I told Wade this once, I said, 'Son, the journey getting there isn't going to be as hard as what happens once you arrive. The hard part is going to be staying levelheaded. Keeping your feet planted. The world needs heroes. You're going to be one of those heroes, but it's an awesome responsibility.' "

Which, of course, raised the question of whether he could handle that responsibility. I told her about my conversation with Garth's mother, who had said backstage in Atlanta that if she could tell her son anything, it would be that he was what he had wanted to become. And that he should be happy and enjoy it. Trish nodded knowingly.

"If I could tell my son anything," she said, "it would be: 'Don't be too serious. Enjoy where you are.' Wade is real business-minded. He is obsessed with getting in there and getting his security. I understand that: It's still shaky ground out there; there are no guarantees. But I told him that recently, when he was in L.A. for the ACMs and called to say, 'I hope you're not disappointed that I didn't win.' I said, 'Wade, let me tell you something. We couldn't be prouder of you. We are honored that you are even there. I want you to stop and look around at who you're *with*. You are with the people and in the places that most people only dream about. If you walk away tomorrow, you've lived the dream that most people never get. Go out and enjoy being young tonight. Have a ball.' "

"And did he follow your advice?"

Trish looked up at the portrait of Wade on the wall. For the first time all day, there were tears in her eyes. "Not really," she said.

"And why is that?"

"Because his dad was that way. When Don was that age, he was obsessed too. The music came first. It takes special women to stay with these men. I've told people before: 'I don't care how hard it got. I would never ask Don to leave music.' That's who they are. Wade is the accumulation of centuries of musicians who have been put into his one body. He *is* music. And he is that age when music is more important to him than life. Yet he's also at the stage where he's competing with so many other talented people. It's very intense out there. We all can't stay on the highest plateau." She sighed. "As long as he pays attention to God, though, he'll be okay. We've all got to have pain in our lives. I wish a parent's pain could take care of a kid's pain. But it can't. The only way to learn is through the struggle."

A few hours later, Wade arrived at the Wal-Mart. He was running late. His flight from Houston had been delayed. He had to be driven to a local television station in Oklahoma City for an interview, then drop by two radio stations, and finally pay a brief visit to a local distributor. By the time Wade made the drive to Bethel Acres, he got caught up in rush hour traffic and was forced to divert directly to the store. "Where have you been?" he snapped at his mother, who had been waiting for him at home. She looked at him coolly, then reached to embrace him. As soon as they separated, Wade was hustled into the showroom.

By the time he arrived, an enormous crowd had gathered in front of the makeshift signing booth—a stack of paper towel boxes covered in denim and decorated with bales of hay. They whooped when Wade appeared. Close to six hundred people were crushed together in a three-person-wide line that stretched from in front of the music department, through the ladies' underwear, past the popcorn maker and fifty-cent airplane ride, out into the mall, and through the front doors into the parking lot, where all three country radio stations from the area were blasting music. Wade took a step back and blinked his eyes when he realized the magnitude of his draw. The mayor, Chris Harden, appeared from the huddle and read a proclamation in honor of Wade Hayes Day that few outside her immediate presence could hear. Wade stooped down and gave her a hug. It was over so quickly, Trish didn't even have time to snap a picture. Presently, Wade squeezed behind the table and began to greet his fans.

He would stand there, by himself, for the next six hours, until well after midnight. He would wait there, smiling, until the last fan had brought one of the six hundred CDs or four hundred cassettes that Wal-

Mart had bought for the occasion. And he would stay there, alone, until so close to his flight time the following morning that he'd have to go back to his hotel when he finished and wouldn't even have time to stop off at home for a piece of his mother's banana bread.

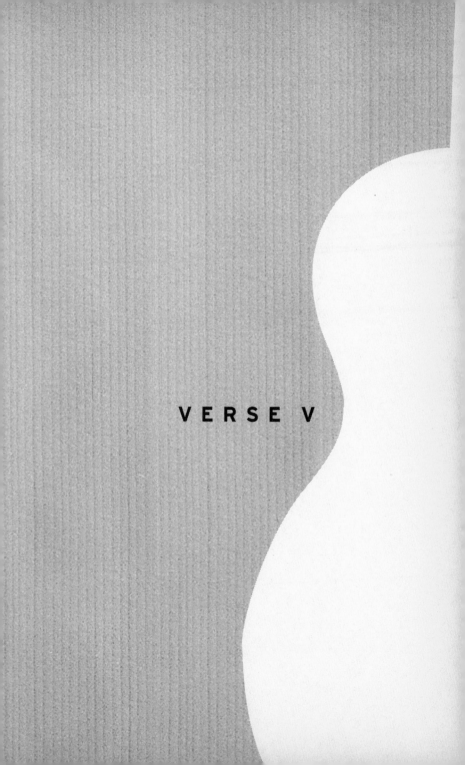

VERSE V

SEVENTEEN

THE SHOW

The only thing visible onstage is the giant light rigging, which lies dormant on the ground. Crumpled, it looks like those futuristic tangles of catwalks and robot arms that litter the ends of Arnold Schwarzenegger movies. Bursts of steam spew from the sides. Yellow spotlights spurt from the center. Presently, as a space-alien wail emanates from the rigging and a "Wrestlemania" roar radiates from the crowd, the Cyclops begins to quiver. The outer arms are the first to move, lifting themselves off the ground in a slow, enticing galactic come-hither. Next the body itself lifts into the air—pink and white with grasping lights. Finally, as the Cyclops reaches its destination overhead, a white grand piano rises from the base of the stage. A man is perched at its bench, looking remarkably like Garth Brooks. Then from a different place comes a voice: "Oh, I said a little prayer tonight." And suddenly, from the belly of the piano, like Venus from the shell—or, it seems unavoidable, Christ from the grave—the real Garth Brooks emerges, his legs pressed together, his arms spread apart, his head thrown back toward the sky. "With all due respect to God," he announces, "tonight we're going to raise some hell!"

By fall, Garth Brooks had reconquered the stage. From the moment he leaped off the piano in Atlanta, it became clear that he did still have the hunger, the drive, the all-consuming "it" that Garth quoted Katharine Hepburn as mentioning and that Garth feared he had lost. GB, in other words, did make his return. "But it wasn't from me," Garth explained. "That's what was so cool." In fact, at the first show in Atlanta, the show Garth said was his worst in ten years, GB originally did *not* show up. "But the crowd lifted us and took us across the finish line," Garth said, "and GB showed up because they brought him, and if that makes me sound egotistical or crazy, then so be it. But he wasn't there at the time we went on. That's because he's out *there*. The people bring him out in me."

It's a feeling, he said, of being comfortable—*more* comfortable on-stage—than he is in other arenas of his life. "I've often described it as being sick as a dog—things can be bad at home or at work—but once you hear that molecular sound, the show starts. When the lights go down, when that crowd goes up, when you feel that first wave of stage smoke across your face, it's like reaching down, grabbing a doorknob, and comin' in. I run in, strip down to my underwear, jump on the couch, and sit there for two hours, just tell jokes and sing."

If anything, the stage became the one place Garth felt totally free. There, Garth was the person he'd always hoped to be. He was humble. He would lumber around, remove his hat, muss his hair, and just stare, wide-eyed, at the unabashed joy of twenty thousand people singing the words to "Friends in Low Places." "Man, I just can't believe you all *came*," he said, awestruck, in Atlanta. He was raucous. He would dart around his playground like a hyperactive Pac-Man, gobbling up the gifts from the audience and screaming in his drummer's ear, locked in a mad race to get through his most spirited hits like "American Honky-Tonk Bar Association." "This is country music with *muscle*," he said, defiant, in Birmingham. Above all, he was fun. He would strut like an NFL receiver, slap high fives with his fiddle player, and dash up and down the Plexiglas drum pod with reckless abandon. "You don't know how much this means to me that you still love my music," he said, teary-eyed, in Los Angeles.

And naturally the fans were teary-eyed themselves. His concert tour was the biggest in North America, with 2.55 million people paying to see him in the first year alone. They got up early, waited in line, and set records in almost every city he played: three shows in Calgary, four in Indianapolis, five in Detroit, six in Jacksonville. The critics mostly loved him as well. "The opening is *2001: A Space Odyssey* meets Billy Bob," wrote the *Virginian-Pilot* in Norfolk. "Imagine Garth Brooks songs done by James Taylor acting like Billy Joel on Van Halen's stage," said

The Cincinnati Enquirer. "It would be easy enough to find fault with the show," commented *The Baltimore Sun*, saying Garth's voice was weak, his microphone was malfunctioning, and the sound was badly mixed. "But if you think all that added up to a bad or even a disappointing performance, you'd be wrong. What Brooks does isn't about perfection—it's about heart."

And, of course, Garth was prepared with a series of media stunts to stimulate publicity. In Denver, he presented the one-millionth ticket buyer with roses, T-shirts, hats, and the keys to a green Chevrolet Z-28 Camaro convertible. "We couldn't do this for everyone," he said at the backstage ceremony, "so we wanted to make you their surrogate." In Charleston, West Virginia, he presented his two-millionth customer with a choice of two cars, a Chevrolet Tahoe or a Pontiac Grand Prix (the fan, Melinda Huffman, chose the Tahoe), as well as a Sony video camera, an Olympus still camera, beach towels, snorkels, underwater goggles, and a four-day vacation to the U.S. Virgin Islands. "What we're doing for you we'd love to do for the other 1,999,999 people who came out to see us on the tour."

But all the excitement and adulation masked the larger truth. Garth Brooks was deeply unhappy. His concert sales were robust (in the music business, conventional wisdom holds that people will always pay to see you sing your old hits), but his records sales had plummeted. "It's a tough time right now," he confessed. "To me, ticket sales and record sales are the direct voice of the people. One of our legs is strong; the other is weak." The press, with the exception of local reporters, had started ignoring him. Plus, since leaving his AMA trophy at the podium, no one dared give him another award. "Maybe I need to do something to right the course," he said. So naturally he tried. In his quieter moments onstage, surrounded by the love he had craved for so long, bathed in the lights focused only on him, he concocted his most daring plan yet to resurrect his falling star. In a way, what he imagined was every child's dream, that secret fantasy that boys and girls have when they feel the world has turned against them: Garth would have himself killed.

From the moment he came of age as a boy, Garth Brooks was fascinated with movie stars. His childhood heroes were Hollywood legends: John Wayne and Marilyn Monroe. His earliest musical idols were artists who were the most theatrical: Kiss, Styx, and Queen. His entire persona was built around being a *showman*. "At Thanksgiving dinner," he told me, "when that fight breaks every year at the table about something stupid:

Dad gets choked up, veins start popping out, and people start screaming about the best live concert they've ever seen. I want somebody at that table to yell out our name." Millions, of course, did. "You know me, I don't take credit for many things," Garth told me, "but this is one thing where I look at the crew and band and say, 'Boys, step up and take your bows. You brought a whole new meaning to entertainment in country music.' "

That meaning had to do with taking country music off the front porch and moving it into the bedroom. As he told Alanna Nash, Garth viewed music as making love to the audience. "And like good sex," he said, "the wilder and more frenzied it gets, the quicker you turn that around and get gentle, tender, and slow, keeping your partner off balance, then smacking it again with something wild and crazy, and just doing that over and over until one of you drops dead . . . That's great physical sex. That's also great, physical music." Like sex itself, though, performing became both a source of pleasure for Garth and a way for him to act out his demons. On the one hand, he told me, he craved the intimacy. "The gig is," Garth said, "the people come to a place and fall in love with the artist. The problem with our gig is, the artist comes to a place and falls in love with the people. It tears me up because I can't stop everything, walk over, and say, 'You wanna talk? Tell me what you're feeling right now.' It's an ego thing. I'm sorry, but when you see that look in someone's eyes, that's very attractive to me." On the other hand, he also craved the dangerous part. At times he would lose himself onstage, take physical risks, or disappear into a delirium. "I mean, that's what you want," he said. "You want to feel invincible. You want to feel scared to death."

In general, Garth's show was built around the idea of making him seem larger than life. "We like to have big special effects at the beginning," he explained of his strategy. "Then take the people down. Then we pump it up again for 'Shameless.' At the end of the song, when the crowd is on its feet and cheering, the lights will pan down and hopefully blind everybody on the floor. Then suddenly everything will go black and this O-ring will be tilted so that it looks like a halo. The last thing you'll see is this halo shooting down from above and illuminating Garth, in the dark, all alone."

Inevitably, he decided to transfer this role of savior to the screen. Several years into his career, Garth did what he'd always secretly wanted to do: He opened a film company in Hollywood. Called Red Strokes, the company was affiliated with Twentieth Century-Fox. On screen,

Garth said he wanted to play darker characters, "polar opposites" from his country persona, "like a priest by day who, at night, turns into a psychopathic killer." The first picture he planned to make would capture that grim side and, he hoped, also stimulate his lagging sales. Called "The Lamb," after the sacrificial lamb in the Bible, the film would tell the story of a major artist who fought with his record label. When the artist's sales start to slip, wreaking havoc on the company's bottom line, the label hatches a surefire plan to resuscitate his career. They would have the artist assassinated, then sit back and watch his posthumous sales soar. Even though Garth said he would not play the artist, the symbolism was still unavoidable, made more so when Garth said he would tape his concerts for the soundtrack. Further, he arranged for the climactic scene of the film to be shot in Central Park at the culminating event of his three-year tour. After less than a decade in the public eye, Garth Brooks had become so consumed with his own mythic downfall that he decided the best way to revive his reputation was to turn himself into a martyr.

All of which raised perhaps the ultimate question about him: How did this happen? How did the genre's biggest star, a man beloved by millions of Americans, a man who had achieved greater success in a shorter period of time than any recording artist in American history, become a sacrificial figure in his own mind?

The answer comes back to the basic tension at the heart of Garth Brooks, which is also, not surprisingly, the fundamental battle at the center of country music: the struggle between art and commerce, between the desire to express yourself (to create genuine, heartfelt music) and the desire to be successful (to create a product and an image that Americans will buy). In the early part of his career, Garth Brooks embodied the first half of that equation: He and his music were fresh, humble, and indisputably sincere. Nashville, sensing that sincerity (and reveling in his success), embraced him. Garth Brooks was what everyone on Music Row wanted to believe *they* were: genuine, self-sacrificing, and interested in the *music*. Also, he was a good citizen. When I moved to town, Bruce Bouton told me a story. Bruce, a steel guitar player, worked on all of Garth's records, as well as Wade's. Bruce's wife became pregnant with their first child at the same time Sandy Brooks got pregnant with hers. In the middle of his wife's pregnancy, Bruce lost his insurance. When Garth heard this, he called Bruce from a gas station in Arizona and offered to pay the cost of delivery. "I've worked on over

two hundred albums in this town," Bruce told me. "I can tell you something shitty about every artist I've worked with. I've never even heard
one bad thing about Garth Brooks."

A year and a half later, I ran into Bruce again and we both realized
how long ago that seemed. In the intervening time, it was hard to recall
hearing one *good* thing about Garth Brooks. The reason: As Garth's
career progressed, he began to reveal his alternate side—less genuine,
more contrived; less neighborly, more cutthroat. Garth was not the first
superstar to make this transition, nor will he be the last. But he was the
most visible. And Nashville, sensing the shift from a man who was humble because he felt that way to a man who was humble because he liked
how it looked, turned on him. At first this change in attitude toward
Garth surprised me. After all, Garth merely did what everyone else on
Music Row did (try to make more money; try to manipulate the media),
only he did it better. Eventually I realized there was more to it: Nashville
turned on Garth Brooks in part because they didn't like what his transformation said about them. Music Row would like to think of itself as
being genuine and pure, while in fact it's often conniving and ruthless,
too.

Indeed, the wholesale disengagement of Garth from Nashville—and
Nashville from Garth—only served to focus attention on one of the
unspoken lessons of country music in the 1990s: The veneer of values
that lies at the heart of the music, its artists, and the community that
promotes them can be astoundingly fragile. If anything, the foundering
of Garth Brooks was merely the latest chapter in a timeless story, where,
for the hero, hubris triumphs over humility and, for many of his followers, disappointment replaces infatuation. As it turned out, this was the
dramatic denouement of the "Gone Country" era: The king would fall.

Garth Brooks's fall from grace took place slowly and over time. First it
happened with his show. All during his career, of course, Garth had
gone out of his way to litter his show with Barnum-esque pyrotechnics
that he hoped would gain him entrance into the Showmen Hall of Fame.
One of those stunts, though, backfired severely. In the three performances that were taped at Irving, Texas, in 1992, Garth prepared a
special ending for "Friends in Low Places." Standing in the center of
his plateaulike stage, he and Ty England, his roommate from Oklahoma
State who became his background guitarist and later went on to an ill-
fated solo career, strutted around the stage waving their acoustic guitars
above their heads. After awhile, they lined up like dueling baseball slug-

gers, counted to three, and smashed their guitars together, sending splinters shooting in the air. When they had finished, they stomped the remaining pieces to shreds and cavalierly flung the debris around the stage. Though the crowd roared its approval, all across the country people were horrified. Letters of disapproval flooded radio stations, and Garth was forced to issue an explanation that the gesture was merely for visual effect. This comment only made matters worse. As one industry executive and self-described former fan told me: "First, it was offensive because he said it was just for effect. For someone who says he's such a 'normal' person, it showed he was really calculating. Second, it showed his disrespect for tradition. The guitar is the most sacred symbol of American music; to destroy it diminished us all."

Later, when Ty England's smashed guitar went on display at the Country Music Hall of Fame, it became the most controversial exhibit in the museum—with more people saying they disapproved of it than any other item in the collection.

Second, the change happened within his label. After Jimmy Bowen left Liberty Records, a new team was hired, led by Scott Hendricks, a tall, handsome, soft-spoken Oklahoman who had produced successful albums by Alan Jackson, Brooks & Dunn, and his onetime fiancée Faith Hill. Scott (unlike Bowen, he was referred to by his first name) also hired several executives from MCA, including Walt Wilson, who became general manager of the renamed Capitol Nashville. Garth was originally furious that Scott was hired. "I would've loved to see Joe Mansfield with Capitol Records," he told me, "simply because in my belief, the guy did everything that he should've to sell a ton of records, and he never got the chance to finish what he started." Also, as he noted, "Mansfield was a friend." Scott, by contrast, was a neophyte. "He's a super-sweet guy," Garth said, "who just—and he'll be the first one to tell you—doesn't want anything to do with the business side of music." At Garth's first meeting with the new team, he expressed his disappointment to their faces. "I don't believe in you guys," he said. "And from what I understand, you're not very good." (As to why he felt the need to make such a remark, Garth told me: "If I'm going to become a partner, I owed it to them to say what I thought, then they could sit there and go 'Oh, we can take his mouth.' ")

Following that dreadful introduction, relations between Garth and Capitol actually improved. After not setting foot in the building for two years, Garth, before the release of *Fresh Horses*, began showing up for regular Thursday afternoon meetings. "I was so happy that I brought my baby—my new album—back here, away from New York to be marketed here," he told me. Almost immediately, though, Garth began flexing his muscles. He chose his singles without consultation. He demanded a $5 million marketing budget. He insisted on shipping 6 million records, even though none of his previous records had sold that quickly. Over objections, Garth got what he wished. Where Capitol did voice concern, however, was the release date of the record. To capitalize on the start of the Christmas season, Garth wanted to release the album on the Tuesday before Thanksgiving, the same date EMI, Capitol's parent, was releasing the first *Beatles Anthology* album. Capitol was concerned.

"We were sitting in my office," Walt Wilson told me, "and I said, 'Garth, you know the Beatles are going to have an ABC special, and the company's going to spend about fifty million dollars promoting the hell out of the album. Everybody thinks the Beatles are going to kick your ass, including people in the company.'" Garth got up and started pacing, eventually demanding that Walt get Charles Koppelman, the head of EMI, on the phone. "Whatever you want to do is okay with us," Koppelman told Garth. Once again, Garth started pacing. "His whole thing," Walt recalled, "his whole drive in life is that he wants to outsell the Beatles. So this was a head-to-head challenge with his particular dream. And he wanted it. He wanted it bad." Finally, after several minutes, Garth spoke. "Let's go for it," he said. The result: The *Beatles Anthology* set sold 900,000 units in its first week in stores; *Fresh Horses* sold 480,000. The media declared the Beatles champion. Within weeks, further weakened by poor reviews and faltering singles, millions of unsold Garth Brooks albums were shipped back to Capitol Nashville.

Following the collapse of *Fresh Horses,* relations between Capitol and Garth returned to the chill of the Bowen Cold War. Again, Garth could have spread responsibility around. He could have blamed the fans. "I never blame the fans," he told me. He could have blamed the music. "The music's fine," he told me. "It's there." Or he could have blamed the label. Garth chose that option. "When you've got six records that average [8.5 million], and you've got one record that comes in very short of that, then you've got to sit there and go, 'Is it the music?' or 'Is it the first album that this company took over and ran with?'" The label, he told me, even admitted that it withdrew marketing support for the album. "Has my label sat down across from me at the table and apologized for dropping the ball on *Fresh Horses?* Yes, they have. Am I out there fighting an uphill battle with the record? Yes, I am. Am I out

there doing it because I know the record is worthy of it? Yes I am."
Scott Hendricks, faced with such bombast, smiled politely and said he
was honored that Garth Brooks was on his label.

Garth was in no mood for politeness, though. By spring, he began
putting out the word that he planned to get Scott and Walt fired. (Walt
eventually was fired, which raised the inevitable question: Did Garth
have a hand it? "Let's just say that it didn't work out," Garth told me.)
Inside the label, meanwhile, Garth was even more aggressive. He ignored deadlines, pushed frantic staffers to the limit, and, when faced
with pleas of exhaustion, sent lawyers slogging through his contract to
back him up. As one executive put it to me: "Dealing with any artist of
that stature is difficult, and when you have to deal with him on every
little stinking problem, it's unusual; it's bizarre." Plus, Capitol believed
Garth had spies in the building. "He was very paranoid," one executive,
afraid to speak on the record, told me, "scared that he was going to get
ripped off, that somebody was going to take advantage of him. When
you're working with an artist, you want to embrace your artist. But if
you have an artist that threatens you and continually reminds you that
he can get rid of you and just has this constant feeling that it's a conspir-
acy against him, how can you deal with him?"

Eventually Garth shocked Capitol Records—and Nashville in general—by going public with his feud. "The Nashville's office handling of
Fresh Horses pisses me off," he told *Billboard* in a story that was so
unprecedented in the tight-knit community of country music, it was
reported the following day on page one of *The Tennessean.* "The label
gave up on the album after it had sold 2.3 million records. They called
it a dead album." In the future, he noted, all Garth Brooks marketing
would be handled by EMI's offices in New York. This new posture
pointed up the change that Garth had undergone: He believed he had
outgrown Nashville's ability to help him. "Maybe Garth needs a David
Geffen," Walt speculated, referring to the legendary czar of the L.A.
music scene, "somebody that he can idolize and compare himself to.
'I'm the biggest selling-artist in North America, therefore I want to deal
with David Geffen; he's a billionaire.' I had the feeling that Garth respected that kind of power and wasn't getting it anywhere in the EMI
system."

This fascination with power, Walt believed, helped explain Garth's
desire to be associated with Nike, still a nagging goal of his. "It was the
power that Nike represented for him," Walt said. "There were multimillion-dollar deals coming in from major, major companies: Coke, Pepsi,

Kentucky Fried Chicken, Kellogg's. And he didn't want to deal with them. Of course, when you're sitting on several hundred million dollars in the bank, you can basically do whatever you want. He has that freedom and he uses it." Ultimately, this cavalier attitude turned off the people whose job it was to help him. "Remember," Walt said, "we not only had that problem, but Bowen had that problem as well, which is Garth's whole career. So I'm not sure that it was our personality or the personality of Bowen, as much as it was the personality of Garth."

That personality led to a third major source of Garth's decline, a breakdown within his own camp. One of the central reasons Garth was so successful was that from the start of his career he seemed to solve one of the testiest dilemmas facing any artist: He surrounded himself with a top-notch team. As Pam Lewis, his comanager, described it to me: "Garth made me believe in his dream so convincingly that I wanted to take it to the world. If you had told me after the first year that he was the Second Messiah, I would have believed you. He was so magical. He was such a great coach. He was such a great people motivator—of his band, of his staff, of *my* staff. People loved him."

Slowly, perhaps inevitably, that team began to crumble. Some, like Allen Reynolds, his producer, remained on board; while others—Pam Lewis, his comanager; Bud Schaetzle, who directed his early specials; even his sister, who was in his band—left for one reason or another. Pam Lewis, in particular, was deeply wounded by her experience with Garth. Even after settling her legal dispute with Bob Doyle (who later returned to managing Garth), Pam felt great sadness for her former client and friend. "I think he's feeling like, 'What did I do this for?' " she told me. " 'Is this all there is?' I can remember having a very poignant conversation with him. He had just turned thirty and was really depressed. He said, 'What happens when you're thirty years old and you've reached every goal you've ever set for yourself?' I said, 'The first thing you do is get on your hands and knees and you thank God for the incredible blessings that you've received. Then you find new goals.' "

Garth, she said, had always been so driven, so insecure, that he hadn't planned on achieving success and thus had no plan to enjoy it when it arrived. After all, she noted, what's so bad about selling 4 million records (which is what *Fresh Horses* sold in its first year). Most artists, even Garth Brooks in his early years, would have been thrilled. "It doesn't have to end this way," Pam said. "There are choices. I don't think it's this way for Vince Gill. I don't think it's this way for Alan

Jackson. I don't think it's this way for a lot of rock stars you can name. It's only this way because he has chosen to be the tragic hero. There's no introspection. He's blaming everybody but himself. And when you start blaming, you divide; when you start dividing, you isolate; when you start isolating, you get fearful, paranoid, and angry. Frankly it's heartbreaking to me because there's something I love about him very much."

What was most painful about these observations—coupled with Garth's own expressions of doubt in recent months and his mother's comment to me in Atlanta ("Son, you're there. Be happy and enjoy it . . .")—is that they all seemed to agree on one central point: Garth *had* achieved the goals he had set for himself; he had become an icon. "He is that symbol," Pam Lewis agreed. "But you can't hold on to that. There's no peace in it. It's shallow. It's trite. It isn't real. That's the crux of the problem. It's a hoax. It's all a hoax. He's a talented man, but his image is a meaningless hoax that's been perpetuated on the American people." Look at his eyes, she added. "They are the eyes of an angry man."

She told me a story to illustrate her point. At the height of the post-SoundScan bonanza, Garth presented both Pam and Bob, his managers, with Jaguar automobiles as gifts. For Pam, who had dreamed of having only three things in her life—a Jaguar, a Grammy, and a brick farmhouse—all of which she got because of her association with Garth, the moment should have been a culmination. "At the time there was the big fanfare," she remembered, "there was this big party, and I thought to myself, 'You know, what I really want is to sit down and have a really good talk with him.' Honestly, that would have meant so much more to me than the car. I loved the car. I was very happy driving it. But it was so empty. It was a press thing. It made him look really generous. I know that's a terrible thing to say, but that was my immediate reaction. It was like a kid who never gets any attention, so his dad keeps buying him stuff. 'Don't buy me any presents,' I wanted to say. 'Just listen to me. Can't we just sit down and talk?' "

Finally, and saddest of all, Garth became isolated from the town. On a warm, sunny day some months into his tour, Garth returned to Nashville for what should have been the highlight of his career. Capitol Records was spending a whopping $80,000 to throw a party for six hundred people to celebrate Garth's having sold 60 million records. The party was held at the Sunset Studios, the cavernous television studio where Wynonna had taped her special with Bette Midler. With all the equipment removed, the room was an enormous, windowless black box several hundred feet long in every direction. In homage to the number of records he had sold, the room had been decorated from top to bottom with memorabilia from the sixties: psychedelic banners, spinning black-and-white kaleidoscope wheels, a Volkswagen bus, even go-go dancers in white vinyl boots gyrating atop towering centerpieces. Plus, everyone was required to come in costume. "It was my idea that if you don't come in costume, you're out of here," Garth told me. "We didn't want anybody thinking this was a big industry party. This is a thing among friends, a thing among dreamers." Amid the sea of Deadheads and mop-tops, a few look-alikes had been hired to work the room. Jackie O was there. As was Batman. Even John Wayne himself, who later posed for a picture with Garth in front of the VW bus.

For the first hour or so, guests mingled, ogled the costumes, and groused about the refreshments. (Despite all the guests wearing tie-dye and toting joints, there was no liquor served, an irony that even *Billboard* felt obliged to point out: "For God's sake, why does a man who's richer than God serve no alcohol at his party?") Finally a small ceremony ensued. At Garth's request, the festivities were hosted by Jim Fogelsong, the courtly man responsible for signing Garth in 1988, who was then pushed aside in favor of "Hurricane" Bowen. On this day, Fogelsong was the only one of Garth's bosses with whom he was still speaking. "This is what I wore in the sixties," Fogelsong said of his church-deacon blue suit. "This is also what I wore in the fifties." Garth, standing behind him, was dressed in a cream St. Louis Browns baseball uniform with the unexplained number ⅛ on the back (it turns out the Browns left St. Louis in 1954, but few realized this at the time). Sandy was dressed in a blue sundress with a red fringe vest and floppy leather cap.

"I remember when twenty-five thousand albums was respectable in country music," Fogelsong said, "even ten thousand was acceptable. We've come a long way in Nashville and we all should feel very proud." After Marilyn Monroe came to the stage to sing "Happy Sixty Million to You," a string of dignitaries presented Garth with tokens: a letter from the governor, a plate from the mayor, a plaque from the Recording Industry Association of America. Scott Hendricks, looking awkwardly out of place, unveiled a brand-new Kawasaki Mule, a buggy for Garth to drive around the farm. And Charles Koppelman, visiting from New York, had Scott drive in a $200,000 Ford New Holland front-load tractor. Garth seemed genuinely excited by this. "When you sell one hundred million," Koppelman said, "you can call me 'Bubba'."

Finally Garth took the podium. "I prepared no speech," he said. "I'm going to talk without thinking." He then proceeded to tick through a series of acknowledgments, that under the circumstances sounded more like an apologia than a celebration. To his family in Oklahoma, he said: "Thanks for the patience. Meaning something to you is all I ever wanted to do." To Bob Doyle and Pam Lewis: "I hope we can remember how innocent and fun things were when we started." To Jimmy Bowen: "My hat's off to you. Our hottest time was under your roof." To Scott Hendricks: "Anytime something old gets new, there are going to be fights. If there's another milestone, I hope I can celebrate it with you, and I hope you feel the same way about me." And finally, to his band and crew, to his label, and to all the players: "You need to know what's about to be said. They had to order and ship fifteen million units this week just to reach sixty, so tomorrow morning, back to work. God bless you guys. Here's to the dream."

After his speech, the crowd started humming as Garth stepped down and, as always, began to sign autographs and pose for pictures. The mood in the room was valedictory: eulogizing more than jubilant. "There's so little joy here," said one frequent partygoer. Indeed, the comparison between this party and the one five years earlier when Garth had topped the SoundScan chart and trumpeted Nashville's arrival into the mainstream were unavoidable. Though they had been invited, no other label heads were there. No artists, other than a few novice acts on Capitol, came. Few if any reporters outside the local press turned up. "He burst down the door for all of us," one manager told me, "but then he turned his back on us. He just hasn't been embraced by the rest of the town. They see the lack of sincerity. If you believe in this town, the town will believe in you. If you isolate yourself, the town will isolate you too."

The mood was perfectly captured in an article in the Nashville *Scene* entitled "The Thrill Is Gone." "These days," wrote columnist Beverly Keel, "the superstar's nickname, 'GB,' more likely stands for 'Garth-bashing'—the most popular Music Row trend since ponytails on middle-aged men." The most common complaint about Garth, she noted, was his insincerity. "It's hard to say when the cracks first started appearing in his armor of earnestness. Was it when he left the American Music Award or when he outed his sister on Barbara Walters without her permission? It could have been when he announced his 'retirement,' attacked used CD stores, or told a reporter that he cowrote most of his songs because he couldn't find any others that were good enough." Above all, the most damaging fact about Garth was that he refused to take any blame for his decline. "Brooks should realize that he's neither John Lennon nor John Wayne and that he's become a man the ten-year-old Garth probably wouldn't have liked. Certainly, his contributions are not to be overlooked. His music has touched millions and forever changed the face of country. But unless he refocuses on the music, he's likely to be remembered in the industry not for his successes, but for his shortcomings."

An hour or so later, after the line dwindled down, I went to get my picture taken with Garth. He was still standing in front of his tractor, shaking hands, hugging, and smiling into the camera. It seemed to be rote now—sincere yet methodical. When my turn passed, Sandy called me over. She was sitting in the mule, observing Garth. This was not the dutiful wife I had seen at the Opry, but the concerned best friend. "I just like to sit and watch him," she said. "He has such a knack with people." But from her wistful tone, it was obvious that she too believed Garth had been scarred by his success. His moment, as he had feared, had passed. "You remember you asked me once if Garth was a cowboy?" she said. "Well, I was thinking about that the other day. I realized that he'll never be that cowboy. He's just too tender. He wears his heart on his sleeve." She smiled awkwardly. "He hides it a lot of the time, but he gets hurt. I know he does. This business has hurt him. These people have hurt him. Nashville, especially, has hurt him. And that's the hardest part of all. Look at what he did for them."

At the end of his hour-and-a-half stage show that year, Garth sang "The Dance." He would stand in the spotlight near the front of the stage, staring at the sea of swaying bodies and waving arms, Bic lighters in the air. Now, more than ever, the words seemed prophetic: "For a moment, wasn't I the king / If I'd only known the king would fall / Hey, who's to say you know I might have chanced it all." It was a transcendent moment, and when it was over he would disappear below the stage. Moments later, to waves of applause, he returned for his encore, "American Pie," about a music star who dies prematurely. At the end he sings his grand finale, "Ain't Going Down (Til the Sun Comes Up)." For that number, he would bound once again up the Plexiglas drum pod, which lifted him on its hydraulic piston and delivered him into the sky.

"It's the greatest feeling in the world," he told me of those moments onstage. "That's where I feel the safest. Any pictures you've seen of us onstage you can bet only our heels are left onstage—we're jumping up, or flying, or leaning out into the crowd. I can't get enough. 'Cause, dude, I've got to tell you, it's those ninety minutes when all the bullshit is off. That's when the hype leaves, and you check your sanity at the door. That's where me and my managers fighting over business deals, me and my label fighting over contracts—all those things are just left at the side of the stage. It's just me and them."

Indeed, as was becoming clear, Garth needed his time on the stage because it was the one place where he could still communicate with his fans. They were the one group still in his corner. When, late in the year, I sat with Garth again in his stark, contemporary office on Music Row and asked him about Sandy's comment that Nashville had turned its back on him, Garth seemed unnerved, but agreed: "Those people that vote for the CMAs and ACMs, I don't do them any good. I mean, you're talking about your fair buyers: We don't play fairs no more. Talking about your agencies: We don't have an agency. We do our own booking." Add it up, he said: Few people in Nashville stand to gain from his winning awards, doing well, or being on top. "I'm sure they're saying Garth's doing his own thing and that's fine. You make yourself an island, get ready to be treated like an island."

Which is why he seemed so removed from Nashville: Garth Brooks, the team player, had completely given way to Garth Brooks, the lone operator. In essence, he had lost the old-fashioned values (cooperation, neighborliness, humility) that had once made him so country. When, in that conversation, I read him the quote from his mother, who believed that he had achieved his goals and should feel happy, Garth looked at me directly and said, "That's sweet. That's very sweet. But she's wrong." He had yet to reach his peak, he insisted. "There's sixty million American sales. There's what, seven billion people on the Earth? You do the math." There were millions he had yet to reach, he said, there were entire countries he had yet to conquer, there were entire generations yet to be born. "Think of all the young people," he said, his eyes gaping at the thought. "Think of the next generation."

That degree of ambition may be understandable, but under the circumstances, it also seemed harmful. Garth Brooks, it turns out, *was* real: He couldn't let go of success. "Do I think it's over?" he asked. "That's only God's call. But I haven't seen a sign yet that says 'Step down.'" The only change, he allowed, was that "now I'm an underdog again. Which is good because the big pressure's off. Now the only pressure left"—he lifted his middle finger to his forehead—"is in here."

It was one of the last conversations we had before he took his tour to Ireland, the first of his overseas stops. Garth had retreated to the sanctuary of his own mind, where he comforted himself with the belief that, having fallen, he might soon rise again. If anything, he seemed to relish the symbolic position, his most iconic achievement. That winter, on the cover of his *Believer* magazine, sent to his most devoted fans, Garth printed the picture of himself he had been saving all along. In the photograph, he was dressed in black jeans, a black long-sleeved shirt, black ropers, and a black Stetson hat. His back was leaning against a white wall. His left leg was bent, his left foot tucked behind his right knee. His arms were outstretched parallel to the ground, his wrists dangling lifelessly. His hat was drooped to his chest. Garth Brooks, on the face of his own publication, had crucified himself.

EIGHTEEN

THE FAMILY

Wynonna was getting ready. She strolled from one end of her hotel suite to another, lofting her daughter onto her hip, rifling through bags of unused diapers, then pausing for a sip of too-sweet iced tea ("I've got to order some more," she says. "Northerners just don't know how to make sweet tea . . ."). She spritzed a bit of raw lemon down her throat, then went to open the curtains, letting the noonday sun pierce her cocoon for the first time all day. Presently she switched on the television, clicking through the channels before finding an old movie to mute. The television was a tiny comfort, a scrap of home, part of the elaborate nest she constructs for herself in every place she goes: the towels on the floor, the socks on the chair, the magazines littered across the table. "Oooh, wait, we must have our daily love offering," she said, darting into the bathroom, peering behind her bed, then returning joyously with a paper sack from which she pulled a package of strawberry PEZ candies and a pink plastic bunny dispenser. She handed the trove to me. FUN 'N' GAMES INSIDE! the wrapping boasted.

Wynonna was preparing for her day. It begins with what she calls "family time." She plopped down on the sofa and grabbed a half-empty bottle of formula from alongside an overturned (completely empty) bottle of Cristal champagne (price: $150). Grace, bald now, with a pink dress and white dimples, squirmed in anticipation of the bottle, then lunged for the rubber nipple. "Look at these socks that my mom bought her," Wynonna said. "Aren't they adorable?" Wynonna lifted her own feet to the edge of the coffee table, dispersing her eclectic devotional of reading material: *Rolling Stone, Us*, the latest issue of *Fortune*, which contained a special report, "Is Your Family Wrecking Your Career?" They sat alongside T. Berry Brazelton's latest parenting book and a small white Bible with the name WYNONNA ELLEN embossed on the front. Just as Wynonna launched into a tirade about the *Fortune* article, her husband called from home. Arch had stayed on the farm with Elijah this weekend while Wynonna performed three sold-out shows in downstate Connecticut at the Foxwood Resort Casino on the grounds of Mashantucket Reservation. "I love you, honey," she said at the end of their conversation.

Wynonna was getting ready for the world. Finally nearing the end of one of the more tumultuous years of her career, Wynonna was inching toward equilibrium. She had a new manager (her stepfather), a new record label (Curb/Universal), and a new producer (Brent Maher). She was also trying to develop a keener sense of where to take her music. "I'm going to do some of the things I've always wanted to do, but didn't think I had the right to because I didn't want people to think I was crossing over," she said, stringing together a list of friends she hoped to work with: Bonnie Raitt, Elton John, Babyface, Bryan Adams, Melissa Etheridge. "Even Meat Loaf wants to sing with me," she said. "Go figure." Yet she knew the one thing she must do was stabilize her family situation. "The general public continues to be interested in Wynonna's personal life," she said. "I guess I can understand that. There are so many dynamics at work in my life that everybody deals with. I have my sister. I've got Arch and all that drama. Then I've got my mother, my typical all-American divorced parent situation. People characterize me as this victim, but I don't see myself as the victim at all."

It was that feeling of misunderstanding that most plagued Wynonna and that caused the greatest transformation in her life. For years, Wynonna had shunned the press and let her mother do the talking, a process that created a shadow Wynonna that the real one never fully understood. Now, having been smothered by that shadow for so long, Wynonna finally decided to open her life and confront her ghosts head-on. Only by doing so could she hope to answer the question that had circled her since childhood: Could she become the legendary figure she'd always seemed destined to become?

°   °   °

The first step was to confront her reputation as the weirdest person in country music.

"You know how some people just attract things?" Wynonna said. It was early afternoon by now and Wynonna was sitting alone on the sofa, spending her "personal time" in reflection. Grace had been handed off to her nanny; the television was still on mute. "I guess I'm one of those people, like that poor kid Culkin. There are plenty of kids just like him in the world. Why does he get the stuff he does?"

Wynonna was wearing black sweatpants, a baggy black zip-up shirt, and a pair of mod black-mesh hightops with the phrase ROCKET SHOES on the sole. Her most conspicuous sign of stardom was the round turquoise sunglasses she wore even though she was indoors and that seemed, in either case, more evocative of Yoko Ono than Barbara Mandrell. Her hair was shorter than it had been in a while, with a touch of blonde at the top, deeper red around her bangs, and a foxy tail reaching just beyond her shoulders. A hint of gray was visible in her bangs. Up close, it was a messy mix. But far away, carefully stroked under her brush, it was a brilliant frame for her face, as beautiful as ever.

"I remember the morning of my wedding," Wynonna said, continuing her theme of media abuse, "when I walked out of my house, the helicopter was so close I could have thrown a rock at it. Later, I was saying my vows and I could hear the helicopter hovering over the building. I thought, 'I have a choice here. Either (a) I can admit to everybody that I'm pissed or (b) I can try to let it go and concentrate on the moment, which is what's important: saying "I do" to my future husband.' "

Letting go, she admitted, had long been her strategy. "My mother is a bulldog when it comes to the press," Wynonna said. "Her attitude is: You control them or they control you. It's always been her number one rule. My attitude has been: You don't need to defend what doesn't need defending." She rolled her eyes and sighed. "I was playing Tahoe recently and one of my ex-boyfriends came to the show with his mother. This is a guy I have known since 1984. He's like a brother to me. He's married and has kids and we exchange photos. We'll always be close. He's my soul mate. Sure enough, two days later someone calls him and says, 'We have a package to deliver from Federal Express. What's your address?' Then they show up and hound his mom." The following week a story about the two ex-lovers appeared in one of the tabloids. "They say we're having an affair," Wynonna said. "Three *blissful* days spent together. Arch is described as having the personality of a donkey. It hurts my feelings."

After enough such incidents, Wynonna began to realize that her strategy of "letting go" wasn't working. The reason had to do with the new reality of American media. In today's atmosphere of "infotainment," the tabloid coverage seeps into the mainstream coverage, which then seeps into the critical coverage, which, in turn, seeps into prime time. In the mid-1990s it was impossible to read an article about Wynonna's music that did not mention her personal life. As a result, the public had an impression of Wynonna—bratty daughter, binge eater, adulterous wife— that utterly undermined any respectability she might have once had as an artist. Ultimately, she concluded she had to change. "I'm not one of those people who puts on Christian blinders and just acts like tabloid stories aren't there," she explained. "So now my new motto is: Bring it on."

The result is a veritable smorgasbord of contemporary plagues. Indeed, the most remarkable thing about being around Wynonna is the sheer volume of issues that clog the air—her children, her marriage, her father, her hair, her weight, even her sexual orientation. With Garth, the biggest issue surrounding him is his psyche: Where is his head? What is he thinking? When will he quit? Why doesn't he enjoy his once-in-a-generation success? Being with him is like attending a master's course in the psychology of American celebrity. With Wynonna, the issues are different, broader, and ultimately more exhausting. Being with her is like riding the wave of a remote control clicker. One moment you're on VH-1, the next the Family Channel. Now let's check in on Oprah, QVC, CMT, E! Entertainment Television. Oh, God, too shallow. We need to drop in on "The 700 Club." There, that was reaffirming. Now how about a dose of Home & Garden TV? Through it all what emerges is the manic sense of a woman trying to stay focused, struggling to discover who she is.

On her weight: "I went to a nutritionist for two years and I came to the realization that I'm not a gorger," Wynonna said. "I'm a condiment freak. My problem is cheese and mayonnaise and white bread. I come from that side of life where we had fried bologna sandwiches. You have a tomato sandwich and it has about six tablespoons of mayonnaise on it. That's my problem. I like potato chips *with* onion dip. Two of the world's worst foods *combined*. I like peanut butter and butter sandwiches. That's a problem. If you take white bread and you put peanut butter *and* butter on it. How many grams of fat is that, Bruce? It's a lot of fat." Her new goal, she said, was to lose thirty pounds.

On her reported liposuction: "This is the truth. I can't even believe

I'm talking about it. I had fascia problems. It happens to a lot of women in pregnancy. Elijah basically tore up my insides. I had to have surgery to strengthen and sew back my muscle wall. The guy who fixed me is a plastic surgeon in Nashville, and he's the guy who does all the breasts. And he was 'Mr. Liposuction' for about a year. He went in, sewed me up, and gave me a line from here to here." She gestured across her abdomen. "He removed one mole and did that surgery, so I was linked to him. At first I thought, 'That's none of your damn business.' This is a very personal thing here. I just didn't feel like going out and talking about what the tabloids said." And why not get liposuction? "Because it's the easy way out. Sure, I'd love to fix about sixteen things, but this is the hard part: accepting who the hell you are."

On her larger-than-life appearance: "I've had hair extensions off and on for years," she said. "Now I have the smallest amount I've ever had. That's me in a nutshell: You have too many because you're insecure. Then you start to learn more how to be yourself." It was the same, she said, with her body. "You make peace with the fact that you're never going to be a size eight. Would I be any happier if I were a size eight? I have been, and I wasn't. As a matter of fact, some very interesting things happened that were overwhelming. We had this party in Hollywood and these actors were coming on to me. I had just met Arch. I remember him flipping out. I walked out of that party feeling so bizarre because it wasn't about the music. The reason that man was coming on to me was because of my appearance. I had this natural ability to be all woman and the attention, frankly, was odd."

Wynonna could go on like this forever. She was locked in her own version of what Garth called his Zone.

On her husband: "I have such a tenderness for who Arch is. He's such a childlike person. Also, he's the most generous person I know. All the times I was sick and nobody called but him. The birthday when he threw me the party he put together himself. Plus, you try being a man coming into this family and keeping both your testicles. The burden on Arch is so tremendous. He puts up with all the speculation, all of the talk. I've never worried about him cheating on me. Not once. As matter of fact, I've had people tell me that when I'm gone all he does is go to church and stay home. He doesn't drink. He doesn't party. What he wants to do is ride the tractor and be a farmer."

On reports of her own infidelity: "God, no. There've been two people in my life I thought were cute, but I couldn't go there. I'm too afraid of going to hell. That's too big an issue for me. Plus, you saw how my road manager just walks through my room. My nanny's always in my room. I couldn't have an affair, even if I wanted to. Sure, in my fantasies. But there's no way. There's just no time."

Finally, on the most persistent rumor of them all, reports that she was a lesbian: "I know people say that about me, but it's not true," Wynonna said. "Maybe it's because I hang around women a lot. The ovary fest around here is quite obvious. It's like being on Rosie's talk show. But the truth is, I've never even had that fantasy. Maybe the fantasy of walking in on someone having sex. But not the dream of being with another woman. I couldn't be more hetero if I wanted to."

All these issues pale alongside the biggest drama in her life. It's the one she's been most reluctant to talk about. But it's also the one that most prevented her from getting where she wanted to go.

On Sunday, January 30, 1994, almost exactly three years after their pay-per-view farewell concert, Wynonna and Naomi Judd reunited for the first time in public at the Georgia Dome in Atlanta to perform during the halftime ceremonies of Super Bowl XXVIII. The event was scheduled to be the emotional close to a collection of country performances, a shining postcard of the New South to be beamed around the new world. As it happened, it also turned out to be a watershed moment for Wynonna, one that unleashed a series of events that resulted in her long-delayed coming of age and, as odd as it seems for a thirty-year-old woman whose raw talent had been sustaining her family for years, a modicum of independence from her own mother.

Though Wynonna had never realized it, Michael Ciminella had always been banned from being backstage with her. Naomi, working first with Ken Stilts and later with John Unger, had conspired to keep Michael completely removed from Wynonna's professional life. "I was sad because I needed my dad," Wynonna told me. "I couldn't figure out why he wasn't invited to things. When he was there, they kept me from being in a room with him by myself. I remember Dad once telling me he wanted to talk, but then, before it happened, Mom and Ken came running to tell me they were worried he was hooked on cocaine again." At the Super Bowl, it was supposed to have been different; Wynonna had invited Michael herself. Yet, again, something went wrong. "I was getting ready to go on," Wynonna said, "and the security's got Dad. They won't let him in. I'm thinking, 'What is this? What's going on here?' " What was going on was that Naomi had outflanked her daughter. "She was in fear," Wynonna said of her mother. "She had John in a room and they were controlling the situation. Of course, I'm oblivious to anything. All I know is that I'm supposed to ride back on the bus with him to Nashville."

That night Michael and Wynonna did ride back to Nashville together, and when they arrived the next morning, Wynonna received a call from her therapist asking her to come to her office at two o'clock that afternoon. When Wynonna arrived, Ashley's car was there, along with Naomi's. The two were seated in a private room when Wynonna appeared at the door. Naomi was crying. Ashley looked stricken. Wynonna was horrified. Something must be wrong. She sat down. Ashley started to speak, but Naomi interrupted, blurting out the news she had concealed from her daughter for over three tumultuous decades: "Michael Ciminella is not your biological father." Even for a woman who had seen her mother extend to staggering extremes of cunning and manipulation, the depth of this deceit was overwhelming. Wynonna sat bewildered as Naomi retreated to the window to cry. "Of course, my whole life was passing before me," Wynonna recalled. " 'No *wonder* Dad was never allowed on that stage. No *wonder* he was never allowed on the bus. Mom was afraid he was going to tell me. I get it.' It put everything in its proper place. Each piece of the puzzle went in. Like the year he told me that he always felt a greater connection with Ashley than with me. '*Now* I get it. *Now* I see.' "

What happened next, though, was even more surprising. "I think everyone was terrified about what would happen," Wynonna said. "Would it mean that Mom and I wouldn't speak for ten years? Would it mean the end of our relationship?" Instead, the news meant a subtle shifting of roles: Suddenly the person who had been kept powerless for so long was holding the power in her hands. For a spiritual woman like Wynonna, a woman who had just found out she was going to become a mother herself, that power was transforming. "I don't want to sound like Mother Teresa here," she said, "but after a while, I got up and I went over and hugged Ashley and Mom. I was so caught up in their pain. This was such a burden for everyone else for so long because they knew. I didn't know. Dad told Ashley when she was in eleventh grade, so she carried it around. Mom carried it around for thirty years. Imagine what she had been through."

As noble as that sentiment was, didn't Wynonna want to strangle her mother? At least for an instant? "No," she said meekly. "She's my mother." What about her father? Why hadn't he bothered to tell her? "It's a really tricky thing," Wynonna said. "It wasn't his place. It was

Mom's place. She orchestrated everything. She picked Dad, and he loved her, and he married her because he felt it was the right thing to do, regardless of whether it was right ten years later." As soon as she left her therapist's office, Wynonna went to see Michael. "It was one of the best talks we've ever had," she said. "He said, 'Do you understand it was not my position to tell you?' I said, 'Yes, I do.'"

In the months that followed, Wynonna's sense of liberation only grew. "For six months, I was in a state of bliss," she said. "The truth was out, which brought Mom and I closer. It also brought Ashley and I more together because we had to work harder at letting each other know that 'This doesn't change our love for each other. Period.' Ashley wept in my arms and said, 'I love you and you are my sister. It doesn't matter how much of your blood is running through my veins.'" As Wynonna observed dryly, "The grace was flowing." Also, Wynonna was struck by the reaction of her biological father, Charlie Jordan, who was married and living in Kentucky and whom Wynonna first saw pictured in a tabloid, not long after the revelation. "I'll tell you when I knew that 'Hey, this guy must really be a good guy . . .' is that he never asked anything of me. I've got a brother; he knows. I've got a grandmother; she knows. The whole town knew. But I kept thinking, 'What a decent guy Charlie must be because he doesn't want anything from his daughter.'" (Contrary to reports, the two have never met: "And don't worry," she cracked, "I'm not going to do it on Oprah.")

Eventually the bliss came to an end. For months after she first learned of her parentage, Wynonna focused on her son, whom she was about to deliver. "I wanted to concentrate not on the shock of finding out about my real father, but on life, touching my belly, speaking to Elijah," Wynonna said. "I realized something that I never had before and that is: 'I'm not a body with a spirit. I'm a spirit with a body.'" It was a personal epiphany, she said, but one that almost ruined her. "By the time Elijah was born," she said. "I was a wreck." Several weeks later, attending church in Brentwood, the collective emotion of her experience surged through her body. "I broke down," she said. "I purged. I wept from the time I walked into church until the time that I left. I just went through this thing where all of a sudden it all just hit me. My grandmother, Michael's mom, had died without telling me. We were very close; I still wear her wedding ring. And what struck me was that she had dug me despite it all—more than she did anybody in that whole friggin' family. She and I had a spiritual connection, and that's what did it for me, brother. More than anything else in my life. Because I knew that I had been chosen by her. She put her sights on me and her affections on me. She *picked* me. And for the first time in my life, it made me feel wanted."

As momentous as that experience was for Wynonna, it was still deeply private. She had yet to confront the public reaction to the new wrinkle in her life. That wouldn't happen until the following year, and when it did, it would further widen her distance from her mother.

In May 1995, NBC planned to air the two-part miniseries *Naomi & Wynonna: Love Can Build a Bridge*, based on Naomi's book. Originally "mortified" at the thought of a miniseries, Wynonna was even angrier when she saw the script. "I was miffed with Mom," she recalled. "She spent more time building the media image of the Judds than I did, but still she let NBC turn our life into this sensational mother-daughter fighting thing. I told her, 'You can go there and sit all day long on the set and watch this happen. I'm not.'" Wynonna never set foot on the set, a gesture Ashley emulated. "This was a real lonely time in my life," Naomi told *People* magazine. "I tried to tell them it was important to support *me* in the emotional struggles I was having."

The big battle, though, was yet to come. In fine Judd punch-and-hug tradition, Ashley ultimately agreed to provide the show's narration, and Wynonna recorded portions of the soundtrack. The two also agreed to appear, yet again, on the cover of *People* magazine. But Naomi wanted more. She wanted the entire family to go on Oprah to promote the series. After initially agreeing, Wynonna backed out, fearing a backlash against her unwed motherhood and a spate of questions about her paternity. She was also concerned about feeding rumors that a Judds reunion was in the works. "I'm trying to work on starting over and not living in the past," she said at the time. In the end, though, she did agree to appear. "The reason I decided to do Oprah," she told me, "is that I had just found out about my dad and I said, 'I would rather go on Oprah and be faced with this as a family than sit back and let Mom speak for me.'" Arch, she said, encouraged her. "He said, 'If you don't ever stand up for yourself again, do it now, so that you'll know what it's like.'"

The night before the Oprah show Wynonna and Ashley gathered on the bed in their mother's suite to watch the first half of the miniseries. Even after all the buildup, Wynonna was stunned at what she saw. "I was really pissed how they portrayed me in the early years as such a snob," she told me. "I remember sitting up in that hotel room crying and being devastated about the family watching the show. I was hurt. I was upset. I would just sit there and go, 'I didn't do that. I don't remember doing that . . .' and thinking, 'This is what America is seeing?' "

The following morning Wynonna, for what she told me was the first time in her life, confronted her mother as an equal. "That was the turning point in our relationship," Wynonna said. "I went to her and said, 'Mom, I want to speak for myself. Anybody asks any questions, I'm talking about this. If you say one word, I mean *one* word, against Dad or anybody else, I will never do another thing with you again.' " Naomi, concerned with her own problems, initially dismissed Wynonna's concerns. "She was worried about herself," Wynonna recalled. "She thought she was going to be stoned in the town square. She kept saying, 'I'm going to get crucified. I'm going to get punished.' " By the time the limo arrived to take them to the studio, the tension had reached its highest point in years, with Naomi worrying that Michael was going to storm the studio and Wynonna threatening to back out of the show if she didn't get certain assurances from her mother. "In the limo that morning was probably the hardest I've ever been with Mom," Wynonna told me. "I stayed on her from the time we left the hotel till the time we went on camera. I remember her hairdresser looking at me and going, 'Holy crap! What's going on?' " Finally, just before showtime, Wynonna heard what she wanted. "Mom said, 'I understand. I will honor your request.' "

It didn't matter. On the May 15, 1995, episode of "The Oprah Winfrey Show," Naomi Judd looked as insecure as she had ever appeared in public. Flanked by her two daughters, who were chilly at best, Naomi had to admit that she had deceived the public on three separate occasions: first in presenting her life story as a fairy tale come true, then in recasting it in her book as the tragic story of a long-suffering woman who triumphs over all the men in her life, and finally in retelling it on television as the heart-wrenching story of a mother who had to choose, Sophie-like, between her two daughters. In one of the more explosive moments on the show, Oprah asked Naomi what scene in the miniseries was the most difficult to watch. "The scene where I had to say goodbye to Ashley so we could chase our dream," Naomi said. "But, Mom!" Ashley shouted, nearly springing from her chair. "That's invention. It's fiction! The producer told me he needed to add a *Sophie's Choice* element to the story to give it more of a dramatic punch." Naomi, faced with such mutiny, meekly retreated.

But the damage had already been done. First, the show further splintered Naomi's relationship with her daughters. "That started a whole family fight for about a week," Wynonna told me. "That's why I don't like doing those things in public." Even worse, it brought the feud into the public eye. "One thing I'll have to say about myself and compliment myself," Wynonna told me, "is that I've never hung my mother in public, nor would I ever." Instead, she opted to let her mother hang herself. "And she did," Wynonna said. "I saw her not only hang herself, but be ridiculed and looked down on. And that was hard because I was watching my own mother be a victim." But, she knew, it was also important. "There was this real balance I had to achieve, which was to honor thy mother, but also to stand up for what you believe in. I stayed focused on the best thing, which was to remain calm. For me, the whole episode was a real awakening."

In the months that followed, a period of time that included her second pregnancy, her wedding, and the birth of her daughter, Wynonna began steadily drawing more discreet boundaries between herself and her mother—everything from keeping her at arm's length at home, to keeping her removed from the recording studio. It wasn't a full break— the "divorce" that Hazel told me she hoped would transpire between the two—but it was a start. "Yes, she's hovering," Wynonna said. "Mothers are like that. It's the dynamic of the relationship. But she's been working really hard at it. She realized she couldn't just come through the gates anymore. Plus, she'll call and say, 'I'm getting ready to do this interview. If you don't want me to say anything about this or that, tell me.' I give her a lot of credit. Some mothers never get that. They think you can always go in their room and go through their stuff."

Ultimately, perhaps the best illustration of Wynonna's emerging independence was her show. For her entire solo career, Wynonna had refused to perform any of the songs she had recorded with her mother as part of the Judds. It was as if that part of her life had been frozen and put into a morgue when her mother retired. Returning from the birth of her daughter, though, Wynonna, for the first time, reintroduced the Judds material into her act. "It was my idea," she told me. "But when I did it, the first night, I was blown away. It was like, 'Whoooaaa . . .' It shifted into this whole other moment. I didn't realize that for so many people the Judds are still such a part of their daily diet." The "Judd medley," as she called it, included everything from "Had a Dream (for the Heart)" to "Grandpa (Tell Me 'Bout the Good Old Days)." It quickly drew the best response of the night. And Wynonna, hip shaking, voice growling, finger pointing, performed it with stunning confidence. "Do I live by that?" she mused. "Partially. Do I grow by that? No. But it's part of me, and it's the essence of who I am. So by God, I better do honor to it because that's where I've come from."

After all the twists and turns of her epic life, this was the ending no one had predicted: Only by embracing her mother in such a public way did Wynonna finally stand alone.

"There's an inner peace in knowing that I can do it on my own," Wynonna told me. "I don't have to draw on that for strength, but I can dive back into it because it's a memory that makes me happy. Just like at Christmas everybody loves to sit and say, 'I remember five Christmases ago. . . .' It's that feeling of comfort that you can love something, but it doesn't have to own you. It's been an interesting passage into realizing that I don't have to have the Judds to be somebody. I paved the way myself. It's like tough love: You say no to somebody, you move forward, then you embrace them again down the road."

Late in the afternoon, as her show draws closer, Wynonna finally shifts her attention from her family to herself. It's the part of the day she calls her "preshow ritual." The sun was reddening outside her window now. The baby was taking a nap. A surreal calm had settled over the room, which still appeared chaotic with its twisted sheets and half-eaten pieces of chocolate cake.

The first thing she did was check in on the Wy-line. As part of her fan club, Wynonna has established a 900-number where fans can call up and receive updates on her life and leave her personal messages. She estimates she gets about two dozen calls a day. One man, she said, calls several times a week reporting news about his car ("I just changed the oil this morning and took it for a drive . . ."), and one woman calls every morning to read her a different Bible verse (on this day, Colossians 3:13: "Be tolerant with one another and forgive one another whenever any of you has a complaint against someone else . . ."). Among the messages this afternoon: A woman in Atlantic City called to offer Wynonna free dry cleaning service when she was next in town and a woman in Michigan whose fiancé had just broken off their wedding called for support. The woman said she had been concerned about what she would do in June, the month her wedding was scheduled, until she heard that Wynonna would be having a fan club party that month: "It's the only thing that could take my mind off the wedding," she said. "I think it's the work of God." The most emotional message was from a woman in Allentown, Pennsylvania. "It's me again," she said, giving her name. "My mom was your wisest fan and your oldest fan club member. She received an award at your fan club party several years ago. She passed away this morning, and you're the first person I wanted to tell. It was one of the highlights of her life, standing up on that stage next to you. I just really wanted to thank you. You made her life happier. On behalf of myself and my entire family, we love you."

Next, Wynonna started pacing the room, drawing strength for her show. " 'Putting on the armor' is what I call it," she said of this ritual walking, talking, sitting, praying, and mumbling to herself. "That protection gives me the openness to be vulnerable onstage. It's one of those things where knowing I'm safe in the arms of God gives me great, tremendous at ease." It was all part of the larger process of learning to believe in herself. "Here's the bottom line," Wynonna said. "I want to go to the mountain and claim what's rightfully mine. I'm worthy. God loves me." Whereas five years earlier, Wynonna had said poignantly in an interview: "My strength was always as Naomi Judd's child. My identity was not Wynonna, child of God; it was Wynonna, child of Naomi Judd." Now she spoke as a different person. "My biggest identity is that I am a child of God," she told me. "Period. I'm not Arch's wife. I'm not Naomi Judd's daughter. I'm not even my children's mother. I am a child of the Lord. My soul is eternal."

At nightfall, Wynonna dresses quietly by herself—black silk pants, a black shirt, and mauve snakeskin jacket with a Nehru collar and jagged black leather Elvis-type cuffs. She puts on her makeup and eyelashes by herself as well and carefully smoothes her hair with a metal-bristle brush. Then she descends the staff elevator and retreats to her backstage dressing room. It's a small, tastefully appointed space, sort of like the first-class cabin in a 747. There are mirrors on the wall and a purple carpet on the floor. The tables are covered with orange cheese trays and dishes of M&M's, dozens of bottles of Evian, Barq's, and Cristal champagne, and a heaping crystal bowl filled with slices of tuna sashimi, chilled stone crab claws, and disks of fresh boiled octopus. Momentarily, Wynonna summons her band and backup singers for their nightly prayer. "It's sort of a tribal thing," she explains. "It's the glue that binds us. We all have a lot going on in our lives. The nights when for whatever reason we don't pray, we're too busy or too hurried, and we forget to ask God to be present. Those are the nights when we all know there's no energy up there, we're lethargic. There's no light."

When the band arrives, the dozen or so members, dressed in elegant black and white, stand in an informal circle holding hands. It's the most integrated band in country music, a sort of traveling gospel choir complete with rock band and horn section: a Nashville version of Earth, Wind and Fire. "I see you all got your invitations," Wynonna says. There is a nodding chuckle and a summons to worship, then Wynonna leads the group in a short prayer. "Thank y'all for coming to celebrate with me tonight," she says. "Just so you know, so I say it in front of everybody, I'm a survivor. I have a lot to be thankful for. Going onstage is my way of celebrating. It's a way of saying to all those people out there, 'Dreams do come true. If I can do it, you can too.'" There is a general rocking in the circle, a squeezing of hands. "But I want you to know, none of us can do this without God." A brief murmuring ensues. The gathering assumes the momentum of a call-and-response invocation. "None of us can do it without God in our lives. He is the source of our light." *"That's right."* "The granter of our gift." *"Yes, ma'am."* "So we must give into our gift tonight." *"Our gift."* "None of us knows what's going to happen. We just don't. So we must go out, relax, and let the music speak through us." *"That's right."* "Remember, we are the music. We are the light." *"Amen, sister. Amen."*

After a few minutes of hand-holding and one last communal hug, the band departs and Wynonna is left alone. She pulls aside a chair and gets down on her knees. The first gleeful applause from the crowd can be heard through the wall as the players take their place onstage. Wynonna quietly lowers herself to the ground. When she arrives, she moves herself into a prone position: her legs pressed together, her arms spread apart, and her forehead touching the floor. "Oh, Lord, let me be humble," she says, her voice pleading and cracking with longing. "Open the door and let me in. Take me into your arms." Worshipful, even trance-like, she murmurs to herself for several more minutes until there's a gentle knocking on her door. Then, as if summoned, she stops, lifts her head, and slowly opens her eyes. A smile drifts gradually across her face. She's ready. But before she rises and steps into the light, she lets her arms reach around her body and utters this hopeful plea to herself, "Love, love, love."

NINETEEN

THE AWARDS

LeAnn Rimes was in control—sort of. Two hours before the thirtieth annual Country Music Association Awards, the fourteen-year-old superstar was standing at the back door of the Grand Ole Opry House giving orders like a wannabe grown-up to her considerably grown-up entourage. "No, I don't *want* to pull my hair back, Mom," she said as her mother tried to tug at her shoulder-length blonde hair. "Wait, I'll stand *here*," she said to the top-hatted chauffeur who was preparing to open the door for her. "Hey, can I go in first?" she chirped to the producer from "Entertainment Tonight" who was busy negotiating with her mother, her father, her publicist, and her agent about the order in which the various members of the party should enter the rented limousine. "Actually, um, we sort of have to hurry," the driver said. "I have to pick up Tim McGraw in half an hour."

Eventually they hit on it: LeAnn would enter first in her blue taffeta and black suede ball gown that under the circumstances seemed more suited for the prom; she would be followed by Mark Steines, the reporter from "ET," dressed in tuxedo trousers and shirt with one of those ubiquitous banded collars. The cameraman, soundman, and segment producer, all in rented tuxes and black tennis shoes, would remain outside and shoot the stars' entrance. Then, once the crew got what it wanted, LeAnn and Mark would turn around and get *out* of the limo again, at which time the producer, cameraman, and soundman would get *in* and shoot the entrance again from that angle. At that point the limo would make a circle (actually, it ended up being more like a ten-point turn) in the cramped parking lot and deposit the entire crew on the meager red carpet—about the size of a large bath mat—that was poised at the building's entrance, about fifty yards from where we were now standing. There they would repeat the double ins and outs again for the purpose of shooting LeAnn's "arrival." "All this for maybe twenty seconds on the air," the producer joked.

In fact, after twenty *minutes*, when the limo had not yet left the starting position, the scene became so clogged that LeAnn's parents, Wilbur and Belinda, plump, proud, and clearly the source of their daughter's Miss Piggy-like *joie de vivre*, decided to jog alongside the limo like two Secret Service agents trying to keep up with their daughter. "I'd love for her to win both awards she's up for," Belinda shouted to a radio reporter who scurried up to her during her run. "But she's already won just by being nominated so quickly as far as I'm concerned. I'm just crossing my fingers and hoping for the best. This is a dream come true."

The CMA Awards are Nashville's annual application for Hollywood-style glamour. Begun in 1967, the three-hour, black-tie, prime-time broadcast takes place every fall on the dolled-up stage of the Grand Ole Opry House and is the centerpiece of a weeklong series of assemblies and awards dinners that comprise the culmination of the year in country music. Because it's one of the few times of the year when the national press turns its gaze however briefly on Nashville, it's also a good indication of what is hot at any particular moment. At this moment it was LeAnn Rimes. "Limousine . . . openin' up the show . . . fourteen years old!" Mark Steines gushed in the back of the car. "Boy, you've got the world by the tail, don't you?" "Yeah," she said with a giggle, "I kinda do."

LeAnn, a bright-eyed, marshmallow-cheeked bundle from Dallas, Texas, arrived in Nashville in the middle of its commercial plateau and rode her phenomenal voice and fresh-faced story to a yearlong lock on the country album charts. Born in 1982 (she's actually younger than the CD itself: "I used to go to my grandmother's and listen to records!" she boasted), the Mississippi native decided at age five that she wanted to be a professional singer. At first she fashioned herself after Barbra Streisand and Judy Garland, then she moved on to Patsy Cline, Reba McEntire, Wynonna. "If I had to model my career after anyone, it would have to be Reba," she told me. "She's made some great business decisions in her career to stay around for twenty years, and my biggest goal is to stay around for a long time."

At fourteen, the challenges of battling the industry were even greater than for her elders. For starters, she ran the risk of seeming like a freak. When we first met, I tried to avoid using industry jargon. Referring to the new custom of releasing an "electronic press kit" on videocassette, I said, "I was just watching this video they sent out on you—" when suddenly she interrupted me, "Oh, you mean the EPK." Later she even dropped the term "rack jobber" in conversation, referring to the wholesalers who stack discount retailers. The only other artist I heard do that was Garth Brooks. "I love to do this," LeAnn said, explaining her facility with the business details, "but it's basically like a job to me. I'm around adults all the time. That's basically who all my friends are. I have no friends my age."

Another problem of LeAnn's age was that her mother had to sit in on interviews. In our first conversation, I asked LeAnn to describe what she looked like. "They say I look like a young Claudia Schiffer," she said. "I'm five-foot-five. I have blonde hair and blue eyes. My hair is right past my shoulder." And then, from across the room, her mother added, "And she's *stacked*."

Perhaps worst of all, LeAnn was being forced to go through adolescence in public. When our conversation turned to her home life, LeAnn and her mother started bickering. "You can't date right now," her mother said, correcting her daughter's response to a question. "We'll let you go out with friends." "But I've done it before," LeAnn muttered. "I don't see what's so bad about *that*." In the limousine with Mark Steines, the subject came up again. "Do you have a boyfriend?" he asked. "Not yet," she answered. Then Steines picked up the telephone. "We need to get a boyfriend back here quick," he said. LeAnn grinned. "I need to get a date for the CMAs!" she echoed. The giggling moment was one of the few that made it onto "ET" the following night.

Once LeAnn disembarked from her limo and disappeared backstage, the crew from "ET" set up the small bank of cameras arranged on the curb to shoot the arrival of the stars. This was the real showtime: Nashville's answer to the runway. But with so many artists demanding starlike treatment and with so many executives trying to flatter themselves, the industry quickly exhausted Nashville's supply of limousines. A few reserves were brought in from nearby, a few more from as far away as Atlanta, but still the demand outstripped the supply, which meant one thing: limo recycling. The drivers would drop off their stars at the red carpet, then hurry the quarter mile or so to the guard booth at the Opry, where other stars would be waiting in their pickup trucks or buses for an empty limo to escort them to the door. All of this made for a quite comical scene on what was supposed to be the most enchanting night of the year. Shania Twain actually showed up in a golf cart. Tim McGraw drove up to the door in his pickup. And John Michael Montgomery, stepping out of his sedan in blue jeans, seemed so surprised to see cameras staring him in the face that he muttered, "Oops, I'm in the wrong place," jumped back in his car, and drove around to the back door.

After an hour, the steady stream of artists in front of the "ET" camera was almost numbing: Sawyer Brown, Ty Herndon, Trisha Yearwood, Faith Hill, Mindy McCready. A few, like Garth and Wynonna, snuck in through the back. But most ran the gauntlet, answered several questions, and were hurried along by their publicists to the next camera, the next interview, then inside the door. After a while, I almost lost track—Neal McCoy, Kevin Sharp, James Bonamy—until suddenly, out of the pack, Wade appeared. Having not seen him in several months, I was startled. He seemed taller and a bit more poised. Nominated for the prestigious Horizon Award, Wade had earned his place in front of the camera.

"Hi, Wade. Nice to meet you," Mark Steines said. "Are you nervous?" Wade thought for a second, as if he had been asked a tough question. He was wearing his black Manuel jacket and a crisp white tuxedo shirt, which made his green eyes stand out. He was sweating slightly under the lights. "Actually, I've been okay," he said. "I was out here awhile ago, went through rehearsal. I thought, 'Gee whiz, I hope I stay cognitive.'" "How about the competition?" The others up for the award were Bryan White, Terri Clark, Shania Twain, and LeAnn Rimes. "I can honestly say that I know all the people in my category very well," Wade said. "I hope they all win." "And so what did you do today?" Mark asked. On this, Wade didn't hesitate. "I rode my hog around," he said, referring to his one indulgence so far in his career, a brand-new Harley Fat Boy.

And just like that it was over. He turned to head inside, but, as he did, he noticed me standing behind the "ET" crew. Since midsummer, Wade had been elusive, keeping mostly to himself. "How ya *doin'*?" he asked. He walked over and took my hand. "Good to *see* you, man." He nodded in a gesture of mock horror and, for him, endearing self-effacement. "Boy, can you believe what happened to me the last few months? Whoever would've imagined."

On a Good Night, the second album by twenty-seven-year-old Wade Hayes, was released by Columbia/DKC Music on Tuesday, June 25 (since SoundScan most albums are released on Tuesdays to give stores an additional high-traffic day along with Saturdays and Sundays). The reviews were mixed, though mostly positive. On the glowing side, *Country Music* magazine likened Wade to his idols: "He's got the sort of guts, attitude, and instincts that you just can't fake. At times he sounds like he sort of sprang full-blown from the Oklahoma dirt. While a lot of his 'new country' contemporaries were cutting their teeth on Queen, Aerosmith, and the Eagles, Hayes was obviously busy soaking up Keith Whitley, Waylon Jennings, and Willie Nelson. Throughout *On a Good Night*, Hayes not only wears his Waylon Jennings influence on his sleeve; he practically sounds like he's got a touch of seventies outlaw fever flowing through his veins."

The Dallas Morning News said Wade's second album was better than the first: "A decade after Randy Travis spearheaded the new traditionalist movement, Nashville's cowboy culture continues to crank out male hat acts that try to balance country roots with radio sensibilities. Unfortunately too many of them opt for the latter—Tim McGraw and John Michael Montgomery, anyone? But every once in a while, authenticity takes a stand. A second album by Oklahoma native Wade Hayes, the twenty-seven-year-old honky-tonker with the booming baritone, proves Merle, Waylon, and Willie are still inspirational." Perhaps the best review came from Michael McCall, writing for the Microsoft Network: "Put Wade Hayes in a lineup with a dozen young male newcomers and his lanky frame, his baby-face good looks, and his oversized Stetson would make him nearly indistinguishable from the rest of the hat pack. However, blindfold a listener and make each of them sing, and Hayes's deep, grizzled baritone would be the voice chosen as least likely to belong to anyone in that young country bunch. In that sense, he's a modern-day Nashville dream. He's a handsome young man with an old man's voice, and he knows exactly how to use it."

The harshest review came from *USA Today*, with David Zimmerman giving the album two and a half out of four stars. "Hayes is a double-threat guitarist/singer who had distinguished himself from the pack, but he stumbles badly on this sophomore album. While he's as appealing as ever on the up-tempo tracks, an overwrought approach diminishes such ballads as 'The Room' and 'I Still Do,' in which Hayes's vocals lose the engaging naturalness that first won him notice with such early hits as 'I'm Still Dancin' with You.'"

The Washington Post was more balanced, with Geoffrey Himes writing: "Hayes's *Old Enough to Know Better* was the only debut country album of last year to be certified gold, and a lot of the credit goes to the young Oklahoman's deep, twang-filled baritone. His follow-up effort doesn't have as many knockout songs, but it confirms that he has a very special voice. Hayes's producer is Brooks & Dunn mastermind Don Cook, and Hayes had the opening slot on B&D's recent tour, so it's not surprising that he imitates the boot-scootin' duo from time to time. He's at his best, though, when the production values fall back and leave room for his voice to grab hold of a vowel and turn it into a note of desire and regret, as on 'Where Do I Go to Start All Over?'"

The album sold 12,268 copies in its first week in stores, placing it eleventh on the country album charts, directly behind Garth Brooks's *Fresh Horses* and in front of Lorrie Morgan's *Greater Need*. By comparison, *On a Good Night* sold one fifth as many as Shania Twain's *The Woman in Me*, which was at number one on the country chart that week, and one twentieth as much as Metallica's *Load*, which was at number one on the pop chart. Perhaps more importantly, Wade sold fewer copies in his first week than Bryan White's 14,000 when his second album, *Between Now and Forever*, debuted several months earlier. On the bright side, though, Wade did have the highest debut that week. "All in all, we're thrilled," said sales VP Mike Kraski.

The second week, which included the Independence Day holiday (and, more importantly for country music, a payday for many country consumers), the album actually ticked up in sales, reaching 13,217 units. It stumbled one place in the rankings, though, to twelfth. The following week, with sales of 12,001, the album fell to fourteenth. The week after it tumbled to sixteenth. By week five, with sales of a mere 8,643 units, *On a Good Night* was on its way to slipping out of the Top 20 entirely. Even more ominously, the album's descent was taking place even as its debut single, "On a Good Night," continued to *climb* the singles chart, a clear sign that the public didn't much care for the song, since customarily as a single ascends the chart it spurs more album sales.

As Debi Fleischer had predicted, Ricochet's second single, "Daddy's Money," went to number one the first week of July. "It's a happy day," Debi said in her office as the Jack Daniel's flowed freely. For all the joy, though, the success of "Daddy's Money" actually presented a problem for "On a Good Night." All the research indicated that Ricochet's single was strong enough to stay number one for a second week. If Debi tried to hold the song at number one, though, she might jeopardize Wade's chance to go to the top three weeks later, since Columbia would be asking for additional favors. In the end she took the chance, and it backfired. "Daddy's Money" dropped to four, while Wade inched forward to six bullet. Though Wade was on every *R&R* reporting station at this point and his record was spinning 6,154 times a week, he was in a testy situation. Above him was George Strait's "Carried Away," a surefire number one, and coming up strong behind him was Brooks & Dunn's "I Am That Man." Debi, knowing she had only one Hail Mary shot, announced that on July 29, the fourteenth week of the record, "On a Good Night" would go for number one. This meant her field reps were asked to tell their stations to give the record its maximum number of plays that week, after which they could drop back if they chose. This was payday for any favors Columbia had done over the previous months.

It failed. "On a Good Night" peaked at number two in the last week of July on both the *Billboard* and *R&R* charts. "I was disappointed," Wade told me. "I like getting number ones. I like the way that feels. And I had hoped the first single off the new album would do that." Considering its uneven history, though, number two was actually a remarkable achievement. "Radio still believes in Wade Hayes," Debi said. "I never believed it was a number one record." She also didn't believe the album would achieve Allen Butler's goal of selling 1 million copies. "It'll be gold-plus," she said, "but I'd be surprised if it took him to platinum. I'd be thrilled, but I'd be surprised. I'd like to go three singles, then move on. That's really the best we can hope for."

As it turned out, they didn't even get that far. After a several-week hiatus, Columbia Records made what would turn out to be the crucial decision for the life of *On a Good Night*. Instead of releasing "The Room," which everyone in Wade's camp believed to be the best song on the album, they decided to release "Where Do I Go to Start All Over?", a brooding ballad about a man wondering how to put his life back together after a breakup. "It's the strongest vocal performance," Mike Kraski said, "and it's the one most likely to get Wade a performance slot on the CMAs." "The Room," he said, was "too dark." "Where Do I Go" also proved to be too dark (and, many radio programmers complained, seemed to be cut in the wrong key). It lingered at radio for several weeks, then died before reaching the Top 30. The album, burdened now by lack of exposure, also suffered, dipping to below 5,000 units a week and plummeting to the bottom of the charts. After licking its wounds, Columbia decided to reverse its fortune by releasing "It's Over My Head," a ditty written by Wade and Bill Anderson. It didn't chart at all.

By early fall, less that six months after its release, *On a Good Night* fell off the Top 75 country album charts entirely and, with Columbia deciding to release no more singles, was officially dead. It had sold fewer than 150,000 units on SoundScan. With record clubs and some generous accounting, Sony was able to have it certified gold, but the fact remained: Wade had fallen into a sophomore slump.

With failure, inevitably, came blame. Wade's managers targeted Don Cook. "All the songs on the album were from Sony/Tree," one of them told me, referring to the publishing company Don helped manage. "He refused to consider songs that he didn't control or benefit from." Don, though he disagreed with that claim, did accept responsibility. "I think there were a lot of factors that came into play," he said. "We probably tried to put it together too fast. We probably didn't have enough choices for the audience and for people who promote records. We probably had some problems in song selection. I'll gladly take the blame for it. It's my job to give them the requisite number of pieces." Even Hazel echoed this point. "I think Wade was undoubtedly not given enough time with that album," she told me. "If he was given enough time in the studio, then he could very well slide into the Ray Price slot. There've been three great voices in country music: Hank Williams, Ray Price, and Wade Hayes. Wade even looks like Ray Price to me. But this album wasn't there at all. Wasn't even close. They've got to study his voice, and work with him, and see what he can do."

The greater problem was Wade's state of mind. By late summer, he had cut off almost all communication with his managers. "Just stop sending me the numbers," he told Mike Robertson. "Don't tell me anything about the album. Don't tell me anything about the singles. I don't want to know." He'd also stopped speaking with Don and went out of his way to avoid running into him around town. "I think he's going through a depression," Don told me. "I don't think it's anything else." The only source of hope in this otherwise bleak time was that in late August, just at the point when Wade's album was reaching its nadir, Marty Stuart and Lorrie Morgan stood outside the front doors of the Grand Ole Opry House and announced the nominees for the CMA Awards. Wade Hayes received his first nomination.

◦　　◦　　◦

Inside the Opry House, the preshow hurry-up was well under way. Since 1970, the CMA show has been broadcast on CBS, and since 1993, it has occupied the entire three-hour prime-time block. The show opens with a coveted performance slot—this year taken by LeAnn Rimes—and culminates with the award for Entertainer of the Year. In between, most of the genre's top performers make an appearance. As a result, the limited backstage area of the Opry is crammed with most of the denizens of Music Row: fidgeting publicists, worrying managers, and frantic wardrobe ladies battling fallen hems and missing sequins with tubes of Krazy Glue and ribbons of duct tape. In Wade's dressing room, for example, while he sat quietly drinking bottled water, Bryan White stood on one leg as a seamstress tried to remove a wrinkle from his Manuel-made faux Armani suit, and Collin Raye sat before a mirror as a hairdresser tried to layer his thinning hair over his scalp. Because such activities are thought to be of interest to the press (who presumably would love a shot of Bryan White's legs or Collin Raye's bald spot), the area is further crowded by a SWAT team of security agents—mostly former high school wrestlers with tight-fitting yellow shirts—that give the space an air of anxiety completely at odds with the stop-by-and-chat friendliness of the Opry itself.

The stage is even more different than the Opry. On this night, the normally quaint red barn backdrop and plain oak floor had been replaced by a high-tech, pink-and-black set that seemed as if it had come straight out of a video game—the Statler Brothers vs. Super Mario Brothers. The floor was covered in giant black and white tesselated tiles; the stage was crowned with enormous yellow jutting beams that looked like overgrown McDonald's arches; and the walls were decked out in bright pink neon sculptures, in the shape of eyebrows, that winked whenever they were on camera. As if to accent the theater's Hollywood metamorphosis, the ceremonial circle of maple that had been transferred from the Ryman Auditorium had been topped with a shimmering gold star. "If anyone still needs proof that country music in the nineties is moving in all kinds of new and exciting directions," Lorrie Morgan said, presenting the first award, "all you have to do is listen to the five nominees for Single of the Year. You'll hear everything from blues to swing, from rock to contemporary hymns. It's all over the map, and it's all country."

Though that last comment—"it's all country"—was subject for debate, the essence of her remark was vividly true. The CMA Awards are a stunning reflection of Nashville in the nineties—its openness to musical innovation, its desperate longing for glitz and assimilation, its complete artistic schizophrenia. One year the awards were dominated by Garth Brooks, with his pop-country hijinks, the next by Alison Krauss, whose unrepentant bluegrass style and downright twangy voice were a throwback to a bygone period. In the nineties alone, the symbolic opening slot has been filled by, among others, Mary Chapin Carpenter, Shania Twain, and LeAnn Rimes, three performers who would be hard-pressed to find a song to sing together, no less find common artistic roots. This, of course, is the great accomplishment of Nashville, but also its biggest source of frustration: What is country music anymore? Is the term even meaningful?

Several years earlier, country music seemed poised to become the dominant form of popular music in America, with an artistic breadth and an audience reach that were not only unrivaled, but also seemingly limitless. Now those ambitions definitely seemed inflated. There is clearly a ceiling on Nashville's appeal. Though its fan base is now national and increasingly international, and though its sound and message have become mainstream, country is still the voice of a set of beliefs and feelings that are rooted in a particular sense of American longing that not everyone experiences. It's not the longing for freedom or rebellion that characterizes much of American culture, most notably rock 'n' roll. Nor is it the frustrated yearning for escape and self-expression that characterizes many forms of African American music, from blues to hip-hop. Instead, it's a yearning for security, for comfort, for family, for happiness, and, especially in our rootless society, for rootedness.

These feelings, besides being expressed most clearly in the lyrics of country music, are also expressed in the music itself. Though the sound of country records has changed over time—higher production value; more cutting-edge guitar licks; powerful, pounding drums—the instruments that jump out of country records and that make people feel most at home in them are the ones that have been around the longest—the raging fiddle, the wailing steel guitar, the plucking mandolin. Even the singers' voices are more familiar than most of pop music. Stop 100 people at Fan Fair and ask them why they like country music and 99 will cite as one of their reasons: "Because I can understand the words." For all these factors—the factors that make country music feel like such a club—the word "country" *does* have meaning today. It's a password that no longer says "This is where I live," but instead, "This is who I am."

This change in country music from an art form associated with a particular region to an art form associated with a particular set of values is also a source of confusion. Most people still associate country music with certain areas—the Appalachian Mountains, the Mississippi Delta, the Texas flatlands. Critics who dismiss contemporary country say it has lost touch with these places and the people who live there. While this may be true, it misses the point: Country has lost touch with those places because the country, itself, has lost touch with these places. And country music itself has never had as its mandate the preservation of American regional life. Its goal has always been to chronicle the lives of Americans and to do it in a way that other Americans will identify with and buy. That country music today is less about place and more about values is the result of changes in America. Specifically, it's a lessening of the importance of place in our lives.

So how did this change come about? Regionalism was always based primarily on economics—people were rooted to the places where they lived in large measure because they were tied financially to the places where they lived. Fifty years ago, when the Grand Ole Opry was at its peak, a majority of Americans lived in rural areas and half of them didn't have electricity. As a result, people were less mobile, less wealthy, more isolated (in 1945, for example, only 46 percent of Americans had telephones), and more inclined to define themselves by their communities. In 1970, when the Outlaws were just beginning to wake Nashville out of its stodginess, jet airplanes, interstate highways, and network television (not to mention rural electrification) were just beginning to free Americans from their regional shackles. With greater mobility and prosperity, people began defining themselves less by their immediate surroundings and more by broader social identifiers: race, gender, youth. The foundations for a national culture were being laid.

Today, regional identity is less important than ever and a new pan-American culture has all but taken hold in the United States. In an era when computers, chain stores, and cable television dominate American life, the sense of isolation and disenfranchisement that were once central to the South have all but disappeared. Instead, they have been replaced by a general sense of well-being and good fortune (down here we've got jobs *and* nice weather), so much so that others have even starting flocking *to* the region. Twenty million Americans have moved to the South since 1970, twice the rate of emigration to other regions. In addition, non-Southerners were suddenly prepared to identify themselves with the region, something unthinkable two generations ago when the dominant images of the South were of oppressed minorities, barefoot, pregnant women, and toothless, racist men. Even if country music had embraced national themes in 1960, a limited number of Americans would have been prepared to embrace it because of the rank connotations still associated with the place from which it came. This change in perception of the South is still the most underdiscussed—and underappreciated—reason's for country's growth in recent years.

Moreover, as the South blossomed in recent decades, not only did it become more like the rest of the country, but it also started to *influence* the rest of the country, making America more like it. As historian (and Nashvillian) John Egerton predicted in the 1970s, the Americanization of Dixie was accompanied by a corresponding Southernization of America. What he meant was that Southern problems—namely race—were becoming national problems. And Southern issues—namely how to raise the prominence of values in American life—were becoming national issues. He was right. The dominance of values in the American conversation of the 1990s—how to preserve our families, our relationships, our communities—is the biggest mark that a Southern aesthetic and value system has taken hold in the country. In place of regional and class rivalry, a generic social conservatism has now taken hold among most middle-class Americans. Ask any Southerner (or Northern transplant) what they like about the South and, after the weather, they are likely to say: "The people are friendly, there's a sense of tradition, and there's an emphasis on families." In short, it's better living. Not coincidentally, it is these topics that country music handles best, and its prosperity during this period has been based primarily on the remarkable synchronicity between its agenda and the agenda of the country.

All of which is fine as a foundation for country's popularity, but what about the music? Is it any good? When I first came to Nashville, I associated country music with everything I hated about the South: Its narrow-mindedness, its unwillingness to change, its attachment to outmoded imagery. The first thing I discovered was that most of my prejudices about the past were incorrect. Country, for all its caution, has tolerated alternative viewpoints, it certainly has changed over the years, and at least some of its artists have tried to punch through the paper sack of Southern clichés. As a result, one curious result of my time in contemporary country music is that I developed more of an appreciation for traditional country and a belief that much of that music does connect to my own deep-rooted sense of being Southern. At the CMA Awards that night, for example, my favorite performance was not by George

Strait or Alan Jackson, but by Ray Price. At seventy years old, he was clearly the most effortless singer on the stage.

As for the music of today, what I feel about it tends to mirror what I feel about the South. When speaking to outsiders—particularly those who dismiss it out of hand—I defend country music fiercely, citing its contemporary edge, its wit, its new crop of more sophisticated artists: Mary Chapin Carpenter, Shania Twain, Kim Richey. I take the same tack with those who claim to like old country but not to like the new. Have you listened to the Mavericks? Trisha Yearwood? Alan Jackson? Garth Brooks has several early songs—"Unanswered Prayers," "Much Too Young (to Feel This Damn Old)," "If Tomorrow Never Comes"— that I would submit for a list of all-time great country songs. Wynonna has a voice and a body of work—both with her mother and without— that is as strong as Patsy Cline's and equally blended between country and pop. To claim that all country music today is merely watered-down pop, full of mindless radio ditties and faceless knock-off artists, is to miss the fact that *all* forms of popular music (not to mention film, television, and books) have only a few originators and swarms of mediocre imitators. Even traditional country music had talentless copycats, only they are now mercifully forgotten, as will be most of today's bad acts. The only difference is that today, because of Nashville's popularity, there is *more* country music in the culture, so there is more *bad* country music. Music Row should not be blamed for making music that is commercial; it has always done so. Also, like every other aspect of the entertainment business, it will continue to do so.

Having said that, when I am around people who are sympathetic to country music—who are "in the club," so to speak—I can be fiercely critical of Music Row. It's the same attitude I take with the South. Like any Southerner who loves the region, I celebrate the decline of racism, poverty, and isolation in the South, but bemoan the crass commercialization, the mallification, the downright homogeneity that seems to be overtaking the area. It's no surprise that the same problems plague country music. As bluegrass prodigy turned amateur historian Marty Stuart memorably told Peter Appelbome, author of *Dixie Rising*: "The best quote I ever heard about country music these days is that the reason Long John Silver's outsells Uncle Bud's Catfish Cabin on the edge of town is because they've made fish succeed in not tasting like fish." For half a century, record executives and radio stations were able to decide what was good music by listening with their ears and trusting their emotions. There must be some way to bring that process into the future without abdicating all creative decisions to one hundred people chosen at random during dinnertime who listen to six seconds of forty songs and are then asked to rank them. No wonder so many songs on the radio sound like commercial jingles; they are being created in exactly the same way.

"It's our job to give radio songs that will get people's attention while they're driving their screaming kids down the freeway," Don Cook told me when I complained that Wade's version of "The Room," one of the best songs I heard that year, was never released. " 'The Room' is not that kind of a song. It's a beautiful vocal performance. It's a real emotional piece of music. But in a minivan full of screaming kids, 'The Room' is a nonevent." Maybe, but why must country music abdicate all its creative energy, as well as its public face, to screaming kids on the freeway? There are plenty of other people, and plenty of other times of day, when we crave the depth, the humor, and, yes, the sadness that the best country music still provides. My plea to Music Row is simple: Please, stop underestimating the intelligence of your audience. You got what you wanted, you lured us in. Now quit insulting us.

Finally, I know from living in Nashville that while it may be growing more like New York and Los Angeles, while there are colossal examples of backstabbing and greed, there is still a feeling of community that is appealing, reassuring, and, yes, Southern. Several steps from the backstage area at the CMA Awards is a large room called Studio A, where the press, managers, and publicists watch the show on closed-circuit broadcast. It was in this room several years earlier that I first eyed Hazel, holding court and leading the cheers or jeers as the winners were announced. Whereas once I had viewed her reactions as unsophisticated—boosterism more than journalism—now I saw them differently. She was, in her own way, the repository of tradition that Nashville needed more than ever in its headlong rush to the future. "Any time there's problems with country music," she told me that night, "when people say it's watered down—'I can't tell this one from that one'—it's because of this: The people are singing for money and not for love. It's not coming from the heart. It's whenever people start having so many meetings, and there's so much business going on, that they're not focusing on the music. The music will always speak for itself. This is a town where the best songs in this world come from, and if we focus on those songs, we'll do all right."

Late in the CMA show, those words seemed to come to life as Vince Gill walked to the center of the stage and, with Alison Krauss, sang a song, the bluegrass-infused "High Lonesome Sound," that he had written in honor of Bill Monroe, who had died several weeks earlier at the age of 84. At the time he was the oldest member of the Grand Ole Opry. At the end of the song, which later was nominated for a Grammy, the two hundred people in Studio A sat still in deference to the gift of the moment, while in the center of the room, sitting alone, Hazel Smith quietly wept to herself.

And at that moment I finally knew which legend she had once loved.

Back in the Opry House, the buzz was building. This year's show was dominated by George Strait, who won three awards: Male Vocalist of the Year, Single of the Year ("Check Yes or No"), and Album of the Year (*Blue Clear Sky*). Vince Gill won two awards, bringing his total number to seventeen, the highest in history. The Mavericks won their second Group of the Year, and Brooks & Dunn claimed Entertainer of the Year.

But on this night, like many, perhaps the biggest anticipation centered around the Horizon Award. Begun in 1981 as a way to reward upcoming stars, the Horizon Award is granted annually to the artist who has demonstrated "the most significant creative growth and development in overall chart and sales activity, live performance professionalism, and critical media recognition." Besides being a perfect expression of the Nashville philosophy (Where else is "creative growth" measured in commercial activity, stage professionalism—whatever that means—and media recognition?), the award has been a consistent harbinger of success. (The one exception was Terri Gibbs, whose career fizzled soon after winning the first award.) Since then, the CMA trophy—a large crystal candle flame that looks vaguely like one of Madonna's cone bras—has been won by Ricky Skaggs, Randy Travis, Clint Black, and Travis Tritt. The Judds won in 1984; Garth Brooks won in 1990.

This year's class of nominees was as eclectic as ever and, for those looking for clues as to what direction country music would take, as confusing as ever. On the one hand, Shania Twain represented a new breed of adult sexuality in country music. Terri Clark, the first female "hat act" in a generation, showed the continuing appeal of Western tomboys. Bryan White appealed to the *Tiger Beat* audience, while Wade had one of the richest, most sad-sack country baritones in years. Finally, LeAnn Rimes was a teenager singing broken-hearted torch songs. Anyone trying to predict the future would have to account for all these branches, as well as a few others. "I think the future is going to be driven by what the past was driven by," Jimmy Bowen said to me in the wisest prognostication I heard about the future, "unique one-of-a-kind talents. If Nashville can continue to produce five or six of those a year, country will do okay."

The future was on everyone's mind as Travis Tritt strode to center stage in a red Manuel jacket to announce this year's Horizon's winner. As was customary, each of the nominees had performed on the show, with Wade strutting through a shortened version of "On a Good Night," taking care to point to his idol during the line, "On a good night, I can drive to the lake / Turn on the radio and find George Strait . . ." It was that shot that the producer chose to use when Wade's picture was shown on the screen full of nominees.

"All of this year's nominees have two things in common," Travis began. "One is talent. You saw them demonstrate that tonight. The other is that they're all new young performers with lots of exciting ideas. If they're the future of country, that means the future is looking bright." After scrolling through the nominees, Travis produced the envelope from his jacket. Wade, sitting on the aisle directly in front of Garth, smiled grimly. A year earlier, this award seemed destined to go to him. Today, after a string of unfortunate breaks, a slew of uneven decisions, and a general avalanche of pressure, he knew it was a long shot. Shania Twain, even though she was on her second album, was the dominant star that year. LeAnn Rimes, though only around for a few months, had experienced the fastest launch of any artist in the 1990s. And Bryan White, following his tearful acceptance speech at the ACMs, had earned himself even more supporters. Still, Wade, thinking of Ricky Skagg's acceptance speech at the CMAs several years earlier that had lured him back to Nashville, kept hope.

"And the award goes to . . ." Travis cracked the envelope. "Bryan White!"

As completely understandable as that award was, it was followed by something unexpected, and, for those of us around Wade Hayes, something magical. As Bryan White climbed from his chair, a wave of goodwill swept through the audience. The wave seemed to catch Wade by surprise. But unlike six months earlier at the ACM Awards in Los Angeles, when he winced and dropped his head in shame when he lost to Bryan, and unlike three months earlier at the TNN/*Music City News* Awards in Nashville when he slumped out of the Opry building when he lost again to Bryan, on this night, in front of a much bigger audience, Wade stood along with the rest of the crowd, lifted his head, and cheered.

°   °   °

Wade was the first to arrive. He parked his brand-new white Dodge pickup in front of the Soundshop Studio and plopped down on the curb, lifting his Styrofoam cup to his lips for a drip of tobacco juice. It was a gray winter day, with the threat of snow in the air, one of those days in the wake of the CMA Awards when Nashville descends into a several-month period of hibernation and self-analysis. "Oh, well, *denied* again," Wade said of his CMA defeat. "But I kind of saw that one coming. I was a little upset, but Bryan White is a great singer and a really nice guy. I'm happy for him. I wasn't at the ACM Awards. I was ticked then. But, hey, I got over it. It took me about six months, but I did get over it."

Wade was wearing brand-new black Wranglers, black Justin boots, a white T-shirt with THE ROAD printed on it, and a Resistol baseball cap with the slogan THE BEST THERE IS. "It's all free," Wade said of his outfit, in a startling echo of Garth's comment to me a year earlier. "I'm not wearing a damn thing I own." He grinned, impressed with himself. He acted older than I remembered. But cleanly shaven this morning, he also looked younger than he had in a year. His nose was even flatter than when we first met. His pale green eyes were wide open, and, once we went inside, they peered out the window for large periods of time—searching, straining a bit, but intermittently laughing. When was the last time I had seen him do that?

"The funnest thing now is coming to the realization that everything is going to be okay," he said. "I've finally learned to let all the bad stuff just kind of float on by and keep my head down." His voice had lost none of its rich Oklahoma air. "I learned that if you let yourself get mixed up in all of that stuff, then there's no way in the world you're gonna write songs. And that's the whole problem I was facing last year. There's just no way to be creative or fully enjoy what I was doing because I was just so worried all the time."

Was there a moment when he came to that realization?

"It was several months ago," he said. "Right before the CMA Awards. I was lying in bed one night and I happened to open up the Bible. As you know, I'm really into theology. That doesn't mean I'm weird or anything, I just enjoy that. Anyway, when I opened the Bible, it happened to fall on this passage that said: 'Worrying won't add one cubit to your stature. Why are you worrying?' And I thought, 'Well, hell, that's exactly what I needed to hear.' And that's the truth. That's kind of what started it. I don't know if it's because I'm getting closer to thirty; I know people change gears about every ten years. Maybe I'm growing up or something, but I have such a more positive outlook on everything. It wasn't anything anybody did. It was just me praying about it *a lot* and coming to the realization that everything happens for a reason."

The reason for his sudden turn of fortune, he believed, is that he needed a dose of reality. As Don Cook had said to me several days earlier, "Wade got real comfortable when he had two number one records in a row. 'This is real easy.' Well, it's not easy. Wade had an incredibly fast start. God knows, I didn't have any start like that in my career. I was here just beating the bushes for six or seven years. It was twenty years before I had the confidence to make records." For Wade, though, the transition was instantaneous. "Wade Hayes came here and within two years he had a number one record. That's phenomenal. It sort of distorts a young, impressionable sense of reality. Truth is, he got through the gate just before it closed—in the sense that all of a sudden everybody just started throwing product at radio. The competition level rose dramatically. In the process of putting together his second album, none of us had any idea what we were running up against. We were just dead wrong about where we needed to position him musically."

On this point, Wade agreed. "I think there were two things working against me last year that I didn't realize," he said. "First, the whole thing with the record. I was iffy about it. There were songs on the album that I liked, but something wasn't jiving. It was like we had to have the album done for billing, or something like that. That really scared me a lot. Second is that we came out with such a big bang. I didn't realize that that wasn't the way things normally go. I kind of got this perception of things that was not true and just naturally assumed, 'Okay, I'm here. It's just going to keep getting better.' But reality caught up with me. And who's to say I didn't need a little reality check? I probably did. I don't think I've ever been mean or snotty to anybody in my life, but maybe I wasn't appreciating things as much as I should have been.

"Lee Roy Parnell told me something," Wade continued. "He came to my first number one party. He told me, 'You better cherish this because they don't come along every day. And you're going to regret it if you don't enjoy this fully.' And you know, I didn't listen to him. I said, 'Okay, this has happened. I'm ready for something else. Give me more.' And he was absolutely right. The next one I get I am going to enjoy it. I can promise you that." His voice became sarcastic. "They're going to see a side of Wade they haven't seen yet."

At the heart of this change, Wade said, was the realization that he had, in fact, achieved the goal he had originally set for himself. He had acquired the security his parents never had. "I drive by every now and then and see this big house that I worked on here in Nashville," Wade said. "It used to be called the elephant house because it had this big long driveway coming out the front of it that looked like an elephant trunk. I can remember being up on that thing when it was twelve degrees below zero in the middle of winter framing that house. It was the coldest I've been in my life. I started at like six bucks an hour and pretty quick got raised up to nine bucks an hour. I thought, 'Damn, I'm doing good.' It was more money than I'd ever made in Oklahoma. Well, I drove by it recently and thought how lucky I am. I have had two number one songs. And I'm making money." He paused. "Just having that piece of mind—that they're not going to come turn your lights off—is pretty exciting."

Then he stopped himself. "But it's not the answer I thought it would be," he said. "I thought if I ever got a record deal and had hits that there would be nothing wrong in my life at all. But that's not the case. You've still got your same old problems. You've still got girl problems. You've still got worries with your career. There's people around you that get sick and pass away. You know, just regular life problems that I never even considered beforehand because I was so worried about money."

And, of course, Wade still faced another big problem: music. "I tell people I want to sing country music," Wade explained. "That's what I write. That's what I listen to. That's what I am, personality-wise. I like Willie Nelson, Waylon Jennings, and Merle Haggard. Their songs *say* something, and they were different. They stirred your heart. One way or another, you either got pissed off or mad right with them or you had your heart broke with 'em. That's what I want to do, man. Connect with people. There's something about the process—creating a song, putting it on a record, playing it onstage, then having somebody tell you they like what you wrote. That's what it's all about. And when you do it right, it's like 'Hey, he's been reading my mail!' "

Which, in the end, is the task of country music: to read people's mail, to go into their empty bedrooms, to probe around the dark corners of their imaginations, to inspire them, to commiserate with them, to enlighten them, to poke fun at them, to dance with them on Saturday night, to wake with them on Sunday mornings, to bring them out of their shadows and into the day, and to do it all with grace and wit.

"You know what I was thinking about just the other day?" Wade said. An hour had passed on the sofa that morning. Two years had passed since we first met. Later that afternoon, he would return to the studio to begin cutting his third album. "I realized that since I had come to Nashville I've been to every state in the country except Hawaii and Alaska. And I was thinking, I had this notion that people up North or out West had nothing to do with country music. I was really apprehensive about playing there. Finally I realized a good song speaks to anybody. Just because they're living in a different part of the United States doesn't mean they don't like the same things I do. It's made me realize that people aren't as different as I thought they were. You've got your basic types of people everywhere you go. There's good ones and there's bad ones. There's people that get it, and people that don't get it. People with soft hearts, people who don't give a care."

"And as a performer?"

"My job is just to go out and entertain 'em. Let 'em see and hear what they came to see and hear. And I try. I try harder now than I ever have to go out and entertain them. I still want it. I want it just as bad, if not worse, than when I first came to town. I've got it caught. I've got it corralled. I just don't have it roped and broke yet. That's my next goal."

"And are you going to do that?"

"I believe so. I didn't know anything when I came to town. It's been a lot to absorb. I got a little sidetracked, but I sincerely feel like my old self again. And I think it's going to work."

Later that day, as the sun peeked its head through the Nashville skyline, Don Cook drove his cherry red Mercedes to the front door of the studio, hopped out, and wandered into the control room, where Wade was waiting. The musicians in their hatchbacks and pickups soon followed. And by just before two, the familiar buzz had returned to Studio A. The men gathered in the soundproof room on Division Street, just east of Music Row, were legends. They had come, on this day, to help create another.

CODA

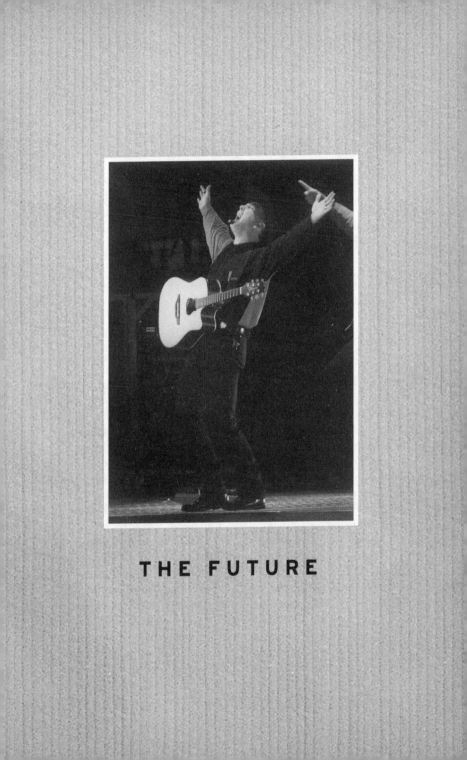

THE FUTURE

The idea of Garth Brooks playing Central Park in New York City had been floating around Nashville for years. And for years Garth Brooks had turned it down. "Politely we told them, 'There's just no reason to be there,'" he said. By early 1997, with his career seemingly stalled, Garth finally had a reason. He needed another of his patented public spectacles to galvanize attention around his new album, *Sevens*, which he hoped would launch his resurrection. Central Park, he said, would provide the ultimate "media event" to initiate that revival.

On Thursday, August 7, 1997, at the start of an unseasonably pleasant evening in Manhattan, several hundred thousand fans crammed into the North Meadow of Central Park to hear a free concert by the person HBO, which broadcast the show live, called "America's most popular performer." (As for the exact number of attendees, Mayor Rudolph Giuliani said there were 150,000 people; Park police placed the number at 250,000; while Garth announced from the stage that the figure was 900,000.) Numbers aside, the evening was an unabashed triumph for Garth Brooks. Though he had given better vocal performances, and certainly ones with more spontaneity (for much of the show, Garth seemed to be apologizing to New Yorkers for invading their city), the content was hardly important. After a remarkably brief span of eight years, Garth Brooks had come to the country's most famous venue in the nation's most unforgiving city and dominated media coverage for weeks on end, obliging the mayor to put on a cowboy hat, local television reporters to seek out barbecue spots, and the *New York Times* to put Garth's photograph on Page A1, above the fold.

More to the point, Garth had finally elevated himself to the pantheon of heroes he had long sought to infiltrate. The show began with an homage to Art Garfunkel, reached its climax with a guest spot by Billy Joel, and ended on an elegiac note with an appearance by the poet laureate of lost innocence, Don McLean, who joined Garth in a version of "American Pie." For a man who had complained for years that he was widely viewed as being more pop than country, Garth's decision to include three pop (and no country) idols as guests seemed to confirm his desire to be considered a pan-American superstar unencumbered by regional roots.

As much as Garth had become that icon, though, his activities leading up to Central Park proved that he was still a restless icon, one haunted by his own ambitions. As with so many high points in his career, the concert was almost completely overshadowed by turmoil in his business affairs. Two months before the show, the head of Thorn-EMI in London summarily dismissed Garth's ally Charles Koppelman and his deputy from the stewardship of EMI North America. For Garth, the moment was cataclysmic. Suddenly there was no one whom he trusted to implement the $20 million marketing plan he had demanded. "When this record thing went down, I didn't sleep," he said. "I was on the phone until four in the morning to London, screaming my guts out." When that failed, Garth decided to withhold his record until further notice, thereby undermining one of the central purposes of Central Park. As a result, in the weeks leading up to the biggest moment in his career, Garth spent much of the time in interviews explaining why a concert that was specifically designed to promote an album was not accompanied by an album. As he admitted, perhaps this turn of events proved that he thrived on controversy and was incapable of enjoying his success: "I do love being miserable," he said.

By Christmas, Garth had finally achieved what he had been demanding all along, namely the ouster of Scott Hendricks as the head of Capitol Nashville. For the second time in two years, Garth Brooks had flexed his muscle and demanded that as the best-selling solo artist in American history he deserved to control the hirings and firings at his record label. The new executive would be Pat Quigley, a New York transplant who admitted he knew nothing of country music and cared even less and whose sole priority was helping Garth achieve his goals, a priority the new president signalled by naming his son after his chief corporate asset. With Quigley in place and prepared to spend $25 million dollars (he boasted) to market Garth, *Sevens* went on to become one of the fastest selling albums of the decade, notching close to three million units in less than a month. Though the album plummeted after the holiday season, nobody noticed. Garth went on to release a boxed set in the spring and his much vaunted "live" album that Christmas. With some creative marketing and generous accounting, Garth seemed determined to reach his ultimate goal: selling 100 million records by the end of the millennium, thereby surpassing the Beatles, leaving Madonna, Michael, and Elvis, in the dust, and assuring his place in retailing history. As an institution he seemed more on par with the corporations he so effortlessly bandied about—McDonald's, Disney, Wal-Mart. Garth had certainly become an icon, but was there anything real about him anymore? After all these years, the question seemed to answer itself.

While Garth grew more and more distant with every gesture, Wynonna suffered a much different fate. Her much ballyhooed comeback album, *The Other Side,* produced a few hits and a few more sassy numbers, but it failed to ignite any support in the increasingly limited realm of country radio. The album dropped quickly from the charts, leaving Wynonna in the homeless state she had been slipping toward for years: too brash for country, too sweet-tempered for rock, too unfocused (and unfashionable) for pop. Her professional limbo, naturally, pushed her back into the waiting arms of her mother. At Tammy Wynette's funeral, Wynonna was introduced from the stage of the Ryman to sing, but it was Naomi who appeared from the wings, dressed in black, and claimed the microphone to give an unsolicited speech about her friend, her daughter, and, inevitably, herself. Though Wynonna might never go back to the browbeaten days of her past, she would never quite escape them either. By Fall 1998 she was even prepared to admit publicly that her mother had never liked Arch; weeks later Wynonna surprised no one by filing for divorce. In yet another stunning echo of her mother, Wynonna Judd, at age 34, would be a single mother of two. In fact, at this point it seemed almost predetermined that by decade's end, less than ten years after Naomi had retired, the Judds would be making their long-awaited reunion—first in the studio, then on the talk shows, and, finally, on the road. As had always been the case, the two needed—desperately needed—each other to survive.

While Wynonna had the escape hatch of her mother, Wade Hayes was not so lucky. When his career skidded, he had much more difficulty regaining control. Wade's third album, *When the Wrong One Loves You Right,* was a much stronger album than *On a Good Night,* and its reception was stronger as well. The album made its debut in the country top ten, a first for Wade, and it produced several hits on country radio. But none of the songs had any particular staying power and, with listenership at country radio down dramatically, a mid-level artist like Wade had even more problems creating a sense of excitement around his career. Wade Hayes, once the most promising star of his class, was suddenly having difficulty standing out in a crowd of young hunks in hats. What seemed remarkable about Wade, though, was how loyal his supporters were (almost as loyal as Garth's and Wynonna's), throughout this fallow period. In 1997, to everyone's surprise, fans voted him "Male Star of Tomorrow" at the TNN/Music City News Awards. And the following year, in what seemed like a typically moving example of the familial relationship that still pervaded the world of country music, Wade's fans on the Internet stood up between his concerts at the Oklahoma State Fair and presented Trish and Don Hayes with a gift for their new dream home: a new computer. Since the money collected from the fan-written, electronic newsletter, "A Word on Wade's World," exceeded the cost of the computer, the fans donated the balance, $1019, to Wade's favorite charity, Habitat for Humanity.

Whatever else they may have in common, the ups and downs in the lives of Garth, Wynonna, and Wade seemed to embody the state of country music at decade's end. On the up side was the legacy of Central Park. The appearance of the best-selling country artist in history in the heart of New York City was the exclamation point at the end of a ten-year growth cycle that reached its apex with the arrival of country music at the forefront of American popular culture. For Music Row, there could be no better symbol of this achievement than the silhouette of a cowboy hat-wearing, guitar-toting country singer inside a picture of a bright red apple hanging from every lightpost and screaming from every bus stop from Wall Street to Yankee Stadium. For Nashville, a community that had long viewed itself as the underappreciated stepsister of American entertainment, this was not only receiving tickets to the ball, but having the ball thrown in its honor.

At the same time, the Central Park event coincided with a moment of great fear in Nashville. Country sales, reflecting a sluggishness that infused the entire music industry, continued to tick downward, a trend that Central Park was unable to stanch. In a way, this chilliness could be seen as a positive sign, a wake-up call to Music Row. The greed and conservatism that seemed to inflict the town would have to be replaced with greater risk and more innovative music. Nashville had captured the country's ear during the previous decade, but in order to keep the country's heart, it would have to continue to listen and change.

That spirit of change was the talk of Music Row—in new crossover music from Shania Twain, Faith Hill, and Deana Carter, and innovation from such stalwarts as Alan Jackson, George Strait, and Vince Gill. That spirit even seemed to infiltrate the chilliest of relationships. In a clear sign that Garth and Nashville might be inching toward an uncomfortable detente with each other, for example, the members of the Country Music Association, following a four year freeze in which they had completely ignored the best-selling artist in country history, awarded him Nashville's highest honor, "Entertainer of the Year," for two years in a row. In both cases, Garth, fearing embarrassment, hadn't even shown up at the Opry House for the show. But it didn't matter. The two camps expressed themselves with a quiet, respectful tip of the hat that, under the circumstances, might have been the best outcome of all. After years of acting like a vigilante, Garth Brooks could not so easily escape the shadow that he still cast over the town. Nor could Nashville so easily erase its resentment over the reach of that shadow. But as the sun set on country's brightest era, everyone had reason to hope that the town and the hero could ride alongside each other into the future—with their heads held high, their dignities intact, and their minds—and ears—open to whatever came next.

NOTES

THE PEOPLE

In acknowledgments, as in award acceptance speeches, brevity is preferred, though rarely achieved. With that caveat, I would like to thank
the myriad of people who helped in the creation of this book.

First, a note on technique. This book was researched and written
over a three-year period in the mid-1990s, during which time I lived
and worked in Nashville. The book was completed with the cooperation
and foreknowledge of all the participants, though none asked for, nor
was given, prior approval over the manuscript. The research included
travel in over twenty-five states and three countries, as well as hundreds
of hours of direct observation with the principals, and dozens of hours
of on-the-record interviews. With appreciation for their extraordinary
generosity (and patience), I would like to pay special tribute to the
following people: Garth Brooks, Sandy Brooks, Colleen and Troyal
Brooks, Mick Weber, Scott Stem, Karen Byrd, Maria Thompson, Lorie
Lytle, and Judy McDonough; Wynonna Judd, Arch Kelley, Naomi Judd
and Larry Strickland, John Unger, Pam Mathews, Roach, Tony Brown,
Renée White, Jules Wortman, Toni Miller, Paula Batson, and Liz Thiels;
Wade Hayes, Trish and Don Hayes, Mike Robertson, Carol Harper,
Ronna Rubin, Allen Butler, Paul Worley, Scott Siman, Janet Bozeman,
Bill Johnson, Mike Kraski, Debi Fleischer, Scott Johnson, and John Mul-
lins. A special thanks to Don Cook for his friendship, support, and un-

usual openness in allowing me to sit in for the entire recording of *On a Good Night*.

I would also like to thank the dozens of other people who appear by name in this book and to the many others who welcomed me into their lives, specifically Hazel Smith, Betty and Raul Malo, Trisha Yearwood and Robert Reynolds, Paul Deakin, Nick Kane, Jerry Dale McFadden, Frank Callari, Ken Kragen, LeAnn Rimes, Belinda and Wilbur Rimes, Rod Essig, John Huie, Cleve Francis, Charley Pride, Steve Small, Waylon Jennings, Willie Nelson, Steve Earle, Travis Tritt, Schatzi Hegeman, Marcus Hummon, Deana Carter, Lon Helton, Len Ellis, Clay Smith, Connie Baer, Bob Titley, and Beth Middleworth. Also, Chuck Aly, Marilyn and Brooks Arthur, Jimmy Bowen, Woody Bowles, Tim DuBois, Neenah Ellis, Olivia Fox, Kevin Lane, Pam Lewis, Judy Massa, Mike Martinovich, Sandy Neese, Jim Ed Norman, Greg O'Brien, Janie Osborne, Nicki Pendleton, Jeff Pringle, Lisa Wahnish, Janet Williams, Marion Williams, Mandy Wilson, Walt Wilson, and Tom Wood. Special thanks to Bill Ivey and the staff at the Country Music Foundation, including Bill Davis, Kent Henderson, Paul Kingsbury, and Ronnie Pugh. (One note: in the interests of preserving their privacy, the names of the tabloid reporter and her sources were changed.)

Additionally, this book would not have been possible without the deep commitment of the following people: for their hospitality, Carol and Gary Duncan; for their open arms, Ellen and Bill Pryor; for his knowledge of Nashville, Clark Parsons; for her knowledge of Music Row and relentless good cheer, Beverly Keel; for her daily fellowship and keen editorial judgment, Karen Essex; for her incomparable spirit and unflagging devotion, Nancy Russell; and for her constant support and unrivaled insight, Susan Levy. If only for the friendship of these talented individuals, my time in Nashville would have been the most rewarding experience of my career.

Portions of this book have appeared in other places. For their editorial input and support, my thanks to Tina Brown and Susan Chumsky at *The New Yorker;* Andrew Sullivan and Margaret Talbot at *The New Republic;* Connie Rosenblum, Nora Kerr, and especially Fletcher Roberts at *The New York Times;* Bill Tonnelli at *Esquire;* John F. Kennedy, Jr., and Rich Blow at *George;* Phil Zabriskie at *Icon;* and special thanks to Annie Gilbar, Steve Root, Charlie Holland, and my friends at *Live!* Also, I am indebted to Sean Collins, Jeff Rogers, Noah Adams, and the entire staff of "All Things Considered" at National Public Radio.

David Black is the fount of counsel and friendship I always sought in an agent. Susan Raihofer is a daily source of affirmation and devotion, as is Gary Morris. Lou Aronica has been a tireless champion of this book. Trish Grader, who inherited the project, has improved it immeasurably with her editorial input and heartfelt enthusiasm. At Avon Books, additional thanks to Joan Schulhafer, Mark James, Anne Marie Spagnuolo, Robin Davis Gomez, Andrea Sinert, and Laura Richardson. Karen Lehrman was a central part of the development of this idea and an unswerving cheerleader throughout its completion. Also, my everlasting thanks to Tina Bennett, Ruth Ann and Justin Castillo, Andy Cowan, Cathy Collins and Doug Frantz, Jessica Korn, David Shenk, Dana Sade, Ellen Silva, Karen Gulick and Max Stier, and Jane von Mehren.

As always, my family has provided a boundless supply of encouragement, good humor, and blunt editorial opinions. I am grateful for their continued presence in my life and work. In particular, I pay special tribute to my brother, Andrew Feiler, for mastering yet another field in his exhaustive quest to be the perfect editor, no matter what subject I throw at him. Finally, being in Nashville and living in the world of country music has reminded me time and again of the importance of family in the creative process. Two members of my family, Ellen Abeshouse Garfinkle and George Alan Abeshouse, have passed in the course of my writing life, and this book is dedicated to their memory.

THE SOURCES

"Writing about music is like dancing about architecture." That comment—variously attributed over the years to Elvis Costello, Frank Zappa, Laurie Anderson, and Martin Mull—has been in my mind continually during the process of writing this book. With all due recognition of the difficulty of capturing music—and musicians—in print, I would like to pay tribute to the many writers who have written so lucidly about Nashville and country music in recent years and whose writing has informed and improved this book.

First, one of the joys of researching this project was reading the autobiographies of many current and legendary country music stars. In particular, I recommend the following: *Coal Miner's Daughter*, Loretta Lynn with George Vecsey (Warner, 1976); *Man in Black*, Johnny Cash (Zondervan, 1976); *Minnie Pearl*, Minnie Pearl with Joan Dew (Simon & Schuster, 1980); *Pride*, Charley Pride with Jim Henderson (Morrow, 1994); *The Storyteller's Nashville*, Tom T. Hall (Doubleday, 1979); and *Waylon*, Waylon Jennings with Lenny Kaye (Warner, 1996). Also, I have drawn from several other autobiographies, including *Love Can Build a Bridge*, Naomi Judd with Bud Schaetzle (Ballantine, 1993); *Crook & Chase*, Lorianne Cook and Charlie Case (Morrow, 1995); and *Rough Mix*, Jimmy Bowen and Jim Jerome (Simon & Schuster, 1997).

The best general purpose history books I consulted were *Country*

Music, U.S.A., Bill Malone (Texas); *Country: The Music and the Musicians*, the Country Music Foundation (Abbeville Press); *The Grand Ole Opry History of Country Music*, Paul Kingsbury (Villard); *Finding Her Voice*, Mary Bufwack and Robert Oermann (Crown); *The Nashville Sound*, Paul Hemphill (Ballantine); *Nashville: Music City, U.S.A.*, John Lomax (Abrams); *Country: The Twisted Roots of Rock 'n' Roll*, Nick Tosches (Da Capo); and *Singing Cowboys and Musical Mountaineers*, Bill Malone (Georgia). Also, a special thanks to Don Doyle for his two-part history of Nashville, *Nashville in the New South* and *Nashville Since the 1920s*, both published by the University of Tennessee.

As for books about individual artists, I have drawn from a wide variety of sources, including the following: *Garth Brooks*, Michael McCall (Bantam); *Garth Brooks*, Edward Morris (St. Martin's); *The Judds*, Bob Millard (St. Martin's); *The Outlaws*, Michael Bane (Dolphin); *Waylon*, R. Serge Denisoff (Tennessee); *Get Hot or Go Home*, Lisa Rebecca Gubernick (Morrow); *Honky-Tonk Angel*, Ellis Nasour (St. Martin's); *Your Cheatin' Heart*, Chet Flippo (St. Martin's); and *Last Train to Memphis*, Peter Guralnick (Little, Brown). Also, I have benefited immensely from several collections of writing, including *Behind Closed Doors*, Alanna Nash (Knopf); *Everybody Was Kung-Fu Dancing*, Chet Flippo (St. Martin's); and *Lost Highway*, Peter Guralnick (HarperCollins).

In addition, several notes of acknowledgment. For their insight into the music business, I pay tribute to the following: *Hit Men*, Frederic Dannen (Vintage); *Moguls and Madmen*, Jory Farr (Simon & Schuster); *The Rise and Fall of Popular Music*, Donald Clark (St. Martin's); and *Stiffed*, William Knoedelseder (HarperCollins). For their insight into the South: *The Americanization of Dixie: The Southernization of America*, John Egerton (Harper's); *Dixie Rising*, Peter Applebome (Times); and *A Turn in the South*, V. S. Naipaul (Vintage). And for their insight into Tennessee, a special nod to the novels of Peter Taylor, particularly *A Summons to Memphis* (Ballantine).

Three reference books in particular have worn holes through my desk. They are *The Comprehensive Country Music Encyclopedia* by the editors of *Country Music* magazine; *Definitive Country*, edited by Barry McCloud; and *Sing Your Heart Out, Country Boy* by Dorothy Horstman. I can't imagine having done this book without them.

In addition, I am profoundly indebted to the limited number of publications and the even more limited number of writers who cover contemporary country music in a serious and thoughtful way. This book draws extensively from work that has appeared in *The Tennessean*, the *Nashville Banner*, *USA Today*, *People*, *Entertainment Weekly*, *The Washington Post*, *The New York Times*, the *Los Angeles Times*, *Country America*, *Country Music*, *New Country*, *Billboard*, *Music Row*, *Radio & Records*, the Nashville *Scene*, and the *Journal of Country Music*. In particular, I have benefited from the work of a small cadre of dedicated writers, notably David Browne, Patrick Carr, Daniel Cooper, Nicholas Dawidoff, Robert Hilburn, Geoffrey Himes, Jack Hurst, Beverly Keel, Brian Mansfield, Michael McCall, Bob Millard, Ed Morris, Jay Orr, James Patterson, Neil Pond, Jim Ridley, Tom Roland, David Ross, Brad Schmitt, Rob Tannenbaum, Mario Tarradell, and Kay West. Many of these writers are now my friends, for which I am deeply thankful. In addition, eternal gratitude for their kinship and support to Peter Applebome, Chet Flippo, Peter Guralnick, Alanna Nash, Melinda Newman, and David Zimmerman.

Finally, my heartfelt appreciation to James Hunter, who not only opened up his treasure chest of music knowledge to me, but also improved this book immeasurably with his painstaking edits and intellectual camaraderie. For this and so many other measures of support from the community of writers in and around Nashville, Tennessee, I am forever grateful.

THE MUSIC

"Hank Williams You Wrote My Life." That classic song, written by Paul Craft and performed by Moe Brandy in 1975, still captures the feeling that many people have when they hear a well-written country song, whether it's by Jimmie Rodgers, Hank Williams, or Wade Hayes. Even in an era of increased competition and bottom-line pressures, Nashville in the 1990s continues to produce music that is thoughtful, emotional, and wickedly clever ("If I can make a living out of lovin' you / I'd be a millionaire in a week or two . . ."). What follows is a selective list of some of the many albums I have enjoyed during my time in Nashville, with particular emphasis on the artists referred to in this book.

The best collection of Garth Brooks's music can be found on *The Hits* (Liberty), a gathering of eighteen of his biggest singles, from "Ain't Going Down (Til the Sun Comes Up)" to "The Dance," that covers the years 1989 through 1995. To ensure that the collection would be available for a limited time only, Garth buried what he claimed were the masters of the album underneath his star on the Hollywood Walk of Fame. In 1997, the final copy of the collection was shipped to stores. As for his studio albums, the two records that best capture his mix of fun-loving, honky-tonk dance numbers and warm, sensitive-guy ballads are 1990's *No Fences* (Liberty), which contains his masterpieces "Friends in Low Places" and "Unanswered Prayers," and 1991's *Ropin' the Wind* (Liberty), which contains "Shameless" and "The River."

Wynonna (Curb/MCA), released in 1992, is a perfect pop-country album and best captures Wynonna's blend of country, R&B, and rock; it contains two of her signature hits, "She Is His Only Need" and "No One Else on Earth." The hits from that album, along with her other two Curb/MCA solo albums, *Tell Me Why* and *revelations*, are gathered together on a single disc entitled *Collection*, released in 1997. As for Wynonna's work with her mother, twelve of the hit records recorded by the Judds are presented together on a single album called *The Number Ones* and available on Curb/RCA.

It would be hard to think of a better traditional country debut album this decade than Wade Hayes's *Old Enough to Know Better* (Columbia), which contains four hit singles (including two number ones), along with several unreleased gems, including "Family Reunion" and "Kentucky Bluebird," written by Wally Wilson and Don Cook. Also, despite its limited commercial success, *On a Good Night* (Columbia) contains five or six outstanding songs with exemplary vocal performances, including the dark ballads "Hurts Don't It" and "The Room" and Wade's gentle tribute to his parents, "Our Time Is Coming."

Other albums I have enjoyed that highlight the breadth and depth of contemporary country music include the following: Mary Chapin Carpenter's *Come On Come On* (Columbia); Deana Carter's *Did I Shave My Legs for This?* (Capitol); Alan Jackson's *Who I Am* (Arista); the Mavericks' *What a Crying Shame* (MCA); Kim Richey's *Bitter Sweet* (Mercury); George Strait's *Blue Clear Sky* (MCA); Shania Twain's *The Woman in Me* (Mercury); Trisha Yearwood's *The Song Remembers When* (MCA); and Dwight Yoakam's *This Time* (Reprise). Twenty years from now, these albums will be on the shelves of music lovers of all generations, and critics will once again be complaining that contemporary country music does not hold up to the past. What they will mean is Nashville in the nineties.

INDEX

391

THE PHOTOS

The Players: Garth Brooks performing at the Rose Garden on July 25, 1996, in Portland, Oregon. *Courtesy AP/Wide World Photos*

The Opry: Grand Ole Opry, Ryman Auditorium, c.1956. *Courtesy Country Music Hall of Fame.*

The Studio: Wade Hayes and Don Cook during the recording of Wade's new CD. *Copyright © Edward A. Rode Photography.*

The Tabloids: Wynonna Judd arriving at her wedding reception on Sunday, January 21, 1996, in Nashville. *Courtesy Bill Steber Photography.*

The Town: Minnie Pearl checking her well-worn purse, 1953. *Courtesy Photofest.*

The Hat: Garth Brooks, 1989. *Courtesy AP/Wide World Photos.*

The Cover: Wade Hayes in a Manuel jacket, 1996. *Photo by Tamra Reynolds.*

The Face: Wynonna and Naomi Judd receiving their trophies for top vocal duet at the 26th annual Academy of Country Music Awards. *Courtesy AP/Wide World Photos.*

The Legends: Willie Nelson and Wylon Jennings. *Courtesy AP/Wide World Photos.*

The Interview: Garth Brooks, 1995. *Courtesy Photofest.*

The Party: Wynonna Judd, 1995. *Photo by Randee St. Nicholas.*

The Single: Wade Hayes, 1997. *Photo by Matthew Barnes.*

The Politics: The Mavericks, 1996. Left to right: Nick Kane, Raul Malo, Robert Reynolds, Paul Deakin. *Photo by Norman Roy.*

The Stage: Garth Brooks performing at the Sioux Falls Arena on October 29, 1997, in Sioux Falls, South Dakota. *Courtesy AP/Wide World Photos.*

The Money: Nashville skyline at night. *Courtesy Phyllis Picardi/Stock South/PNI.*

The Fans: Garth Brooks at Fan Fair. *Courtesy Beth Gwinn Photography.*

The Launch: Wade Hayes as a young boy. *Courtesy Trisha Hayes.*

The Show: Garth Brooks in concert. *Courtesy Beth Gwinn Photography.*

The Family: Wynonna Judd in concert. *Copyright © Steve Jennings/ Corbis.*

The Awards: LeAnn Rimes backstage at the 1997 CMA Awards. *Courtesy Country Music Association.*

The Future: Garth Brooks performing at Central Park's North Meadow on August 7, 1997, in New York City. *Courtesy AP/Wide World Photos.*